Understanding Nanomaterials

Second Edition

T0418614

TEXTBOOK SERIES IN PHYSICAL SCIENCES

PUBLISHED TITLES

Concise Optics: Concepts, Examples, and Problems
Ajawad I. Haija, M. Z. Numan, and W. Larry Freeman
ISBN: 978-1-138-10702-1 • 2018

Understanding Nanomaterials, Second Edition
Malkiat S. Johal and Lewis E. Johnson
ISBN: 978-1-4822-5322-1 • 2018

Understanding Nanomaterials

Second Edition

Malkiat S. Johal
Lewis E. Johnson

CRC Press
Taylor & Francis Group
Boca Raton London New York

CRC Press is an imprint of the
Taylor & Francis Group, an **informa** business

CRC Press
Taylor & Francis Group
6000 Broken Sound Parkway NW, Suite 300
Boca Raton, FL 33487-2742

© 2018 by Taylor & Francis Group, LLC
CRC Press is an imprint of Taylor & Francis Group, an Informa business

No claim to original U.S. Government works

Printed on acid-free paper

International Standard Book Number-13: 978-1-4822-5322-1 (Paperback)
International Standard Book Number-13: 978-0-8153-5438-3 (Hardback)

Library of Congress Cataloging-in-Publication Data

Names: Johal, Malkiat S., author. | Johnson, Lewis E. v.d.L., author.
Title: Understanding nanomaterials / Malkiat S. Johal, Lewis E. Johnson.
Description: Second edition. | Boca Raton : CRC Press, Taylor & Francis Group, 2017. | Series: Textbook series in physical sciences
Identifiers: LCCN 2017047524| ISBN 9781482253221 (pbk. : alk. paper) | ISBN 9780815354383 (hardback)
Subjects: LCSH: Nanostructured materials. | Nanochemistry.
Classification: LCC TA418.9.N35 J64 2017 | DDC 620.1/15–dc23
LC record available at https://lccn.loc.gov/2017047524

Visit the Taylor & Francis Web site at
http://www.taylorandfrancis.com

and the CRC Press Web site at
http://www.crcpress.com

Printed and bound in the United States of America by Sheridan

To my lovely wife, Luisa, and my daughters, Simi and Bianca.

M. S. J.

To Renee, Wayne, and Bruce, who have inspired me in both chemistry and teaching.

L. E. J.

Contents

Detailed Contents

Series Preface

This textbook series offers pedagogical resources for the physical sciences. It publishes high-quality, high-impact texts to improve understanding of fundamental and cutting edge topics, as well as to facilitate instruction. The authors are encouraged to incorporate numerous problems and worked examples, as well as making available solutions manuals for undergraduate and graduate level course adoptions. The format makes these texts useful as professional self-study and refresher guides as well. Subject areas covered in this series include condensed matter physics, quantum sciences, atomic, molecular, and plasma physics, energy science, nanoscience, spectroscopy, mathematical physics, geophysics, environmental physics, and so on, in terms of both theory and experiment.

New books in the series are commissioned by invitation. Authors are also welcome to contact the publisher (Lou M. Han, Executive Editor: lou.han@taylorandfrancis.com) to discuss new title ideas.

Preface to the Second Edition

Since the first edition of *Understanding Nanomaterials* was released, the fields of nanotechnology and nanoscience have continued to grow and evolve at an incredible pace. The major national-scale programs funding user facilities, such as the Molecular Foundry in Berkeley, California, and Environmental Molecular Sciences Laboratory in Richland, Washington, and nanotechnology centers at large research universities have been augmented by increased interest in nanotechnology across the spectrum of postsecondary education. According to the National Nanotechnology Initiative program directory,[1] at present, there are 33 undergraduate programs, 22 master's programs, and 20 doctoral programs in nanotechnology, nanoscience, or molecular engineering in the United States alone. A community college level program was even piloted at North Seattle College.

As nanotechnology and nanoscience evolve into larger, more mainstream fields, undergraduate instruction in nanoscale systems is growing in importance. Programs are introducing nanoscale science earlier in the curriculum and drawing students from an increasingly diverse variety of academic programs. To address this need, we have greatly expanded *Understanding Nanomaterials* to include a wider range of background material covering essential elements of physical chemistry and classical physics for understanding nanosystems as well as expanding to a larger number of materials and applications. The contents of the first edition

---

[1] https://www.nano.gov/education-training/university-college.

have also been restructured into shorter and more digestible chapters, improving fluidity and organization. The second edition is arranged such that it has a twofold use. On the one hand, it may be used as an introductory text, starting with the new unit on the fundamental physical chemistry of nanosystems (Chapters 2–4) to provide a transition from general chemistry/physics to more specialized material on synthesis, characterization, and applications of nanomaterials. On other hand, it may be used in advanced courses by starting with Chapter 5 and transitioning to primary literature midway through the course. The new material in the early chapters as well as more connections to terminology more often used for solid materials are also intended to improve accessibility outside of chemistry departments.

Finally, this second edition is the result of an unusual long-term collaboration. Dr. Lewis Johnson, who joined the writing team for this edition, was a student of Professor Johal at Pomona College in 2006–2007 while the curriculum that led to the first edition of this text was being developed. After graduating, Dr. Johnson was drawn to the University of Washington in part by the strong nanotechnology program, and continued to collaborate with Professor Johal, eventually returning to Pomona College as a postdoctoral lecturer where he taught the nanomaterials course in the chemistry department in 2013. Renewed collaboration and discussion about the course led to the development of the present text.

Malkiat S. Johal
Lewis E. Johnson

Preface to the First Edition

To the Student

Nanoscience is a rapidly changing field where new innovations and discoveries are being made every day. To write a book that captures even a fraction of what the scientific literature has produced over the last few years would be a monumental task. With this in mind, the topics presented in this book are carefully selected to provide a basic understanding of the field. Many important topics, such as computational chemistry and solid state physics, have been given limited coverage, largely because I want this book to be accessible to any student who has taken introductory college-level science courses.

This book is written for a full- or half-semester course in nanoscience with an emphasis on understanding nanomaterials. The stress on "understanding" is the key behind the objective of the text: to provide fundamental insight into the molecular driving force underlying self-assembly processes, as well as to explain how to characterize the resulting nanomaterials. Knowledge of self-assembly and characterization is essential for an understanding of these interesting systems.

It should be noted that this book does not draw heavily from scientific papers; rather, it should be used in conjunction with the primary literature.

To the Instructor

I have drawn relevant material from many scientific disciplines, assuming only a basic level of competency in physics, chemistry, and biology. Mathematical rigor has been limited to presenting key results and simple proofs. Instructors should use their discretion in placing emphasis on or "filling holes" in areas that may seem somewhat inadequate or limited in scope. The half-course model is suggested for teaching material directly from the book and solving the end-of-chapter problems. For a full-semester course, the book should be used in a course that requires students to refer to the primary literature. The latter may be more suitable for intermediate to advanced-level classes, although I strongly suggest training students to read papers early in their careers.

The approach taken in this book is to focus on preparing the student to read papers in this field, and so I have limited specific examples to landmark papers. This book should provide the necessary background to enable the student to comprehend articles from scientific journals. I teach the material from this book in conjunction with a student seminar series in which the students select interesting papers for presentation and class discussion. This approach combines discussion-based and problem-solving skills, and provides exposure to a highly interdisciplinary field of study.

Malkiat S. Johal

Acknowledgments

Writing this book would not have been possible without the support of friends and family. I am grateful to the support of my colleagues in the chemistry department at Pomona College, in particular Professors Cynthia Selassie and Daniel O'Leary, who have been a constant source of support. My research students and members of my Soft Nanomaterials class throughout the years have been crucial in helping me to develop a textbook that meets the needs of students interested in this field. I would like to express sincere gratitude to my good friend and coauthor, Dr. Lewis Johnson, whose input and dedication was crucial in developing the second edition. I would also like to thank Dr. Paul Davies (University of Cambridge) and Dr. Jeanne Robinson (Los Alamos National Laboratory) for providing invaluable opportunities throughout my career. I am grateful to Professor Bruce Robinson (University of Wahsington) for his useful comments on Chapters 2, 3, and 4. I would like to thank my executive editor, Lou M. Han, for his relentless commitment from the very beginning, and his faith in a second edition. Last but not least, I would like to thank my wife, Luisa Lucia Johal, for her love, support, and encouragement throughout the years.

M. S. J.

I am immenensely grateful to Dr. Malkiat Johal for the opportunites to collaborate on this textbook and to teach at Pomona and for over a decade of mentorship and friendship, as well as to our executive editor, Lou M. Han, for his advice and patience during the writing process. I also greatly appreciate the advice and mentorship of many colleagues, including but not limited to Dr. Bruce Robinson, Dr. Larry Dalton, Dr. Delwin Elder, Dr. Charlie Campbell, and Dr. Tom Engel (UW); Dr. Mukesh Arora, Dr. Cynthia Selassie, Dr. Wayne Steinmetz, and Dr. Richard Hazlett (Pomona); Dr. Simone Raugei, Bojana Ginovska, Dr. Peter Sushko, Dr. Niri Govind, Dr. Chris Mundy, Dr. Greg Schenter, Dr. Shawn Kathmann, and Dr. John Jaffe (PNNL), and Dr. James Farr (NOAA). I also thank Dr. Nicholas Bigelow (Intel) for many years of friendship, scientific collaboration, and for advice regarding sections related to lithography and semiconductor fabrication, as well as Dr. Stephanie Benight (Exponent), Dr. Andreas Tillack (ORNL), Dr. Jason Kingsbury (Cal Lutheran), and Dr. Jason Sellers (REC Silicon) for useful discussion and many years of collaboration and friendship. I am also grateful to my students and interns at both Pomona and UW, especially Hannah Wayment-Steele, Conner Kummerlowe, and Dr. Luke Latimer. I also thank Katie Lane for useful discussion, production editor Jonathan Achorn and his staff at MTC for their patient and meticulous work typesetting the book, and Wayward Coffeehouse in Seattle for providing a great place to write. Finally, I am grateful for my family and many other friends for their patience during the ups and downs of the writing process.

L. E. J.

Authors

Malkiat S. Johal is a professor of physical chemistry at Pomona College, Claremont, California. He obtained a first-class honors degree in chemistry from the University of Warwick, UK. After earning his PhD in physical chemistry from the University of Cambridge under the guidance of Professor Paul Davies, Dr. Johal joined Los Alamos National Laboratory as a postdoctoral research associate with Dr. Jeanne Robinson, where he worked on the nonlinear optical properties of nanoassemblies. His research laboratory at Pomona College focuses on using self-assembly and ionic adsorption processes to fabricate nanomaterials for optical and biochemical applications. Professor Johal's laboratory also explores fundamental phenomena such as ion-pair complexation, adsorption, surface wettability, and intermolecular noncovalent interactions in materials at interfaces. He has published more than 60 research papers, mostly coauthored by his undergraduate research students. He teaches courses in physical chemistry, general chemistry, and soft nanomaterials.

Lewis E. Johnson is a research scientist at the University of Washington, Seattle, Washington. He obtained his bachelor's degree in chemistry at Pomona College, Claremont, California, where he conducted research under the guidance of Professor Malkiat Johal, as well as research as an intern at NOAA with Dr. James Farr. He earned his PhD in chemistry and nanotechnology from the University of Washington under the guidance of Professor Bruce Robinson. He has taught at Pomona College (his undergraduate *alma mater*) as a postdoctoral lecturer and worked as a postdoctoral research associate at Pacific Northwest National Laboratory, where he conducted research on glass formation in calcium aluminate electride semiconductors with Dr. Peter Sushko and on allosteric modulation of electron transfer in nitrogenase with Dr. Simone Raugei. His current research involves designing and characterizing new nonlinear optical dyes and modeling the formation and structure of complex noncrystalline materials, among other projects.

CHAPTER 1

A Brief Introduction to Nanoscience

1.1 THE SCOPE OF NANOSCIENCE

Nanoscience—the study of systems with components on the scale of a billionth of a meter—is full of possibilities and presents us with the potential for significant technological breakthroughs in the near future. Nobel laureate Richard P. Feynman realized the importance of this field almost six decades ago. In his legendary speech, "There's Plenty of Room at the Bottom," he stated:

> This field is not quite the same as others in that it will not tell us much of fundamental physics in the sense of, 'What are the strange particles?' But it is more like solid state physics in the sense that it might tell us much of great interest about strange phenomena that occur in complex situations. Furthermore, a point that is most important is that it would have an enormous number of technical applications. (Full text available at http://www.zyvex.com/nanotech/feynman.html)

Nanotechnology, the application of nanoscience toward developing new technologies, is defined as the engineering or manipulation of functional systems at the molecular scale. A functional system is used to describe a material, or interface between materials, that has a well-defined responsibility and performs that responsibility efficiently and while minimizing undesired consequences. Although the term nanotechnology was popularized in the 1980s, scientists have been studying nanostructures for well over a century. As early as the mid-1800s, Michael Faraday investigated the properties that a ruby-colored solution forms when an aqueous solution of $NaAuCl_4$ is treated with a reducing agent. Faraday concluded that fluid contained very finely divided metallic gold dispersed in the aqueous solution. A century later, electron microscopy showed that these solutions were indeed composed of colloidal gold particles (gold nanoparticles) with average diameters of around 6 nm.

Over the last few decades, nanotechnology has focused largely on the use of colloidal systems, polymers, and nanometer-sized particles (nanoparticles) in coatings and materials. For example, silver nanoparticles have found use in hundreds of products because of their antimicrobial properties. More recently, nanotechnology has been used to explore biologically active materials as novel biosensors and targeted drug delivery vehicles for the treatment of diseases. The field is also impacting electronics, including development of new transistors, amplifiers, and adaptive structures. The smallest features in the integrated circuits in computer central processing units, which were over a micron in 1985, are now on the order of 14 nm, with devices on scales of 10 nm or smaller at the prototype stage. The next few decades will inevitably move nanotechnology to the point where we will be able to fabricate complex nanosystems and molecular devices by design on an industrial scale. Feynman's speech continued to discuss the transformative potential of nanotechnology:

> I want to build a billion tiny factories, models of each other, which are manufacturing simultaneously. ... The principles of physics, as far as I can see, do not speak against the possibility of maneuvering things atom by atom. It is not an attempt to violate any laws; it is something, in principle, that can be done; but in practice, it has not been done because we are too big.

One example of how nanotechnology is impacting our lives involves silver **nanoparticles**. Each of these spherical particles is made up of hundreds or thousands of Ag atoms. The precise method of how these particles are prepared will be discussed later. These particles range in size from 1 nm to 100 nm, and their outer surface is usually comprised of silver oxide. Figure 1.1 shows some electron microscope images of silver nanoparticle samples of various sizes.

Among many applications, silver nanoparticles show a remarkable ability to kill bacteria. Thus, they are currently being used as antibacterial and antifungal agents in a host of industries including biotechnology, textile engineering, and water treatment. Some companies are even developing coatings containing silver nanoparticles for household products and medical equipment. While use of silver nanoparticles in consumer products has clear benefits, there are also associated health concerns.

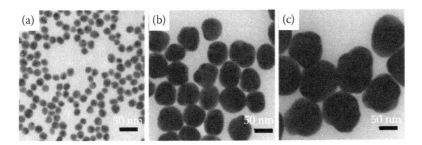

Figure 1.1 Transmission electron microscopy (TEM) images of silver nanoparticles with diameters of (a) 20 nm, (b) 60 nm, and (c) 100 nm, respectively. Scale bars are 50 nm. (Image from Sigma-Aldrich.)

These particles can accumulate in the liver if ingested, and exposure has shown them to be toxic to several organs, including the brain.

While nanomaterials offer tremendous benefit, their small scale and high activity due to large surface area provide some challenges, especially concerning human health and the environment. Figure 1.2 shows a number of nanostructures that will be discussed extensively in this book, along with applications currently in use or being developed.

1.2 THE NEED FOR NANOSCIENCE EDUCATION

Over the last decade, nanotechnology has been one of the fastest growing areas of research globally. Estimates had concluded that by 2015, over 10 million jobs worldwide would be affected by nanotechnology. In light of this, it is important to recognize that significant growth in areas of scientific research inevitably impacts education, first at the graduate level, and then at the undergraduate level. In the last decade, a healthy flow of publications has addressed the needs of graduate students and trained professionals in the field of nanoscience. Rapid research advances in areas such as soft matter, supramolecular science, and biophysical chemistry have fueled the recent surge in the number of professional journals in nanotechnology. Over 50 journals are currently publishing research in nanotechnology, and over 30 are devoted solely to nanomaterials. The impact in education is being felt as evidenced by the steady stream of articles on nanoscience education published in the *Journal of Chemical Education*, a monthly journal published by the American Chemical Society (ACS). These articles are usually aimed at the undergraduate level and cover new concepts in nanoscience, classroom demonstrations, and advanced laboratory exercises.

Figure **1.2** Examples of nanostructures and their applications. Many of these structures will be discussed in later chapters.

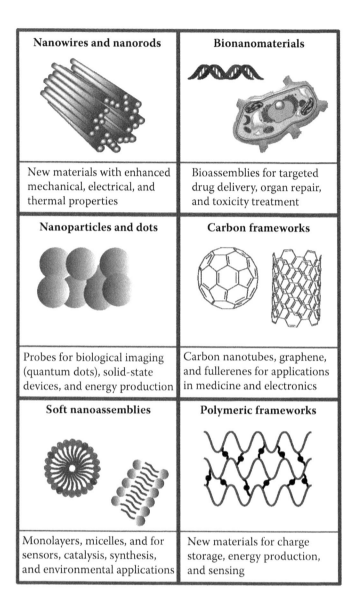

Nanowires and nanorods	Bionanomaterials
New materials with enhanced mechanical, electrical, and thermal properties	Bioassemblies for targeted drug delivery, organ repair, and toxicity treatment
Nanoparticles and dots	Carbon frameworks
Probes for biological imaging (quantum dots), solid-state devices, and energy production	Carbon nanotubes, graphene, and fullerenes for applications in medicine and electronics
Soft nanoassemblies	Polymeric frameworks
Monolayers, micelles, and for sensors, catalysis, synthesis, and environmental applications	New materials for charge storage, energy production, and sensing

Examples of well-established nanoscience publications include the Royal Society of Chemistry's *Soft Matter* and *Nanoscale*, which cover the interdisciplinary science underpinning the properties and applications of soft matter at the nanoscale, as well as the ACS journals *ACS Nano* and *Nano Letters* and other high-profile journals such as *Nature Nanotechnology* (Nature Publishing Group) and *Advanced Functional Materials*

(Wiley-VCH). Open access platforms, which do not require a subscription fee for readers to access full-text articles, have also seen a recent surge in journals devoted to nanoscience. Examples include *Applied Nanoscience* (Springer), *The Open Nanoscience Journal* (Bentham Open), and the *International Journal of Smart and Nano Materials* (Taylor & Francis Group). In addition, there has been steady growth in funding for nanoscience from both private sources and government agencies such as the National Science Foundation (NSF), Department of Energy (DOE), and National Institute of Standards and Technology (NIST). The National Nanotechnology Initiative (NNI) was first implemented in 2001 by the U.S. federal government. The aim of the initiative was to advance nanotechnology research and development, promote transfer of this technology to the consumer, support the safe and responsible development of the technology, and to develop and sustain nanoscience education. The 2014 federal budget provided more than $1.7 billion for the NNI, reflecting steady growth in the NNI investment. Since its inception in 2001, the cumulative NNI investment now totals almost $20 billion. The 2014 NNI budget supports nanoscience research at 15 agencies, with the largest investments from the DOE, NSF, National Institutes of Health (NIH), Department of Defense (DOD), and NIST.

Together with the growth in research activity and sharp increase in professional publications, this rise in funding commitment provides a compelling reason to begin serious training of our future workforce in this area. Thus, the present textbook has been developed to be accessible to undergraduate students in chemistry, physics, materials science, and engineering, as well as professionals in related fields.

1.3 THE NANOSCALE DIMENSION

The word "nano" is derived from the Latin word *nanus*, meaning "dwarf," and is often used in the context of miniaturization. It is given the abbreviation n. In the international systems of units, nano is the prefix used when multiplying a unit, such as a given length, by 10^{-9}. Thus, one can speak of a nanometer (1 nm = 1×10^{-9} m), a nanosecond (1 ns = 1×10^{-9} s), and even a nanogram (1 ng = 1×10^{-9} g). The nano term is typically used to refer to objects with length scale approaching the order of 10^{-9} m. Thus, one can speak of nanotubes, nanofossils, nanowires, and nanofilms as materials in which at least one dimension is on the order of 10^{-9} m. The term nanofabrication refers to the procedure used to construct materials with nanoscale dimension. Nanotoxicity is now a term describing the negative health consequences of some toxic nanomaterials.

To put the nanoscale in perspective, consider the size of a hydrogen atom. You may have learned from introductory physics or chemistry that the Bohr radius (the distance from the 1s electron to the central proton in hydrogen) is about 52.0 pm, or roughly 0.05 nm. This distance arguably represents the lowest limit with respect to atomic distances. Atoms and their ions vary in size between the Bohr radius and about 0.3 nm. This range represents the atomic scale. The hydrogen molecule, H_2, has a proton-proton distance (bond length) of about 0.07 nm. The much larger I_2 molecule has a bond length of about 0.3 nm, and the diameter of the benzene ring is about 0.5 nm. The size of molecules increases rapidly with structural complexity. The diameter of a DNA double helix is about 2 nm. The polyatomic molecule dodecanol, $CH_3(CH_2)_{10}CH_2OH$, also has a length approaching 2 nm. If 48 such molecules were stacked together as shown in Figure 1.3, then it is conceivable that an aggregate of length ~10 nm and height ~10 nm could be formed. Nanostructures may be comprised of thousands of molecules, resulting in aggregates on the scale of tens to hundreds of nanometers. Furthermore, macromolecular systems such as polymers and proteins have average sizes (hydrodynamic radii) approaching 10 nm; the aggregation of such molecules may result in structures on the scale of micrometers.

There is no clear boundary between what one considers the molecular scale and the nanoscale that characterizes aggregates of molecules. The distinction depends on the size of the smallest dimension of the discrete units (molecules or aggregates thereof) within the system; a nanosystem could consist of a large collection of small molecules or a few larger ones. It is sufficient to say that the field of nanotechnology or nanoscience deals with the manipulation and control of structures of a length scale with at least one dimension that is 1000 nm or smaller. The same definition holds for solid materials without a discrete covalent structure (e.g., metals and inorganic crystals). The science is fascinating because physical and chemical phenomena at these scales are markedly different from those observed in bulk (macroscopic) matter. Sometimes the difference is just a result of the much larger surface-area-to-volume ratio as particles shrink in size.

Consider a soccer ball, which has a radius of 11 cm and a C_{60} fullerene (also known as a buckyball), a spherical molecule composed entirely of sp^2 hybridized carbon atoms, which has a radius of 0.5 nm. Given that the

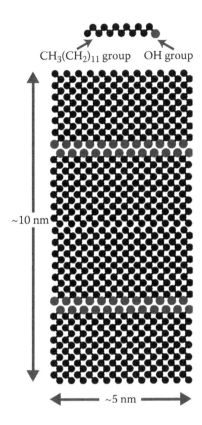

CH$_3$(CH$_2$)$_{11}$ group   OH group

~10 nm

~5 nm

Figure 1.3 An assembly of 48 dodecanol molecules forming a hypothetical aggregate of length 10 nm and width 5 nm. The filled circles represent carbon atoms and the open circle represents the oxygen of the terminal hydroxyl group of the molecule. Hydrogen atoms are not shown for convenience. This two-dimensional assembly is held together by a combination of hydrogen bonds and hydrophobic interactions (discussed further in Chapters 2 and 3).

area and volume of a sphere are, respectively

$$A = \pi r^2$$
$$V = \frac{4}{3}\pi r^3 \tag{1.1}$$

and applying simple algebra, the ratio of surface area to volume for the buckyball is over five billion times larger than for the soccer ball. This massive increase in surface area means that a far larger fraction of a nanoscale system is able to participate in chemical reactions or physical processes such as charge or energy transfer than a corresponding bulk system and that the system is heavily defined by its surface properties. Therefore, surface science plays a central role in understanding nanomaterials.

1.4 DIMENSIONALITY AND ITS IMPLICATIONS

Nanosystems are principally defined by their size (spatial dimensionality). As established in the prior section, if a system has at least one dimension in which the size is under 1000 nm, it is nanoscale. However, nanosystems can be further divided by the number of dimensions that are within the nanoscale (their spatial dimensionality). Silver nanoparticles and buckyballs are both within the nanoscale along all dimensions. Since none of their dimensions exceed 1000 nm, they are examples of zero-dimensional (0D) nanomaterials. Correspondingly, a carbon nanotube (a long tube composed of sp^2 hybridized carbon) is within the nanoscale along two dimensions (the girth of the tube) but may exceed it along the length of the tube; they are one-dimensional (1D) nanomaterials. A sheet of graphene—an atomically thin layer of graphite composed entirely of sp^2 hybridized carbon atoms—is within the nanoscale along a single axis, but may stretch for distances large enough to be visible to the naked eye along its other axes. This an example of a two-dimensional (2D) nanomaterial.

By this definition, a three-dimensional (3D) material may seem impossible. How could a material have all of its dimensions exceeding the nanoscale but still be considered within the realm of nanomaterials? This contradiction can be resolved through the same mechanism by which single molecules can be considered parts of nanomaterials. A 3D nanomaterial is a bulk material composed of other nanostructures arranged in a regular pattern, where the distance between the components is within the nanoscale. For example, a collection of nanoparticles linked together by nanotubes or strands of DNA into a solid could be considered a 3D nanomaterial. Examples of nanomaterials of different dimensionalities are shown in Figure 1.4.

The concept of spatial dimensionality is related to the broader concept of **degrees of freedom**, or ways in which a particle can move. The spatial dimensionality of a system has a strong influence on its electronic properties, as discussed in Chapter 4. However, in addition to spatial dimensionality, nanomaterials can also be characterized by their internal degrees of freedom. Translational degrees of freedom involve movement through space; for example, an atom within a nanotube is only able to

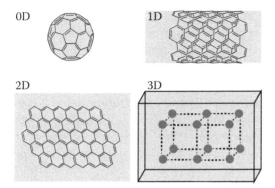

Figure 1.4   Examples of nano-materials of different spatial dimensionalities, including a fullerene (0D), a carbon nanotube (1D), a sheet of graphene (2D), and a nanoparticle lattice (3D). The gray rectangles extending beyond the system indicate the directions in which the system repeats to form an extended structure.

move along the long axis of the nanotube and exhibits 1D translational dimensionality. Similarly, rotational dimensionality, or rotational degrees of freedom, quantifies the extent to which a molecule or nanostructure may change the direction in which it is oriented. For example, a molecule bound to a surface or confined between two surfaces is only able to rotate within a plane and exhibits 2D rotational dimensionality, with the latter also exhibiting 2D translational dimensionality. The concept of 1D rotational dimensionality is less intuitive; it refers to a structure that may only be pointing one of two directions along a single axis (e.g., up or down). Finally, systems have vibrational degrees of freedom, which can store significant amounts of energy and which can be examined using techniques such as infrared and Raman spectroscopy.

1.5 SELF-ASSEMBLY

In general, there are two strategies for designing nanoscale systems. A "top-down" approach describes constructing a material of interest by breaking down a larger structure in an organized manner or by using larger tools to organize molecules. Nanoscale materials can be carved into shape by physical methods such as lithography or placing molecules on a surface using the tip of an atomic force microscope (AFM, discussed in Chapter 8) or similar instrument. In contrast, a "bottom-up" approach describes forming a nanosystem from smaller subunits (e.g., molecules in solution), driven by the physical and chemical properties of the subunits themselves. While nanomaterials can be formed by both top-down and bottom-up approaches, nanotechnology more often refers to being able to synthesize and manipulate materials using a bottom-up approach. The bottom-up approach may take advantage of specific chemical reactions or may involve intermolecular interactions between molecular fragments.

Self-assembly, which is one of the important terms used in nanoscience, describes a process in which a collection of disordered building blocks (molecules or nano-objects) come together to form an organized structure. One common example is the crystallization of small ions into a definite lattice structure. Techniques such as x-ray diffraction show that such highly organized structures are comprised of a repetition of smaller "unit cells" of nanoscale dimension. Self-assembled structures may also result from just a few molecules, but the cooperative interplay between these molecules within the structure may impart a specific function or property.

It is also worthwhile to note that self-assembly may lead to discrete or extended entities. A discrete entity is well defined in terms of the number of molecules it contains. A simple example is the dimer that is formed when two nucleic acids interact through hydrogen bonding to form a DNA base pair (Figure 1.5). Another example, which is discussed in detail in later chapters, is a micelle—a stable, often spherical aggregate of a few hundred amphilic molecules, which have one polar/hydrophilic end and one nonpolar/hydrophobic end; the hydrophobic ends cluster together to minimize energy and surface area. Extended nanosystems are undefined in scale in at least one dimension; they can be either 1D, 2D, or 3D nanosystems. Polymers (1D) and thin films (2D) are examples of extended nanosystems. Nanotechnology exploits both discrete and extended systems to perform specific functions.

Self-assembled systems are commonly found in nature as well as in engineered materials. The folding of proteins and other biomacromolecules and the formation of lipid bilayer cell membranes are examples of natural self-assembly in biology. Artificial methods may involve covalent building blocks based on chemically robust starting materials or may

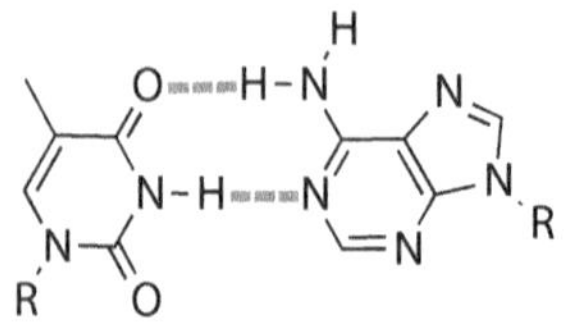

Figure 1.5   A dimer of the nucleic acid bases adenine and thymine, two of the key components of DNA. The dashed lines represent hydrogen bond interactions (discussed in Chapter 5) between the carbonyl oxygen and amine hydrogens. The dimer is an example of a discrete entity, which could form a larger nanostructure as part of a DNA double helix.

exploit the shapes, sizes, and intermolecular interactions of these building blocks to direct the formation of nanostructures. The interactions that drive self-assembly processes will be discussed in detail in Chapter 5.

Self-assembly processes can be static or dynamic. Static self-assembly describes the irreversible formation of a stable structure. Examples of static self-assembly include the growth of nanotubes and nanoparticles as well as covalent polymerization. Dynamic self-assembly describes a reversible process such as the weak noncovalent adsorption of molecules onto a surface and oscillating chemical reactions. Other reversible processes include the formation of the double helix of DNA and the folding of polypeptide chains into a protein molecule. The next two chapters discuss these two processes in more detail in the context of thermodynamics and kinetics.

1.6 SUPRAMOLECULAR SCIENCE

Self-assembly in nanomaterials falls into the broader field of supramolecular science. Supramolecular science refers to the branch of science that focuses on systems composed of a discrete number of molecular subunits (molecular building blocks), such as large, complex molecules, clusters of molecules, or small molecules bound to cavities, pockets, or other active sites in larger molecules. Figure 1.6 illustrates examples of both a single entity and aggregated supramolecule. Generally speaking, the spatial organization of the building blocks is influenced by reversible weak interactions such as hydrogen bonds, van der Waals interactions, and electrostatic forces (Chapter 5). Although irreversible interactions such as covalent bonds may also play a vital role, supramolecular chemistry is concerned mainly with noncovalent interactions.

Supramolecular science is important in a host of processes such as protein folding, molecular recognition, self-assembly, and host-guest chemistry. The hybridization of single-stranded DNA in solution to the double-stranded form is driven by hydrogen bonds being formed between base pairs—this process results in an extended supramolecular entity.

Supramolecular science and the study of noncovalent interactions touches every scientific discipline from biology (e.g., biological cell structure, protein-protein interactions, drug delivery using nanovehicles), chemistry (e.g., colloid stability, micellar nanoreactor synthesis), and physics (e.g., organic photovoltaic systems, holography, optical coatings,

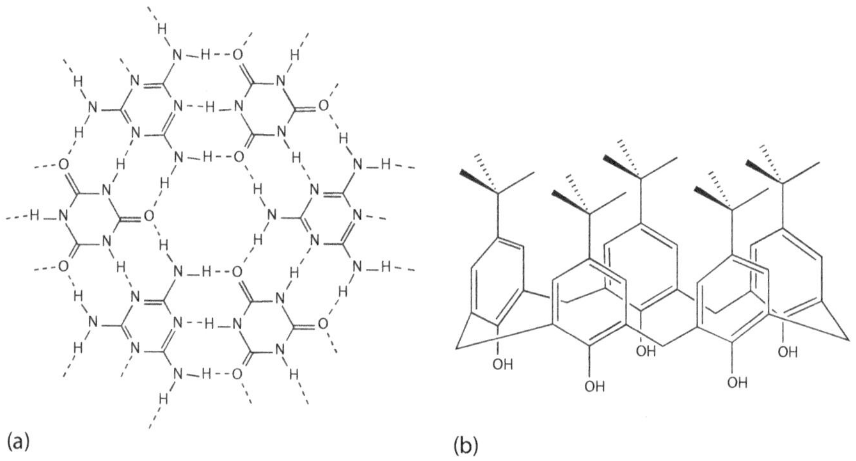

Figure 1.6 Example of an aggregated (a) and single-entity (b) supramolecular structure.

(a)   (b)

data transmission and storage), to engineering (e.g., tertiary oil recovery, large-scale synthesis of nanowires) and environmental science (e.g., remediation on nanopores, detection of hazardous materials on nanofilms). For example, an extended entity such as a thin film containing an antigen on the surface may be used to detect the presence of a specific antibody. The complexation of the antigen and the antibody, driven by noncovalent interactions, may result in a supramolecular "bilayer."

1.7 OVERVIEW OF THE TEXT

This introductory book is composed of four units. The first unit encompasses fundamental physical chemistry required for understanding nanomaterials, including thermodynamics (Chapter 2), chemical kinetics and transport (Chapter 3), and basic quantum mechanics (Chapter 4). The second unit focuses on the interactions that determine the structure of nanomaterials, including noncovalent interactions (Chapter 5) and methods for characterizing nanomaterials in bulk environments (Chapter 6). The third unit concerns interfaces, including the interaction of molecules with surfaces (Chapter 7) and instrumental techniques for characterizing nanomaterials on surfaces (Chapter 8). The final unit consists of two chapters focusing on synthesis and applications of nanomaterials, further focusing on nanomaterials for optical and electronic applications (Chapter 9) and on thin films (Chapter 10). The book captures the interdisciplinary nature of this field and attempts to provide a well-balanced approach to teaching nanoscience, though with a perspective based in physical chemistry and a particular focus on "soft" nanomaterials composed of noncovalent assemblies of organic molecules. The book also showcases nanomaterials research originating from the chemistry, biology, physics, medicine, and engineering communities. In particular, the book emphasizes, wherever possible, topics of current global interest (energy, the environment, and medicine). Recommended preparation includes at least one year each of college-level chemistry, physics, and calculus; prior introductory coursework in biology is also useful.

1.8 SOURCES OF INFORMATION ON NANOSCIENCE

Each chapter of this book ends with a short list of relevant further reading. These books, review articles, and papers have been selected in part due to their clarity, depth, and manageable mathematical rigor. They should be accessible reading material once the student has completed relevant chapters in this text. However, some advanced books are also listed for graduate students, professionals, and ambitious undergraduates.

One of best sources of up-to-date information on nanoscience can be found by visiting the NNI website (www.nano.gov). The site posts some of the most groundbreaking developments in nanotechnology and provides funding information and educational resources. Readers are also encouraged to browse scientific journals for up-to-date information on nanomaterials. The following journals published by the American Chemical Society are highly recommended: *Nano Letters*, *Langmuir*, *Journal of Physical Chemistry*, *Biomacromolecules*, *ACS Nano*, *Applied Materials and Interfaces*, and *Chemistry of Materials.* Other useful journals include *Advanced Materials*, *Advanced Functional Materials* (Wiley-VCH), *Thin Solid Films*, *Nano Today*, *Nanomedicine* (Elsevier), *Soft Matter*, *Nanoscale* (Royal Society of Chemistry), *Nature Nanotechnology*, and *Nature Materials* (Nature Publishing Group).

The following books are additional recommended starting points for students beginning their education in nanoscience.

Deffeyes, K. S. and Deffeyes, S. E. *Nanoscale: Visualizing an Invisible World*, 2009, Massachusetts Institute of Technology, Cambridge, MA. This is a beautifully illustrated book containing many examples of nanomaterials. The book describes the local structure of materials at the nanoscale. It is an excellent read for those beginning their studies in nanoscience.

Jones, R. A. *Soft Machines: Nanotechnology and Life*, 2008, Oxford University Press, New York. Although not mathematically rigorous, this book does an excellent job of presenting fundamental physical laws governing nanoscience.

Pacheco, K. A. O., Schwenz, R. W., Jones, W. E. Jr., Eds. *Nanotechnology in Undergraduate Education*, 2009, ACS Symposium Series 1010. This is an excellent book describing course and curriculum innovations as well as nanoscience laboratory experiences. It is highly recommended for those interested in developing an undergraduate course in nanoscience.

Ratner, M. A. *Nanotechnology: A Gentle Introduction to the Next Big Idea*, 2002, Prentice Hall, Upper Saddle River, NJ. This book focuses on the technical and business aspects of the field. It provides a broad perspective on the subject, from science and economics to ethics.

Understanding Nanotechnology from the editors of *Scientific American*, 2002, Warner Books, New York. This is a popular science book that does a good job of describing the technological implications of nanoscience. The mathematical and scientific background is limited in this book, so it is an accessible overview of the field and nanoscience terminology.

END OF CHAPTER QUESTIONS

1. A material is composed of rodlike discrete structures that have a length of 5.25×10^{-6} m and diameter of 4.50×10^{-8} m.

   a. Express both of these numbers in nm. Determine the cross-sectional area and the volume of the structure in nm and nm, respectively.

   b. Is the material a nanomaterial? If so, what is its dimensionality? Explain your reasoning.

2. Determine the surface-area-to-volume ratio of spherical nanoparticles with diameters 10 nm, 100 nm, and 1000 nm. Do you see a pattern?

3. Consider a sample of silver nanoparticles with an average diameter of 100 nm per particle.

   a. By what factor does the surface area per particle increase if the diameter doubles?

   b. How many Ag atoms make up a typical nanoparticle? Ag has an atomic radius of 144 pm.

   c. Calculate the total surface area of a sample comprised of 100 nanoparticles.

   d. Calculate the total surface area of a 5 g sample of this powdered nanomaterial.

e. Compare the surface area you calculated in part (d) with that of 5 g of solid silver. The density of silver (and other materials) can be obtained from many reliable online databases.

4. Consider the following materials. What is their spatial dimensionality? Write a sentence justifying your assessment of

   a. A molecule of double-stranded DNA.

   b. A ZIF-8 metal-organic framework, which can be used for adsorbing different kinds of gas molecules.

   c. A self-assembled monolayer of dye molecules anchored by thiol groups to a gold surface.

   d. Tobacco mosaic virus, a RNA virus that has been used as a structural model for ordering of molecules and considered as a possible nanomaterial component due to its unusual structure.

   e. A lead sulfide quantum dot.

5. The vast majority of science publishing now occurs online, whether through traditional subscription-based journals or online-only journals that presents papers as soon as they have been peer-reviewed and accepted. The change in publication model from paper to online and the sheer volume of publications has led to the development of new methods of quickly finding publications or works by particular researchers. The digital object identifier (DOI) is now ubiquitous and provides an unambiguous way of finding papers or other works that is independent of bibliography style systems and can be generated before a paper has been assigned a volume and page number. Try looking up the following DOIs at http://dx.doi.org and briefly summarize (1–2 sentences) what journal they are from and what they are about.

   a. 10.1021/nl0347334.

   b. 10.1016/j.nanotod.2012.08.004.

   c. 10.1126/science.1083842.

CHAPTER 2

Thermodynamics and Nanoscience

OVERVIEW

Before we delve into the structure and function of nanomaterials, we need to understand the fundamental **thermodynamics** and **kinetics** that control their formation and reactivity. Thermodynamics is concerned with heat and work and explains, on a macroscopic level, relative stability and the feasibility of change. Kinetics, on the other hand, deals with experimental factors governing reaction rates. Changes in rate with respect to the concentrations of different materials involved in the reaction often give insight into the mechanism of a chemical process. In order to understand the kinetic and thermodynamic factors governing processes such as self-assembly, phase behavior, and chemical reactivity in nanosystems, we need to review some of the well-established ideas pertaining to macroscopic bulk phase matter. This chapter provides a concise review of equilibrium thermodynamics, while Chapter 3 discusses kinetics. These crucial topics serve as an important prelude to material covered in subsequent chapters, where these ideas are extended to materials on the nanoscale. It should be appreciated that thermodynamics and kinetics are enormous areas of study, and therefore we have been selective in covering concepts most relevant to the study of nanomaterials.

2.1 TERMINOLOGY IN THERMODYNAMICS

2.1.1 The system and surroundings

In thermodynamics, the specific region of interest (often the material sample) is always denoted as the **system**. Everything else is referred to as the **surroundings**. Energy, such as heat and/or matter, may transfer between the system and surroundings. We say the system is **isolated** if no such transfer of heat or matter can occur. An **open** system is one that exchanges both energy and matter with its surroundings. However, if transfer of energy but not matter is allowed, then we refer to the system as being **closed**. The earth is an example of a (nearly) closed system, where electromagnetic energy is exchanged between the earth and its surroundings but very little matter is exchanged. Actual systems are rarely *completely* open or isolated, but can be approximated as such to simplify the theoretical treatment of the system and distinguish systems with greatly different degrees of heat or matter transfer. For example, coffee in even a very well designed thermos (which can be approximated as an isolated system) will eventually cool to room temperature, but it will happen *far* slower than in a paper cup (open system).

Dynamic processes, such as a change in physical state or a chemical reaction, can be described by the type of interaction that occurs between the system and the surroundings. If such a process occurs with no heat transfer between the system and surroundings, we call that process **adiabatic**. An **isothermal** process is one where there is no change in temperature of the system or the surroundings. A process occurring under conditions of constant pressure is an **isobaric** process. Again, these are often ideal situations and, for example, an adiabatic process may in reality occur with a slight transfer of heat to or from the surroundings, such as when a gas cylinder becomes cold to the touch when vented rapidly. We use terminology such as open, closed, isolated, isothermal, adiabatic, and isobaric to carefully define the state of a system and/or delineate the nature of the process involving the interaction between the system and its surroundings.

2.1.2 Some thermodynamic variables

The thermodynamic state of a system is specified by a set of variables. For example, the change in the internal energy of a system can be expressed in terms of volume and entropy changes. The most important variables in our treatment of thermodynamics are temperature (T), pressure (P), enthalpy (H), entropy (S), internal energy (U), and volume (V). Some of these variables are **extensive** quantities (i.e., the property is proportional to the size of the system. Examples of extensive variables include mass, volume, energy, entropy, Gibbs energy (G), heat capacity (C), and number of moles (n). All these depend on the amount of material present. In contrast, **intensive** variables do not depend on the amount of material present. Examples of intensive variables include density, concentration, temperature, and molar heat capacity. Dividing an extensive property by a different extensive property will generally yield an intensive value. An important example is the intensive quantity molar volume ($\bar{V}$), given as volume (extensive) divided by the number of moles (extensive) (Equation 2.1):

$$\bar{V} = \frac{V}{n} \qquad (2.1)$$

Example 2.1   The Molar Volume of Gold Nanoparticles

The density of a 100-nm gold nanoparticle is 19.30 g/cm^3. Determine its molar volume in liters per mole.

Solution   The units of molar volume are L/mol. Therefore $\bar{V}$ is related to the inverse of density:

$$\frac{1}{19.30 \text{ g/cm}^3} = 0.052 \text{ cm}^3/\text{g}$$

The relative atom mass of Au is 197 g/mol:

$$\therefore \left(0.052 \text{ cm}^3/\text{g}\right) \times (197.0 \text{ g/mol}) = 10.24 \text{ cm}^3/\text{mol}$$

Since there are 1000 cm^3 in 1 L,

$$\bar{V} = \frac{10.24 \text{ cm}^3/\text{mol}}{1000 \text{ cm}^3/\text{L}} = 0.010 \text{ L/mol}$$

Before moving on, it is worth discussing the thermodynamic variable **internal energy** (or the corresponding intensive variable *molar* internal energy). Internal energy is the energy contained within the system and results from the thermal motion of molecules; for an ideal gas, it depends only on temperature. More complex molecules and nanostructures have additional contributions from rotational and vibrational motion depending on their number of degrees of freedom. An increase in the temperature of a system results in a larger internal energy of the system.

As a simple example of a thermodynamic system, let's consider a simple piston such as that shown in Figure 2.1. The piston encloses a gas at some initial pressure P_i, which comprises the system. Let's say the system absorbs heat from the surroundings. The absorption of heat will increase the temperature and therefore the internal energy of the gas by increasing the thermal motion of the gas particles. The increase in temperature results in an increase of the pressure within the cylinder. This process leads to a new state having a final pressure P_f. In Section 2.3 we will examine the energy changes that occur due to this expansion process.

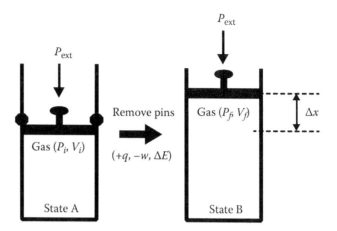

Figure 2.1   The isothermal irreversible expansion of a gas inside a piston. When the pins are removed in State A, the gas expands and does work against an external pressure P_{ext} until it reaches the final State B. In State B, the gas pressure inside the piston is equal to P_{ext}.

2.1.3 Reversible and irreversible changes

Changes to matter may be either reversible (proceeding through a series of tiny changes) or irreversible (proceeding through large, sudden changes). Let's consider a process when one phase (e.g., ice) changes to another phase (e.g., water). Under certain conditions phase changes are irreversible. The melting of a 200-nm-wide gold solid nanoparticle at 1100°C is an example of an irreversible process. The particle's melting point is around 1064°C, and above this temperature it spontaneously melts to the liquid phase. An irreversible change is one that proceeds in one direction and the final state does not revert back to the original. Examples of these processes include the sublimation of solid iodine at 30°C, the mixing of two gases, and the expansion of a gas into a vacuum.

Returning to our Au nanoparticle, we observe that at the melting point (1064°C) the two phases (solid and liquid) coexist with constant amounts of each phase. This is an example of physical equilibrium. At the molecular level, Au atoms from the solid nanoparticle are constantly entering the liquid phase, and Au atoms from the liquid are constantly entering the solid phase. Therefore, at the microscopic level the system is dynamic and reversible, but at the macroscopic level nothing appears to be changing. A chemical equilibrium reaction is another type of reversible change. As an additional example, let's consider an important reaction leading to the formation of silica (SiO_2) nanoparticles. Under certain conditions we have the reaction in which silicic acid converts to silica nanoparticles:

$$Si(OH)_4(aq) \rightleftharpoons SiO_2(aq) + 2H_2O(aq)$$

The double arrow indicates the reaction is reversible. When equilibrium is reached we have constant concentrations of reactants and products despite the fact that the forward and reverse reactions are still occurring on the microscopic level. In general chemistry you learned the effect of concentration of each species in the reaction, and external factors, such as pressure and temperature, on the position of **chemical equilibrium**.

In general, a system reaches equilibrium when the system has no effect on the external environment despite not being isolated from it. For example, a system may reach mechanical equilibrium once there's no change in pressure between the system and its surroundings. The most common types of equilibria include thermal equilibrium (the system and its surroundings are at the same temperature), chemical equilibrium (the chemical composition of system does not change with time), and phase equilibrium (the amount of each phase does not change with time). We will return to physical and chemical equilibria in Section 2.6.

2.2 TEMPERATURE AND NANOMATERIALS

2.2.1 Thermodynamics and size effects

In classical thermodynamics the system of interest is macroscopic and composed of a huge number of atoms. Temperature, T, is defined for the entire system based on its average kinetic energy. The temperature of individual (or a small group of) atoms and molecules within the system is not defined. The equations of thermodynamics are generally valid for systems composed of a large number of particles. What about defining the temperature of nanomaterials, which may be composed of only hundreds or thousands of atoms? Is there some limit below which temperature becomes undefined?

Let's first consider the relative temperature fluctuations in a cubic particle of length L composed of N atoms per unit volume (Equation 2.2):

$$\frac{\Delta T}{T} = \sqrt{\frac{1}{NL^3}} \qquad (2.2)$$

We define the relative temperature fluctuation at T as $\Delta T/T$, where ΔT is the range of temperature that deviates from T. We will consider classical thermodynamics to be valid, when relative temperature fluctuations are below 1% (0.01), (Wautelet and Shirinyan, 2009).

Example 2.2   Relative Temperature Fluctuation and Nanosystems

Silver has a density of 10.5 g/cm^3 and has an atomic mass of 107.86 g/mol, yielding a number density of 6×10^{30} atoms/m^3. Consider a nanomaterial of length L composed of $N = 10^{30}$ atoms per cubic meter. Assuming we accept a relative temperature fluctuation less than 0.01, determine the minimum size of the nanomaterial for thermodynamics to be accepted.

Solution $\Delta T/T$ = 0.01. Rearranging Equation 2.2 for L gives

$$L = \left[\frac{1}{N} \left(\frac{T}{\Delta T} \right)^2 \right]^{1/3} = \left[\frac{1}{6 \times 10^{30}} \left(\frac{1}{0.01} \right)^2 \right]^{1/3} = 1.19 \times 10^{-9}$$

In other words, L must be larger than about 1 nm for thermodynamics to be valid.

Example 2.2 illustrates that our treatment of thermodynamics is suitable for nanoscale systems as long as the particle number is sufficiently large. However, there are some other considerations. Since $\Delta T/T < 0.01$ for classical thermodynamics to be valid, at room temperature ($T = 290$ K), $\Delta T < {\sim}3$ K. Melting processes occur over this range (ΔT), and since the range is relatively large, this implies that such transitions may disappear in some nanoscale systems.

2.2.2 Melting temperatures of nanomaterials

In Section 2.1.2, we stated that volume is an extensive property, but that density is an intensive property since it does not depend on the number of particles or the size of the system. These definitions do not strictly hold at the nanoscale, where the molar volume (and its inverse, density) depends on the size of the nanoparticle. This arises from the fact that nanosystems have large surface-area-to-volume ratios and this ratio decreases sharply as the size increases. Thus, an important ratio affecting thermodynamic and other properties is N/n, where N is the number of surface particles and n is the total number of particles comprising the nanomaterial. Figure 2.2 shows the small density changes for Au nanoparticles going from 50 nm to 150 nm. For comparison, the density of bulk Au is 19.32 g/cm^3.

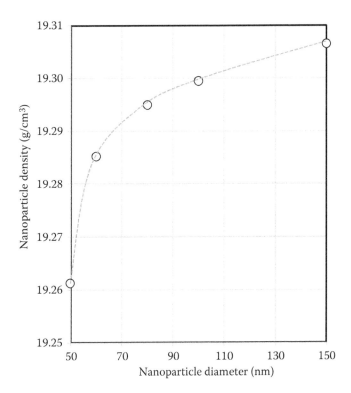

Figure 2.2 The change in the density of Au nano-particles as a function of particle diameter.

It should also be noted that the gradual increase shown in Figure 2.2 is not observed below 150 nm, where the variation shows no discernible trend.

Another very interesting observation is that the melting temperature of nanoparticles increases with size and eventually levels off to a value corresponding to the bulk phase melting temperature. This is illustrated in Figure 2.3 for gold nanoparticles, where beyond 15 nm the particles have a melting point corresponding to that of the bulk phase. To understand this behavior, we will consider the total *cohesive* energy holding n number of atoms in solid material together. The theoretical treatment given in this section is adapted from (Qi, 2005).

If N is the number of surface atoms, then the number of bulk interior atoms is n-N. We begin by defining E_o as the cohesive energy per atom of the bulk material. Since n-N is the number of bulk atoms, the contribution of the bulk atoms to the total cohesive energy of the nanomaterial (E_{total}) must be $E_o(n$-$N)$. Now let's consider the surface atoms. As shown in Figure 2.4, the number of atom-atom interactions on the surface is one half of the number atom-atom interactions in the bulk. We say that half of the total bonds (or interactions) of each surface atom are *dangling* bonds.

Figure 2.3 The melting temperature of Au nanoparticles as a function of particle diameter.

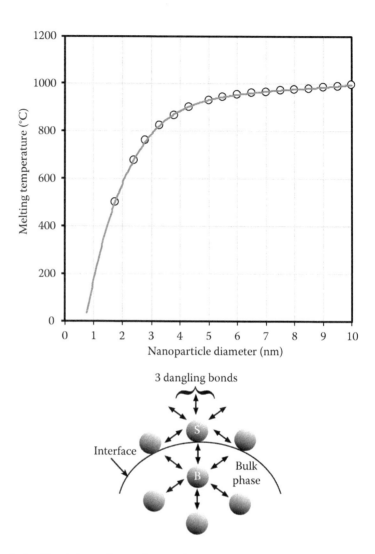

Figure 2.4 The interactions of a surface atom (S) and a bulk atom (B). The surface atom has half the number of neighboring interactions compared to the bulk phase atom. The number of dangling bonds in S is equal to half the interactions of B with its neighboring atoms.

As a result, the surface contribution to the total cohesive energy is $\frac{1}{2}NE_o$.

Therefore, the total cohesive energy of the nanomaterial can be described by Equation 2.3:

$$E_{\text{total}} = E_0(n - N) + \frac{1}{2}E_oN \tag{2.3}$$

We now multiply both sides of Equation 2.3 by N_A/n, where N_A is the Avogadro number, giving the cohesive energy per mole $(\bar{E}_n)$ of the nanomaterial. Doing this separately for the left and right sides of Equation 2.3 gives Equations 2.4 and 2.5:

$$\text{LHS:} \quad \frac{N_A}{n} E_{\text{total}} = \bar{E}_n \tag{2.4}$$

$$\text{RHS:} \quad \frac{N_A}{n} E_0(n - N) + \frac{N_A}{2n} E_0 N = N_A E_0 - \frac{N_A N}{n} E_0 + E_0 \frac{N_A N}{2n}$$
$$= N_A E_0 \left(1 - \frac{N}{2n} \right) \tag{2.5}$$

We see that $N_A E_0$ is the cohesive energy per mole of the bulk material, which we will denote as $\bar{E}_b$. Thus, the cohesive energy per mole holding the bulk solid together $(\bar{E}_b)$ will differ from that holding the nanoparticle together $(\bar{E}_n)$. The relationship between E_n and E_b is then captured in Equation 2.6:

$$\bar{E}_n = \bar{E}_b \left(1 - \frac{N}{2n} \right) \tag{2.6}$$

It should be noted that both the cohesive energies $\bar{E}_n$ and $\bar{E}_b$ measure the strength of interaction between the atoms, and both of these increase linearly with the melting temperature. Furthermore, melting temperature is also a measure of how strongly the atoms are bound to each other. As a result, Equation 2.6 can be written in terms of the melting point of the nanoparticle (T_{mpt}^n) and the melting point of the bulk material (T_{mpt}^b) (Equation 2.7):

$$T_{\text{mpt}}^n = T_{\text{mpt}}^b \left(1 - \frac{N}{2n} \right) \tag{2.7}$$

In Equation 2.7, we see that the ratio N/n is crucial in determining how different bulk and nanomaterial melting temperatures are. The ratio N/n depends on the geometry of the system. Table 2.1 summarizes this ratio for different nanomaterial geometries.

Table 2.1
The *N/n* Ratios for Various Nanostructures.

Nanosolid	*N/n*
Spherical nanosolids	$\dfrac{4d}{D}$
Disclike nanosolids	$\left(\dfrac{4}{3}\right)d\left[\left(\dfrac{1}{h}-\dfrac{2}{l}\right)\right]$
Nanowires	$\left(\dfrac{8}{3}\right)\dfrac{d}{l}$
Nanofilms	$\left(\dfrac{4}{3}\right)\dfrac{d}{h}$

Note: *D* is the diameter of a spherical nanostructure and *d* is the diameter of the corresponding atom. For non-spherical nanostructures, *l* is the diameter (or length) and *h* is the height (or thickness).

Example 2.3   Determining the *N/n* Ratio for a Spherical Nanoparticle

The volume and the corresponding surface area of a sphere is $\dfrac{\pi D^3}{6}$ and πD^2, respectively, where D is the diameter. Show that $\dfrac{N}{n} = \dfrac{4d}{D}$ for a nanophere, where d is the diameter of an individual atom in the nanoparticle.

Solution   The volume of an atom is $\dfrac{\pi d^3}{6}$. The total number of atoms can be obtained by dividing the nanoparticle volume by the volume of a single atom. Thus

$$n = \frac{\pi D^3/6}{\pi d^3/6} = \frac{D^3}{d^3}$$

Each atom's contribution to the surface area of the nanoparticle is simply the area of its great circle (i.e., $\pi\left(\dfrac{d}{2}\right)^2 = \dfrac{\pi d^2}{4}$). To determine N, the total number of surface atoms, we divide the total surface area of the nanoparticle by the area of the great circle of the atom. Thus

$$N = \frac{\pi D^2}{\pi d^2/4} = \frac{4D^2}{d^2}$$

Therefore,

$$\frac{N}{n} = \frac{4D^2/d^2}{D^3/d^3} = \frac{4d}{D}$$

Using the expressions in the table and Equation 2.7, one can determine the predicted melting temperatures of a variety of nanostructures. In order to calculate nanostructure-melting temperature, the corresponding bulk melting temperatures are required. For example, Figure 2.5a and b show

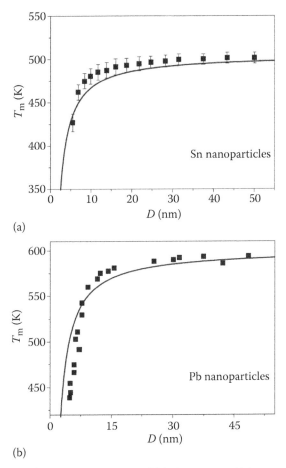

(a)

(b)

Figure 2.5 The melting temperatures of (a) Sn nanoparticles, and (b) Pb nanoparticles, as a function of particle diameter. The symbols represent experimental values and the solid lines are the calculated melting temperatures based on Equation 2.7 and Table 2.1. (Reprinted from *Physica B: Condensed Matter*, 368, Qi, W.H., Size effect on melting temperature of nanosolids, 46–50, Copyright 2005, with permission from Elsevier.)

the experimental and calculated melting temperatures of spherical Sn and Pb nanoparticles, respectively. For nonspherical nanostructures, Figure 2.6a shows the melting temperature of indium nanowires as a function of wire diameter and Figure 2.6b shows the melting temperature of an indium nanofilm as a function of film thickness.

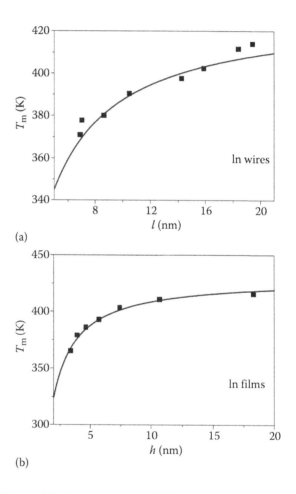

Figure 2.6 The melting temperatures of (a) In nanowires as function of wire diameter, and (b) In nanofilms as a function of film thickness. The symbols represent experimental values and the solid lines are the calculated melting temperatures based on Equation 2.7 and Table 2.1. (Reprinted from *Physica B: Condensed Matter*, 368, Qi, W.H., Size effect on melting temperature of nanosolids, 46–50, Copyright 2005, with permission from Elsevier.)

Example 2.4   Predicting the Melting Point of Spherical Nanoparticles

The atomic diameter of silver is 320 pm. Predict the melting point of a sample of Ag nanoparticles of size 100 nm. The bulk phase melting point of silver is 961.8°C.

Solution   320 pm = 0.320 nm. Using the appropriate equation for N/n from Table 2.1 (see Example 2.3), we have

$$\frac{N}{n} = \frac{4d}{D} = \frac{4 \times 0.320}{100} = 0.0128$$

Using Equation 2.7 gives us the melting temperature of the nano-particle:

$$T_{mpt}^{n} = T_{mpt}^{b}\left(1 - \frac{1}{2}\frac{N}{n}\right) = 961.8\left(1 - \frac{1}{2} \times 0.0128\right) = 955.6°C$$

2.3 THE FIRST LAW OF THERMODYNAMICS

The flow of energy is a vital characteristic of any process. Intuitively, one assumes that processes occur when the energy of the system will be lowered. For example, a rolling ball with a given amount of kinetic energy will eventually come to rest (zero kinetic energy) as it converts its kinetic energy to other forms such as heat due to friction with the surface and the heat is transferred to the surroundings. Even though the ball loses kinetic energy, the total energy of the system (ball) and surroundings (surface) remains constant. The first law is a statement of the **conservation of energy**. We will focus on three kinds of energy transfer: mechanical work, heat flow, and electrical work. In particular, we will see how exchanges among these three forms allow us understand processes in nanomaterials.

2.3.1 Work

Mechanical **work** (w) is a result of unbalanced forces. To understand this, let's consider a system comprised of a gas compressed in a cylinder with volume V_i and pressure P_i (Figure 2.1). The surrounding environment has an external pressure P_{ext}. The piston (of cross-sectional area A) pushing against the gas is held in position by pins. When the pins are removed the gas expands to its new volume V_2. The system has reached mechanical equilibrium because the new final pressure (P_f) equals the external pressure (i.e., $P_f = P_{ext}$). We say that work has been done by the system on the surroundings.

It is convention to assign this work a negative magnitude (i.e., $-w$). Conversely, when the surroundings do work on the system (a compression process) it is convention to assign this work a positive magnitude (i.e., $+w$).

We can use this system to derive an important relationship between work and pressure. From basic physics, work is defined by Equation 2.8:

$$w = -\int f(x)\,dx \tag{2.8}$$

The negative sign in the above equation is deliberately included to ensure getting the proper sign for w. In our system the work done on the surroundings during the expansion process is a product of the displacement of the piston and the constant external force (Equation 2.9):

$$w = -f_{ext}\Delta x \tag{2.9}$$

The external force is related to the external pressure and the area of the piston (A) by Equation 2.10:

$$P_{ext} = \frac{f_{ext}}{A} \tag{2.10}$$

Substituting Equation 2.10 into Equation 2.9 and using the fact that the change in volume $\Delta V = A\Delta x$ yields Equation 2.11:

$$w = -P_{ext}\Delta V \tag{2.11}$$

Example 2.5   Irreversible Expansive Work

Determine the work done when 1 mol of a gas expands from 2 L to 10 L against a constant external pressure of 1 atm at 298 K.

Solution   The change in volume is 8 L. The work done,

$$w = -P_{ext}\Delta V = -(1\,atm) \times (8\,L) = -8\,L\,atm$$

Recognizing that 1 L atm = 101.325 J, we have $w = -810.6$ J.

The above result provides the work done when the gas expands in a single-step **irreversible** process as illustrated in Figure 2.1. What would happen if we carried out the process reversibly? In order to do so, we need to understand exactly what it means to carry out the process reversibly from state A to state B.

A **reversible** expansion can be devised by first recognizing that at any point between the initial and final state, the gas must always be in equilibrium with its surroundings. The equilibrium condition in this case is that the internal gas pressure must equal the external pressure during the entire process. The reversible process must be done such that this condition is met in going from state A to state B. To understand the reversible expansion better, let's do this process in a series of steps, as illustrated in Figure 2.7. Initially (state A) the gas pressure within the cylinder is 2 atm. Tiny weights added to the top of the piston exert an external pressure of 2 atm. The system is in equilibrium. Now imagine that one of these tiny weights is removed such that the external pressure drops to 1.999 atm. The piston moves up and the internal pressure drops to 1.999 atm (State B). Again, this system is in equilibrium with its surroundings. The removal of an additional weight further reduces the external and internal pressure to 1.998 atm (State C). These steps can be repeated until the final state is reached in which the external and internal pressure is 1 atm. In true reversible processes we would have an infinite number of steps in which the pressure is reduced infinitesimally in going from state A to the final state.

Purely reversible processes are hypothetical, but like the types of ideal systems and processes discussed earlier in the chapter, they are useful approximations to real systems and allow derivation of important relationships in thermodynamics. Processes can be carried out slowly and

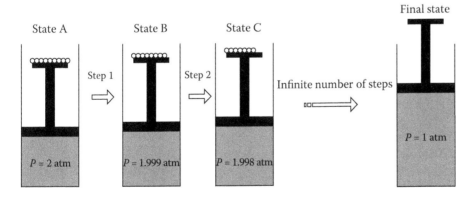

Figure 2.7   The hypothetical concept of reversibility. The removal of weights from the piston in infinitesimal amounts takes the system from its initial to final state. Conversely, infinitesimal weights could be added onto the piston to reverse the process. Each step is essentially in a state of equilibrium with the external pressure exactly equal to the internal pressure of the gas in the piston.

carefully to ensure the system remains close to equilibrium during the process. For example, the slow compression of a collection of nanoparticles over a period of hours approximates a reversible process. Let's use the reversible process to obtain the work done for gaseous expansion and compression.

For a reversible process the internal and external pressures are the same. If we approximate the gas in the cylinder as an ideal gas, then Equation 2.12 gives the external pressure.

$$P_{ext} = \frac{nRT}{V} = \frac{RT}{\bar{V}} \tag{2.12}$$

Note that pressure in Equation 2.12 is not constant but depends on V and so in order to determine work, we substitute Equation 2.12 into Equation 2.8 to yield the integral Equation 2.13.

$$w = -nRT \int_{V_i}^{V_f} \frac{1}{V} dV \tag{2.13}$$

Upon integrating Equation 2.13 from V_i to V_f we obtain the work done for the reversible process (Equation 2.14).

$$w = -nRT \ln \frac{V_f}{V_i} \tag{2.14}$$

Example 2.6   Reversible Expansive Work

Determine the reversible work done when 1 mol of a gas expands from 2 L to 10 L against a constant external pressure of 1 atm at 298 K.

Solution   Substituting the appropriate values into Equation 2.14 gives

$$w = -nRT \ln \frac{V_f}{V_i} = -(1 \text{ mol}) \times (298 \text{ K}) \times \left(8.314 \text{ JK}^{-1} \text{mol}^{-1} \right) \ln \left(\frac{10}{2} \right)$$

$$= -3987 \text{ J}$$

We see that the reversible work of expansion is much larger than the irreversible case shown in Example 2.5. In fact, the reversible process will always yield the maximum possible work for an expansion. This is illustrated in Figure 2.8 where the gray-shaded area represents the total work for the reversible process; this is the integrated area under the P–V curve as described by Equations 2.13 and 2.14. In contrast, the rectangular striped area shown in Figure 2.8a represents the irreversible work as described

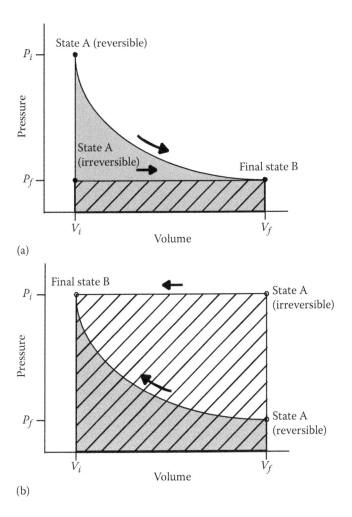

Figure 2.8 Graphical representations comparing the reversible and irreversible path for (a) expansion and (b) compression processes. The shaded areas represent the total work done for the corresponding process. Reversible expansion gives the maximum work, while reversible compression gives the minimum work.

by Equation 2.11. As illustrated in Figure 2.8b, the opposite is true for the compression process; the reversible process requires the minimum work.

As an illustration of a nanoscale scale process involving a reversible compression, let's consider a two-dimensional monolayer composed of insoluble molecules (such as dodecanol) floating on the surface of water. The two-dimensional monolayer can be compressed using barriers (Figure 2.9). In this case, we replace volume by area (A) and pressure

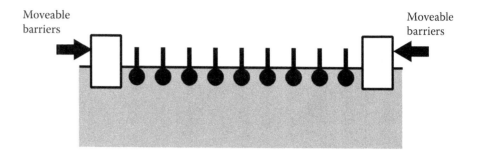

Moveable barriers   Moveable barriers

Figure 2.9 A monolayer of molecules self-assembled at the liquid–air interface. The moveable barriers can be used to compress or expand the monolayer. Doing so very slowly approximates a reversible process.

by its two-dimensional analog, *surface pressure* (Π). Thus, the corresponding equation of state is given by Equation 2.15:

$$\Pi A = nRT \qquad (2.15)$$

For a reversible compression from A_i to A_f, the work is given by Equation 2.16:

$$w = -nRT \int_{A_i}^{A_f} \frac{1}{\Pi}\, dA = -nRT \ln \frac{A_f}{A_i} \qquad (2.16)$$

In practice, the process can be performed reversibly by compressing the monolayer very slowly. We discuss this system more completely in Chapter 7.

The usefulness of work has been applied to a number of nanoscale systems. This usually involves measuring the force required to distort a system. For example, studies have been done where a strand of DNA has been stretched and the corresponding force measured (Cluzel et al., 1996). Figure 2.10 illustrates the process. The naturally extended strand of DNA (its contour length, l_o) has an end-to-end separation of about 15 nm. The distance r is defined as x/l_o, where x is the length of the stretch strand. Thus, $r = 1$ at the contour length. Figure 2.10b shows how measured force changes with r. At (i), $r < l_o$ and the force is zero, corresponding to the unstretched (or coiled) form of DNA. At (ii), stretching begins and the force increases. The flat region (iii) corresponds to a structural transition, beyond which the strand is irreversibly damaged (iv). By using our general expression for work (Equation 2.8), we can derive an expression for the work done in stretching the DNA strand from r_1 to r_2 (Equation 2.17):

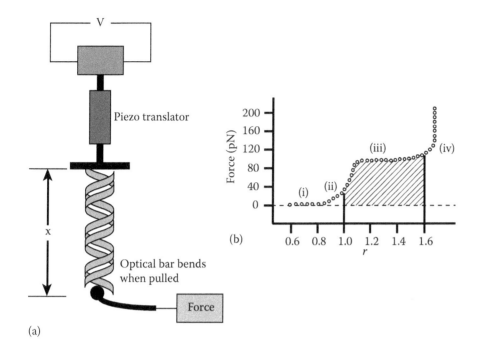

Figure 2.10 (a) The stretching of a DNA strand using a piezoelectric translator. (b) A plot of the force required to stretch the strand versus length of the strand. A detailed experimental on the use of optical tweezers to stretch DNA can be found in Smith, S. B., Cui, Y., and Bustamante, C. *Science New Series*, Feb. 9, 1996, 271(5250): 795–799.

$$ w = \int f(x)\,dx = \int f_{ext}(l_o r)\,d(l_o r) = l_o \int_{r_1}^{r_2} f_{ext}(r)\,dr \qquad (2.17) $$

Thus, as long as we can measure the integrated force between r_1 and r_2, we can determine the work done in stretching the DNA strand from r_1 to r_2 (see Problem 1).

At this point it is worth noting that work done depends on the path taken from the initial to the final state. We have clearly seen that the values of w are different depending on whether a process is carried out in a single step (irreversible) or very slowly (reversible). We say that work is a **path function**: its value depends on the path taken. This is in contrast to other functions that only depend on the difference between the final and initial state, such as energy. These are known as **state functions**.

Finally, it is worth noting that we can write the general inequality $w_{rev} < w_{irr}$, for any process. The inequality holds for expansion or compression, being mindful of the sign convention used. For example, in the expansion case, the system is doing work ($-w$), indicating maximum work for the reversible process, while the converse is true for compression.

2.3.2 Heat exchange and heat capacity

The flow of heat (designated as q) is a form of energy transfer that results from a temperature difference between the system and surroundings. By convention, a minus sign is used $(-q)$ to indicate the transfer of heat from the system to the surroundings. A positive value of q indicates the absorption of heat into the system from the surroundings. The heat capacity of a material is defined as the ratio of the heat transferred to the temperature change, or more accurately,

$$C(T) = \frac{dq}{dT} \tag{2.18}$$

The heat capacity does itself depend on temperature, but is usually considered constant within a small temperature range. The heat transferred due to a change in temperature is

$$q = \int_{T_1}^{T_2} C(T)dT \tag{2.19}$$

Heat capacity is both an extensive quantity and a path function. It can be directly measured, and like any extensive quantity, dividing by the number of moles gives the corresponding intensive quantity. It is important to recognize that molar heat capacity is a path function and values depend on whether they are measured under constant volume (C_V) or constant pressure (C_P) conditions. For solids and liquids, the difference between the two is insignificant. However, for gases the difference between C_V and C_P is approximately equal to the molar gas constant R (or 8.134 J mol^{-1} K^{-1}).

2.3.3 The first law in terms of work and heat

Now that we have discussed work and heat, let's consider processes in which both are exchanged. For illustrative purposes, let's consider the expansion of the gas shown in Figure 2.1. Clearly, the expansion involves work done on the surroundings. If the process occurs adiabatically (i.e., no heat transfer), then the gas suffers a drop in internal energy. This drop in internal energy is manifested by a decrease in the temperature of the gas. Indeed, experience shows us that gases expanding in a canister feel cool at the nozzle.

In an isothermal expansion, where the temperature of the system remains constant, heat enters the system to maintain the constant temperature of the gas. Thus, in a reversible isothermal expansion process, the work done on the surroundings is exactly equal to the heat transferred into the system. Since the temperature of the system remains constant, the change in internal energy of the system must be zero in going from the initial to the final state.

We can use these ideals to formulate a useful version of the First law of thermodynamics. If work and heat are the only forms of energy transfer in a system, then the change in internal energy of the system is given by

$$\Delta E = q + w \tag{2.20}$$

Equation 2.20 is one statement of the first law of thermodynamics. We will use it to obtain other thermodynamic relationships.

To illustrate an application of the first law, let's considers the heating (or cooling) of n moles of liquid water from T_1 to T_2 under conditions of constant atmospheric pressure. The molar heat absorbed is given by Equation 2.21:

$$q_P = n\bar{C}(T_2 - T_1) \tag{2.21}$$

If $T_2 > T_1$, q_P is positive and heat is absorbed. The corresponding work done for this process is

$$w = -P_{ext}(V_2 - V_1) \tag{2.22}$$

The volume of the water will not change very much for this process and so the work done is practically zero. Thus, according to the first law (Equation 2.20), the internal energy for the heat or cooling of water depends only on the molar heating capacity of water and the temperature difference (Equation 2.23):

$$\Delta \bar{E} = n\bar{C}_V(T_2 - T_1) \tag{2.23}$$

Since the process is occurring at constant volume (no expansive work), one only needs to know heat capacity and the temperature difference in order to determine the change in internal energy. Comparing Equations 2.21 and 2.23, we immediately see that heat absorbed or lost is equal to the change in molar internal energy,

$$q_V = \Delta \bar{E} \tag{2.24}$$

Since the heat capacity is defined as the derivative of q with respect to T (see Equation 2.19), we have our formal definition of heat capacity at constant volume (Equation 2.25):

$$C_V = \left(\frac{\partial E}{\partial T}\right)_V \tag{2.25}$$

We will use this fact to introduce another state function called **enthalpy**.

2.3.4 Enthalpy

One of the most important state functions encountered in chemistry is **enthalpy**. It is defined by the Equation 2.26:

$$H = E + PV \tag{2.26}$$

The corresponding change in enthalpy is

$$\Delta H = \Delta E + \Delta(PV) = \Delta E + P\Delta V + V\Delta P \tag{2.27}$$

At constant pressure, this equation becomes

$$\Delta H = \Delta E + P\Delta V \tag{2.28}$$

Recall that Equation 2.24 describes heat transfer in terms of a change in internal energy for a constant volume process. There is an analogous equation describing heat transfer for a process carried out at constant *pressure*. The relationship, shown in Equation 2.29, is used to define the change in enthalpy (ΔH):

$$q_P = \Delta H \tag{2.29}$$

Thus, enthalpy change is simply the heat transferred in a constant pressure process.

Since the heat capacity is defined as the derivative of q with respect to T (see Equation 2.19), we have our formal definition of heat capacity at constant pressure (Equation 2.30):

$$C_P = \left(\frac{\partial H}{\partial T} \right)_P \tag{2.30}$$

Example 2.7   Heat Capacities at Constant Volume and Pressure

Show that the difference $\bar{C}_P - \bar{C}_V$ is equal to the molar gas constant for an ideal gas at a temperature T.

Solution   Start with the fundamental definition of enthalpy:

$$H = E + PV$$

For an ideal gas, we can replace PV with nRT. Thus

$$H = E + nRT$$

Differentiating with respect to T gives

$$\frac{dH}{dT} = \frac{dE}{dT} + nR$$

or

$$C_P = C_V + nR$$

dividing both sides by the number of moles and rearranging gives

$$\bar{C}_P - \bar{C}_V = R$$

2.4 THE ENTROPY STATE FUNCTION: THE SECOND AND THIRD LAWS

2.4.1 The classical interpretation of entropy

Entropy is a thermodynamic state function first introduced by Rudolph Clausius in the mid-1800s based on the working of heat engines. In order to understand entropy, let's consider the reversible isothermal expansion of a gas from some initial volume to some final volume. Since the process is isothermal, the temperature of the system is constant. As a result, heat must flow from the surroundings and into the system to maintain the temperature of the system. Since the process is carried out reversibly, we will call this heat flow q_{rev}. If we divide this q_{rev} by the temperature of the system, we end up with a new state function. We call this new state function **entropy** and describe it by Equation 2.31:

$$\Delta S = \frac{q_{\text{rev}}}{T} \tag{2.31}$$

We know that $\Delta E = 0$ for an ideal gas if $\Delta T = 0$. From the first law

$$\Delta E = q + w = q_{\text{rev}} + w_{\text{rev}}$$

and because $w_{\text{rev}} < w$, we have

$$q - q_{\text{rev}} = -(w - w_{\text{rev}}) < 0$$

The above implies $q < q_{\text{rev}}$, and thus from Equation 2.31 we get

$$\Delta S = \frac{q_{\text{rev}}}{T} > \frac{q}{T}$$

For the isothermal expansion of a gas, we know that $\Delta E = 0$ and so according to the first law (Equation 2.20), $w_{\text{rev}} = -q_{\text{rev}}$, and since w_{rev} is known (Equation 2.14), we have

$$q_{\text{rev}} = nRT \ln \frac{V_f}{V_i} \tag{2.32}$$

and so according to Equation 2.31,

$$\Delta S_{\text{system}} = nR \ln \frac{V_f}{V_i} \tag{2.33}$$

Equation 2.33 gives the entropy change in our system (ΔS_{system}) due to the reversible expansion of a gas from some initial volume V_i to final volume V_f. Since ΔS is a state function, its value does not depend on the path taken. This means that ΔS has the same value if the process occurred as a single-step irreversible expansion.

When we talk about entropy we must consider entropy changes in both the system and the surroundings, because if heat enters the system then it must leave the surroundings and vice versa. Therefore, the total change in entropy must be given by the changes that occur in both the system and the surroundings (Equation 2.34):

$$\Delta S_{\text{total}} = \Delta S_{\text{system}} + \Delta S_{\text{surroundings}} \tag{2.34}$$

Example 2.8   Entropy Change of Expanding Gas

Determine ΔS_{system}, $\Delta S_{\text{surroundings}}$, and ΔS_{total} when an ideal gas expands isothermally and reversibly from 1 L to 10 L.

Solution   Let's consider the system and the surroundings separately.

System: $\Delta S_{\text{system}} = (1\,\text{mol})(8.314\,\text{Jmol}^{-1}\text{K}^{-1}) \ln\left(\dfrac{10\,\text{L}}{1\,\text{L}}\right) = 19.14\,\text{JK}^{-1}$

Surroundings: The surroundings have lost a quantity of heat equal to exactly $-q_{\text{rev}}$. Therefore,

$$\Delta S_{\text{surroundings}} = -\frac{q_{\text{rev}}}{T} = -nR \ln\left(\frac{V_f}{V_i}\right)$$

$$= -(1\,\text{mol})(8.314\,\text{Jmol}^{-1}\text{K}^{-1}) \ln\left(\frac{10\,\text{L}}{1\,\text{L}}\right) = -19.14\,\text{JK}^{-1}$$

Total: $\Delta S_{\text{total}} = \Delta S_{\text{system}} + \Delta S_{\text{surroundings}} = 0$

What is the physical meaning of ΔS? We can examine ΔS to see how it changes for various processes. In the example above, we've already seen that it's zero for a reversible process (for the total system and surroundings). Let's examine some other processes. We will not go through the details here, but rather focus on the findings. It turns out that the entropy change is positive when a gas expands into a larger volume in an isolated system or when two different gases mix in an isolated system. Furthermore, the entropy change is positive when heat flows from a hot body to a cold body (where the two are in contact and together form an isolated system), and $\Delta S = 0$ when the two bodies reach the same temperature (thermal equilibration). In other words, the entropy change is always positive for spontaneous (irreversible) processes, and is zero when the system reaches equilibrium. The entropy change is negative for nonspontaneous processes. We can summarize these facts as

$$\Delta S > 0 \; \textit{Spontaneous} \, \text{process}$$

$$\Delta S = 0 \;\; \textit{Reversible} \, \text{process}$$

$$\Delta S < 0 \; \textit{Nonspontaneous} \, \text{process}$$

If we consider the universe as an isolated system, then any natural irreversible process will tend to an increase in entropy. One common statement of the **second law of thermodynamics** is that the energy of the universe is constant but the entropy is tending to a maximum. A more formal assertion of the second law states that for any change in the thermodynamic state of a system, the entropy change of the system is

$$\Delta S = \frac{q}{T} \; \text{for } \textit{a reversible process}$$

$$\Delta S > \frac{q}{T} \; \text{for } \textit{an irreversible process}$$

2.4.2 The statistical interpretation of entropy

Equation 2.33 tells us that an expansion of a gas will always have a positive entropy change. From quantum theory and statistical mechanics we know that a particle has access to more quantized translational energy states when it occupies a larger volume. We discuss this in more detail in Chapter 4, and the result tells us that a molecule is able to "spread" its energy over a larger volume. Entropy can be interpreted as this spreading of energy. In fact any increase in the degrees of freedom of a molecule or material corresponds to a higher entropy state. For example, H_2O has more ways of vibrating than H_2 and so has more *vibrational* entropy.

As a simple example illustrating the increase in entropy with volume, consider a thin nanofilm on a surface (Figure 2.11a). On top of the thin film we have a layer of firmly attached ions. In the bulk solution we have a large

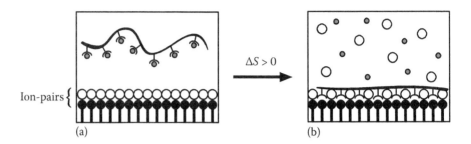

Figure 2.11   The attachment of a polymer chain onto a surface [from state (a) to state (b)] liberates a large number of ions into the solution. The increase in volume available to the ions makes this process entropically favorable.

macromolecule, such as a polymer, which has a tendency to adsorb to the surface of the thin film. When the polymer assembles on top of the film (Figure 2.11b), ions are liberated into the bulk phase where they occupy a larger volume and consequently have many more degrees of freedom. Therefore, this association process is driven by the increase in entropy when ions are displaced from localized positions on the film surface.

From statistical mechanics, we can define entropy in absolute terms rather than as a change, as shown in Equation 2.35:

$$S = k_B \ln W \qquad (2.35)$$

This is one of the most important equations in thermodynamics because it makes the connection between entropy and the disorder displayed in a system. W is the number of ways of arranging the system while keeping the total energy constant. It is, in a manner of speaking, a quantitative measure of disorder. W will always increase with temperature and so a substance's entropy will increase with temperature. Since W increases as the number of degrees of freedom increase, S will be larger for gaseous and liquid phases compared to well-ordered solid phases of the same substance.

To determine W we need to know the number of particles, N, comprising the system, and the **degeneracy**, g, of the system. Degeneracy refers to the number of arrangements that have the same energy. The relationship between W and g is given by Equation 2.36:

$$W = g^N \qquad (2.36)$$

As a simple example, let's consider a linear polar molecule with a slight positive charge at one end and a slight negative charge at the other.

Figure 2.12a shows one of two possible orientations that such a molecule may have. In this example, the "up" orientation has the same energy as the "down" orientation. We say that this system has degeneracy of 2. Furthermore, since $N = 1$, the corresponding entropy of this single particle system is

$$S = k_B \ln W = k_B \ln(g^N) = N k_B \ln g = k_B \ln 2 \qquad (2.37)$$

If we had a larger assembly of these molecules such that they formed a nanofilm comprising 11 molecules arranged as a monolayer (Figure 2.11a), we would expect W to be much larger. This nanofilm can be viewed as an ordered assembly in which some defects exist where some molecules have reversed orientations. Again, $g = 2$ in this system, but $N = 11$. Therefore, the entropy corresponding to the nanofilm is

$$S = k_B \ln(g^N) = N k_B \ln g = 11 k_B \ln 2 \qquad (2.38)$$

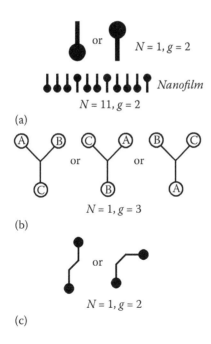

Figure 2.12 Molecular building blocks (a), (b), and (c) that can arrange themselves into well-defined orientations. The allowed orientations for each molecule have the same energy and provide the number of microstates, and hence entropy, available to the system.

In fact we can scale up the number of molecules to 1 mole ($N = N_A$) to arrive at the molar entropy of this assembly (Equation 2.39):

$$\bar{S} = k_B \ln\left(g^{N_A}\right) = N_A k_B \ln g = R \ln 2 \tag{2.39}$$

In the above equation we have recognized that the product $N_A k_B$ is equal to the molar gas constant R (8.314 Jmol^{-1}K^{-1}).

The example described in Figure 2.11a provides the inherent entropy of a material in which two states of equal energy are possible. This twofold degenerate system yields the *orientational* contribution to the overall entropy of the material. Another example of orientational entropy is that shown in Figure 2.11b in which a molecule can have one of three possible orientations. Figure 2.11c illustrates *conformational* entropy in which a molecule can adopt one of two possible conformations of the same energy.

To conclude this section we consider the situation where we have a perfectly ordered system such that $W = 1$. According to Equation 2.35, such a system will have zero entropy. We can imagine a perfect crystal at a very low temperature to approximate such a system. This is the basis of the third law of thermodynamics, which states that the entropy of a perfect crystal is zero at zero kelvin.

2.5 THE GIBBS ENERGY STATE FUNCTION

2.5.1 The direction of spontaneous change

Consider the first law of thermodynamics in which heat and work are the only forms of energy transfer (Equation 2.40):

$$\Delta E = q + w \tag{2.40}$$

The second law of thermodynamics gives the entropy change for an irreversible (spontaneous) process (Equation 2.41):

$$\Delta S > \frac{q}{T} \tag{2.41}$$

Rearranging the above expression for q gives;

$$q < T\Delta S \tag{2.42}$$

Substituting Equation 2.42 into Equation 2.40 and recognizing the fact that $w = -P\Delta V$ for an irreversible process at constant pressure leads to Equation 2.43:

$$\Delta E < T\Delta S - P\Delta V \qquad (2.43)$$

Thus in the above equation, we have deliberately included the condition of spontaneity in terms of entropy into the first law. Rearranging this equation gives

$$\Delta E - T\Delta S + P\Delta V < 0 \qquad (2.44)$$

or

$$\Delta(E - TS + PV) < 0 \qquad (2.45)$$

The variables between the brackets in Equation 2.45 represent a new state function, which we call the **Gibbs energy**, G, which we define by Equation 2.46:

$$G = E - TS + PV \qquad (2.46)$$

Thus, we see that for a spontaneous irreversible process we have the condition

$$\Delta G < 0 \qquad (2.47)$$

Instead of using Equation 2.41 for a spontaneous process, we could have used the ΔS condition for a process at equilibrium; that is

$$\Delta S = \frac{q}{T} \qquad (2.48)$$

Following the steps as before, we end up with the condition

$$\Delta G = 0 \qquad (2.49)$$

for a reversible process. In fact, we can use the sign of ΔG to state whether a particular process at constant pressure will be spontaneous or not under the given conditions. A positive value of ΔG implies a nonspontaneous process.

In examining Equation 2.46, we notice that the term $E + PV$ is the state function H (Equation 2.26). Therefore, we have

$$G = H - TS \qquad (2.50)$$

or for any chemical change at a particular temperature T,

$$\Delta G = \Delta H - T\Delta S \qquad (2.51)$$

The above equation should be familiar from general chemistry. It tells us that both enthalpy and entropy changes must be considered when predicting whether a process is spontaneous or not at temperature T.

In general chemistry, you learned that the standard enthalpy change or Gibbs energy change of a chemical reaction is often obtained from standard enthalpies (or standard Gibbs energies) of the formation of products and reactants. These ideas can be applied directly to reactions involving nanosystems. The standard enthalpy, entropy, and Gibbs energy of reaction is obtained from Equation 2.52:

$$\Delta \bar{X}_r^o = \sum \Delta \bar{X}_f^o(\text{products}) - \sum \Delta \bar{X}_f^o(\text{reactants}) \qquad (2.52)$$

where $X = H$, S, or G.

Example 2.9   Uses of Formation Data

The following reaction is relevant to the synthesis of silica nanoparticles:

$$SiCl_4(l) + 2H_2O(l) \rightarrow SiO_2(s) + 4HCl(aq)$$

Use the data in the table below to estimate the standard molar entropy change at 298 K.

	$\Delta \bar{G}_f^o$(298 K)	$\Delta \bar{H}_f^o$(298 K)
$SiCl_4$	−619.8 kJ/mol	−687.0 kJ/mol
H_2O	−285.84 kJ/mol	−237.19 kJ/mol
SiO_2	−859.4 kJ/mol	−805.0 kJ/mol
HCl	−167.46 kJ/mol	−131.17 kJ/mol

Solution   From Equation 2.52 we have

$$\Delta \bar{G}_r^o = [(-859.4) + 4(-167.46)] - [(-619.8) + 2(-285.84)]$$

$$= -337.76 \, \text{kJ/mol}$$

$$\Delta \bar{H}_r^o = [(-805.0) + 4(-131.17)] - [(-687.0) + 2(-237.19)]$$

$$= -168.3 \, \text{kJ/mol}$$

We can use these values in Equation 2.51 to solve for the standard entropy change for the reaction at 298 K:

$$\Delta \bar{S}_r^o = \frac{\Delta \bar{H}_r^o - \Delta \bar{G}_r^o}{T} = \frac{-168.3 - (-337.76)}{298} = 568 \, \text{J/mol K}$$

2.5.2 The Gibbs energy and surface tension

Now that we've discussed the Gibbs energy, we can use it to describe the thermodynamic state of a nanosystem, such as a nanofilm. Since the surface-area-to-volume ratio in a nanomaterial increases as size decreases, the surface of the nanomaterial becomes more important in determining the thermodynamic state of the material. We've discussed the work done in increasing the volume of a gas, but what is the work done in increasing the area of a surface? Clearly, work must be done to transport a bulk phase molecule to the surface and thus increase the area of that surface. This work represents the Gibbs energy change in increasing the surface area. Strictly speaking, the Gibbs energy of the surface is directly proportional to the surface area. The proportionality constant is known as the **surface tension** of the material in question. We will take a closer look at surface tension and its determination in Chapters 7 and 8. For now, let's define molar Gibbs energy of a surface by Equation 2.53:

$$\bar{G}_{\text{surf}} = \gamma \bar{A} \tag{2.53}$$

In the above expression, $\bar{A}$ is the molar area (i.e., the number of moles per unit area on the surface) and γ is the surface tension (usually expressed in N/m).

Later in this chapter, we will need an expression for the surface molar Gibbs energy in terms of molar volume of a spherical nanoparticle rather than molar area of a planar surface. In order to derive this expression, we need to examine how the volume of a spherical particle increases as the surface area increases. Equations 2.54 and 2.55 give the surface area and volume of a sphere of radius r, respectively:

$$A = 4\pi r^2 \tag{2.54}$$

$$V = \frac{4}{3}\pi r^3 \tag{2.55}$$

We need to know how V increases as a function of time and then compare the result to how A increases with time. To do this, let's first take the derivative of V with respect to time (Equation 2.56):

$$\frac{\partial V}{\partial t} = \frac{\partial V}{\partial r} \times \frac{\partial r}{\partial t} = 4\pi r^2 \frac{\partial r}{\partial t} \tag{2.56}$$

To simplify things, let's set $\partial V/\partial t = 1$ for reference. Therefore,

$$\frac{\partial r}{\partial t} = \frac{1}{4\pi r^2} \tag{2.57}$$

Now let's do the same thing for A by first taking its derivative with respect to t:

$$\frac{\partial A}{\partial t} = \frac{\partial A}{\partial r} \times \frac{\partial r}{\partial t} = 8\pi r \frac{\partial r}{\partial t} \tag{2.58}$$

Substituting Equation 2.57 into the above gives

$$\frac{\partial A}{\partial t} = 8\pi r \frac{1}{4\pi r^2} = \frac{2}{r} \tag{2.59}$$

Therefore, the change in area and volume are related by Equation 2.60:

$$dA = \frac{2}{r} dV \tag{2.60}$$

The corresponding surface molar Gibbs energy in terms of molar volume of a nanoparticle is therefore

$$\bar{G}_{\text{surf}} = \frac{2\gamma \bar{V}}{r} \tag{2.61}$$

where r is the radius of a spherical particle. We will see the importance of this equation in Section 2.4.6.

Example 2.10   Surface Gibbs Energy of a Nanosphere

Calculate the surface Gibbs molar energy of a 10-nm sized water droplet. The surface tension of water is 72.8 mN/m.

Solution $\gamma = 72.8 \times 10^{-3}$ N/m, and $r = 10 \times 10^{-9}$ m.

The density of water is $\rho = 1$ g/cm^3 which is equal to 10^6 g/m^3. The molar volume is found by dividing the molar mass of water by its density:

$$\bar{V} = \frac{M}{\rho} = \frac{18.2 \text{ g/mol}}{10^6 \text{ g/m}^3} = 1.82 \times 10^{-5} \text{ m}^3/\text{mol}$$

$$\therefore \bar{G}_{\text{surf}} = \frac{2\left(72.8 \times 10^{-3} \text{ N/m}\right)\left(1.82 \times 10^{-5} \text{ m}^3/\text{mol}\right)}{10 \times 10^{-9} \text{ m}} = 265 \text{ J/mol}$$

2.5.3 Multicomponent systems and chemical potential

The Gibbs energy of a single component will vary with temperature (T), pressure (P), and the number of moles (n) of the single component. Let's focus on how G changes with n while keeping all the other variables constant. We do this by writing the partial derivative of G with respect to n. This partial derivative is known as the **chemical potential** of the species in question and is the *partial* molar Gibbs energy of the species. It's given the symbol μ and is defined by Equation 2.62:

$$\mu = \left(\frac{\partial G}{\partial n}\right)_{P,T} \quad (2.62)$$

The concept of chemical potential becomes important when discussing a multicomponent system (i.e., a system with n_A moles of species A, n_B moles of species B, n_C moles of species C, and so on). Thus, for a multi-component system, the total molar Gibbs energy is given by the sum of their chemical potentials.

The total Gibbs energy of a system containing components A, B, C,... is given by

$$G(T, P, n_A, n_B, n_C, ...) = n_A\mu_A + n_B\mu_B + n_C\mu_C + ... = \sum n_i\mu_i \quad (2.63)$$

where the chemical potential of each component is given by

$$\mu_i = \left(\frac{\partial G}{\partial n}\right)_{P,T,n_j \neq n_i} \quad (2.64)$$

Chemical potential is a useful concept in thermodynamics in that it helps us understand how multicomponent systems evolve toward equilibrium. Chemical potential is analogous to electrical potential, where charge flows from a region of high electrical potential to a region of low electrical potential. In the case of chemical potential, transfer of matter always occurs from a region of high chemical potential to a region of low chemical potential. As an example, let's consider a nanofilm in contact with a solution containing its constituent molecules. This two-phase system is illustrated in Figure 2.13. The two phases have chemical potentials $\mu(bulk)$ and $\mu(film)$. If $\mu(bulk) > \mu(film)$, then molecules from the bulk phase will enter the film until $\mu(bulk) = \mu(film)$, at which point the two phases are in equilibrium with each other.

We will revisit chemical potential in Section 2.4.3 where we see that its value depends on the composition of the system according to Equation 2.65:

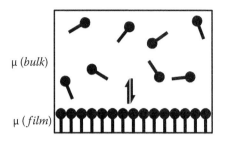

Figure 2.13   A two-phase system. When the adsorption process reaches a state of equilibrium, the chemical potential of the film phase is equal to the chemical potential of the bulk phase.

$$\mu_i = \mu_i^* + RT \ln a_i \qquad (2.65)$$

μ_i^* in the above equation refers to the chemical potential of pure component i, and a_i is the activity of component i.

2.6 PHYSICAL AND CHEMICAL EQUILIBRIA

2.6.1 The equilibrium constant

Consider the general chemical reaction

$$aA + bB \rightleftharpoons cC + dD \qquad (2.66)$$

The extent of the reaction can be described by the ratio of the product concentrations to the reactant concentrations, all raised to the power of the corresponding stoichiometric coefficient. This ratio is called the **reaction quotient**, Q, and is defined by Equation 2.67:

$$Q = \frac{[C]^c [D]^d}{[A]^a [B]^b} \qquad (2.67)$$

The value of Q changes during the reaction, but eventually levels off to a constant value when the reaction reaches equilibrium. Beyond this point the concentration of all species is constant and $Q = K$, the **equilibrium constant** (Equation 2.68):

$$K = \frac{[C]_{eq}^c [D]_{eq}^d}{[A]_{eq}^a [B]_{eq}^b} \qquad (2.68)$$

The concentrations of all species in Equation 2.68 are typically expressed in mol/L for three-dimensional condensed-phase systems. Equilibrium constants for gas-phase systems can be expressed in terms of partial pressures; the corresponding equilibrium constant, K_p, for the reaction occurring in Equation 2.68 is

$$K_p = \frac{P_C^c P_D^d}{P_A^a P_B^b} \qquad (2.69)$$

Treating each gaseous species as an ideal gas, the molar concentration can be related to the partial pressure (Equation 2.70):

$$P = \left(\frac{n}{V}\right)RT = CRT \qquad (2.70)$$

Substituting Equation 2.70 into Equation 2.69 and simplifying yields

$$K_p = \frac{(C_C^c)(C_D^d)}{(C_A^a)(C_B^b)}(RT)^{(c+d)-(a+b)} \qquad (2.71)$$

which provides the relationship between K and K_p (Equation 2.72):

$$K_p = K(RT)^{\Delta n} \qquad (2.72)$$

When expressing concentrations and partial pressures in equilibrium constants, we often divide the values by some reference value. For instance, the partial pressure of component i is referenced to a pressure of 1 atm. Likewise, the concentration of component i is always divided by 1 mol/L. These quantities are known as **activities** (a_i), and being unitless, ensure that K and K_p are unitless (Equation 2.73):

$$K = \frac{a_C^c a_D^d}{a_A^a a_B^b} \qquad (2.73)$$

2.6.2 Heterogeneous equilibria

A heterogeneous system, where more than one phase is involved in the reaction, is of particular importance to nanosystems. For example, solid phase nanoparticles are often synthesized in a liquid phase solvent system. A simple case of a heterogeneous equilibrium system is the vaporization of liquid water to gaseous water at its boiling point.

The vapor pressure of water is constant regardless of the amount of liquid water present. In fact the amount of liquid does not affect the equilibrium, and for this reason it is not included in the expression for the equilibrium constant. It is a general rule that in heterogeneous systems, the activity of pure solids and liquids is always 1. In our example of the $H_2O(l) \rightleftharpoons H_2O(g)$ equilibrium, $K_p = P_{H_2O} = a_{H_2O}$. Another example is decomposition of calcium carbonate to calcium oxide and carbon dioxide,

$$CaCO_3(s) \rightleftharpoons CaO(s) + CO_2(g) \tag{2.74}$$

The amount of calcium carbonate and calcium oxide does not affect the position of equilibrium. Again, $K_p = P_{CO_2} = a_{CO_2}$.

In many instances, nanoparticles are present or functioning in aqueous solutions. A fraction of the particles may dissolve and result in a saturated solution. The remaining solid will remain in equilibrium with the solution, but the amount will not affect the position of equilibrium. For example, barium sulfate nanoparticles may dissolve according to the reaction

$$BaSO_4(s) \xrightarrow{H_2O} Ba^{2+}(aq) + SO_4^{2-}(aq) \tag{2.75}$$

The equilibrium constant in this case is known as the solubility produced, K_{sp}, and is given by Equation 2.76:

$$K_{sp} = a_{Ba^{2+}} a_{SO_4^{2-}} \tag{2.76}$$

2.6.3 The relationship between Gibbs energy and the equilibrium constant

We can use our understanding of the laws of thermodynamics to obtain a relationship between the Gibbs energy and the thermodynamic equilibrium constant. Starting with the definition of the Gibbs and the enthalpy state functions (Equations 2.77 and 2.78),

$$G = H - TS \tag{2.77}$$

$$H = E + PV \tag{2.78}$$

we can write Equation 2.79:

$$G = E + PV - TS \tag{2.79}$$

Taking derivatives yields

$$dG = dE + PdV + VdP - TdS - SdT \tag{2.80}$$

and since

$$dE = q + w = TdS - PdV \qquad (2.81)$$

we have

$$dG = TdS - PdV + PdV + VdP - TdS - SdT \qquad (2.82)$$

Simplifying the above expression gives

$$dG = -SdT + VdP \qquad (2.83)$$

At constant temperature, this equation reduces to

$$dG = VdP \qquad (2.84)$$

Assuming an ideal gas and integrating from some initial pressure P_1 to some final pressure P_2,

$$\Delta G = \int_{P_1}^{P_2} VdP = nRT \int_{P_1}^{P_2} \frac{1}{P} dP \qquad (2.85)$$

gives the relationship between the molar Gibbs energy and pressure (Equation 2.86):

$$\Delta \bar{G} = RT \ln \frac{P_2}{P_1} \qquad (2.86)$$

Let's define $P_1 = 1$ atm (the standard pressure) so that G at P_1 is the standard Gibbs energy. With this in mind we arrive at Equation 2.87:

$$\bar{G}_2(T) - \bar{G}_1^o(T) = RT \ln \frac{P_2}{1 \text{ atm}} \qquad (2.87)$$

Note that we have been careful to recognize that temperature is not considered when defining standard states and so we emphasize that the Gibbs energy is a function of temperature. Equation 2.87 simplifies to Equation 2.88:

$$\bar{G}(T) = \bar{G}^o(T) + RT \ln P \qquad (2.88)$$

This relation can also be expressed in terms of chemical potentials (Equation 2.89):

$$\mu(T) = \mu^o(T) + RT \ln P \qquad (2.89)$$

The molar Gibbs energy change for a chemical reaction is defined as the difference between the sums of the Gibbs energy of the products and the sums of the Gibbs energies of the reactants (Equation 2.90):

$$\Delta \bar{G} = \sum \bar{G}_{\text{products}} - \sum \bar{G}_{\text{reactants}} \tag{2.90}$$

For the simple gas phase reaction $A(g) \rightleftharpoons B(g)$, the Gibbs energy of reaction is

$$\Delta \bar{G} = (\bar{G}_B^o + RT \ln P_B) - (\bar{G}_A^o + RT \ln P_A) \tag{2.91}$$

Rearranging gives

$$\Delta \bar{G} = \bar{G}_B^o - \bar{G}_A^o + RT \ln P_B - RT \ln P_A \tag{2.92}$$

which can be written as

$$\Delta \bar{G} = \Delta \bar{G}^o + RT \ln \frac{P_B}{P_A} \tag{2.93}$$

For our reaction the ratio P_B/P_A is the reaction quotient Q (see Equation 2.67). Thus, Equation 2.93 becomes

$$\Delta \bar{G} = \Delta \bar{G}^o + RT \ln Q \tag{2.94}$$

At equilibrium $Q = K$ and $\Delta \bar{G} = 0$ for the reaction. Thus, the relationship between the standard Gibbs energy change and the equilibrium constant is

$$\Delta \bar{G}_{\text{reaction}}^o = -RT \ln K \tag{2.95}$$

2.6.4 Temperature dependence of the equilibrium constant

We can use Equation 2.95 to see how K changes with temperature. Recall

$$\Delta \bar{G}^o = \Delta \bar{H}^o - T \Delta \bar{S}^o \tag{2.96}$$

Substituting this equation in to Equation 2.95 gives

$$\Delta \bar{H}^o - T \Delta \bar{S}^o = -RT \ln K \tag{2.97}$$

Rearranging gives

$$\ln K = -\frac{\Delta \bar{H}^o - T \Delta \bar{S}^o}{RT} \tag{2.98}$$

which can be written in the form

$$\ln K = \frac{\Delta \bar{S}^o}{R} - \frac{\Delta \bar{H}^o}{RT} \qquad (2.99)$$

Equation 2.99 shows how K changes with T. If we assume that $\Delta \bar{S}^o$ and $\Delta \bar{H}^o$ are temperature-independent, a plot of ln K versus $1/T$ will yield a straight line with a slope equal to $-\Delta \bar{H}^o/R$ and an intercept equal to $\Delta \bar{S}^o/R$. Of course, obtaining a positive or negative slope will depend on whether the reaction is exothermic or endothermic. Alternatively, we write equations for the equilibrium constant at temperatures T_1 and T_2:

$$\ln K_1 = \frac{\Delta \bar{S}^o}{R} - \frac{\Delta \bar{H}^o}{RT_1} \qquad (2.100)$$

$$\ln K_2 = \frac{\Delta \bar{S}^o}{R} - \frac{\Delta \bar{H}^o}{RT_2} \qquad (2.101)$$

Subtracting the above two equations yields

$$\ln K_2 - \ln K_1 = -\frac{\Delta \bar{H}^o}{RT_2} + \frac{\Delta \bar{H}^o}{RT_1} \qquad (2.102)$$

which can be simplified to

$$\ln \frac{K_2}{K_1} = \frac{\Delta \bar{H}^o}{R} \left(\frac{1}{T_1} - \frac{1}{T_2} \right) \qquad (2.103)$$

Equation 2.103 is known as the **van't Hoff Equation**. It is used to determine the equilibrium constant at some temperature T_2 if it is known at some other temperature T_1. The equation assumes that the enthalpy change for the reaction is constant between the limits T_1 and T_2. This is a reasonable assumption if the temperature range is small. If T_1 and T_2 differ by a large amount, the temperature dependence of enthalpy needs to be considered.

2.6.5 Phase equilibria in bulk materials

Figure 2.14 shows the familiar phase diagram of water. It shows the pressure and temperature conditions under which the various phases (solid, liquid, and gas) are stable. The line boundaries indicate the coexistence of two phases at equilibrium, and at the triple point all three phases coexist in

Figure 2.14 The phase diagram of water, showing the various coexistence curves for the solid (S), gaseous (G), and liquid (L) states. the triple point is indicated, along with the critical point beyond which a supercritical liquid is formed.

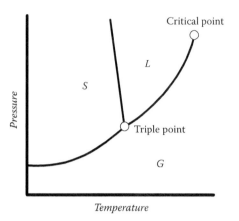

equilibrium with each other. An interesting feature in the diagram is the abrupt termination of the liquid–gas coexistence curve at the critical point. Beyond this region, water exists in a **supercritical** state, a new phase that exhibits the physical properties of both the gas and liquid states.

Recall the total derivative of the Gibbs energy (Equation 2.104):

$$d\bar{G} = \bar{V}dP - \bar{S}dT \qquad (2.104)$$

For two phases (1 and 2), we write the corresponding equations

$$d\bar{G}(1) = \bar{V}(1)dP - \bar{S}(1)dT \qquad (2.105)$$

$$d\bar{G}(2) = \bar{V}(2)dP - \bar{S}(2)dT \qquad (2.106)$$

Let's assume that both phases are in equilibrium with each other. At equilibrium the Gibbs energies will be equal; that is

$$\bar{G}(1) = \bar{G}(2) \qquad (2.107)$$

or

$$d\bar{V}(1)dP - \bar{S}(1)dT = d\bar{V}(2)dP - \bar{S}(2)dT \qquad (2.108)$$

rearranging gives

$$[d\bar{V}(2) - d\bar{V}(1)]dP = [\bar{S}(2) - \bar{S}(1)]dT \qquad (2.109)$$

or

$$\frac{dP}{dT} = \frac{\Delta\bar{S}_{\text{trans}}}{\Delta\bar{V}_{\text{trans}}} \qquad (2.110)$$

Since

$$\Delta \bar{S}_{\text{trans}} = \frac{\Delta \bar{H}_{\text{trans}}}{T_{\text{trans}}} \qquad (2.111)$$

for a liquid in equilibrium with its vapor, the equilibrium constant is equal to the vapor pressure of the vapor phase. Thus, for this system we can replace K_1 and K_2 in Equation 2.103 with the corresponding vapor pressure, giving Equation 2.112:

$$\ln \frac{P_2}{P_1} = \frac{\Delta \bar{H}^o}{R} \left(\frac{1}{T_1} - \frac{1}{T_2} \right) \qquad (2.112)$$

The pressure at the boiling point (T_b) of the liquid is equal to $P_1 = 1$ atm. Setting $T_1 = T_b$ and $P_1 = 1$ atm gives the **Clausius–Clapeyron equation** (Equation 2.113):

$$\ln P = \frac{\Delta \bar{H}^o_{\text{vap}}}{R} \left(\frac{1}{T_b} - \frac{1}{T} \right) \qquad (2.113)$$

2.6.6 Phase equilibria in nanoparticles

In the treatment of phase changes within nanoparticles, we will assume that the particle is made up of a single element, for example, pure gold, silver, or a lead nanoparticle. The total chemical potential of a nanoparticle must be a sum of the bulk chemical potential and the surface chemical potential. The chemical potential was defined in Equation 2.89. Thus

$$\mu_{\text{particle}} = \mu_{\text{bulk}} + \mu_{\text{surf}} \qquad (2.114)$$

We begin by first writing down two expressions for the chemical potential of the liquid and solid phase for the nanoparticle of radius r (Equations 2.115 and 2.116):

$$\mu^L_{\text{particle}} = \mu^L_{\text{bulk}} + \frac{2\gamma^L \bar{V}^L}{r} \qquad (2.115)$$

$$\mu^S_{\text{particle}} = \mu^S_{\text{bulk}} + \frac{2\gamma^S \bar{V}^S}{r} \qquad (2.116)$$

In the above two equations, γ and $\bar{V}$ correspond to the surface tension and molar volumes of the liquid and solid nanoparticle.

At the melting temperature the two phases of the nanoparticle (solid and liquid) will be at equilibrium with each other, and so their chemical potentials will be equal. Thus

$$\mu^L_{\text{particle}} = \mu^S_{\text{particle}} \tag{2.117}$$

or

$$\mu^L_{\text{bulk}} + \frac{2\gamma^L \bar{V}^L}{r} = \mu^S_{\text{bulk}} + \frac{2\gamma^S \bar{V}^S}{r} \tag{2.118}$$

Rearranging

$$\mu^L_{\text{bulk}} - \mu^S_{\text{bulk}} = \frac{2}{r}\left(\gamma^L \bar{V}^L - \gamma^S \bar{V}^S\right) \tag{2.119}$$

Now the change in the chemical potential $\mu^L_{\text{bulk}} - \mu^S_{\text{bulk}}$ is the Gibbs energy corresponding to the phase change (Equation 2.120):

$$\Delta G_{\text{trans}} = \mu^L_{\text{bulk}} - \mu^S_{\text{bulk}} \tag{2.120}$$

From earlier sections we know that $\Delta G_{\text{trans}} = \Delta H_{\text{trans}} - T\Delta S_{\text{trans}}$ and so $\Delta S_{\text{trans}} = \Delta H_{\text{trans}}/T_{\text{trans}}$, in which we can write

$$\Delta G_{\text{trans}} = \Delta H_{\text{trans}} - \frac{T\Delta H_{\text{trans}}}{T_{\text{trans}}} \tag{2.121}$$

In terms of our transition, we have

$$\Delta G_{\text{trans}} = \Delta H_{\text{trans}} - \frac{T\Delta H_{\text{trans}}}{T_{\text{mpt}}} = \Delta H_{\text{trans}}\left(1 - \frac{T}{T_{\text{mpt}}}\right) \tag{2.122}$$

Therefore, from Equation 2.119, we have

$$\Delta H_{\text{trans}}\left(1 - \frac{T}{T_{\text{mpt}}}\right) = \frac{2}{r}\left(\gamma^L \bar{V}^L - \gamma^S \bar{V}^S\right) \tag{2.123}$$

Rearranging the above equation gives Equation 2.124:

$$\frac{T}{T_{\text{mpt}}} = 1 - \frac{1}{r}\frac{2}{\Delta H_{\text{trans}}}\left(\gamma^L \bar{V}^L - \gamma^S \bar{V}^S\right) \tag{2.124}$$

It is should be noted that in the above equation T refers to the melting temperature of the nanoparticle of some radius r, where T_{mpt} refers to the bulk phase melting point. We see that this equation has the exact same functional form as Equation 2.7

The total Gibbs energy change for nanoparticle melting transition can be written according to Equation 2.125:

$$\Delta G_{\text{trans}}^{\text{total}} = \Delta G_{\text{trans}}^{\text{bulk}} + \Delta G_{\text{trans}}^{\text{surf}} \qquad (2.125)$$

For the surface, the Gibbs energy change is given by the difference between the surface molar Gibbs energies of the two phases (see Equation 2.61). Thus

$$\Delta G_{\text{trans}}^{\text{surf}} = \frac{2\gamma^L \bar{V}^L}{r} - \frac{2\gamma^S \bar{V}^S}{r} \qquad (2.126)$$

$$\therefore \Delta G_{\text{trans}}^{\text{total}} = \Delta G_{\text{trans}}^{\text{bulk}} + \frac{2}{r}\left(\gamma^L \bar{V}^L - \gamma^S \bar{V}^S\right) \qquad (2.127)$$

To use Equation 2.127, we need to know the surface tension values of the liquid and solid phases. While the values for the liquid phase are easily measureable, the values for the corresponding solid phase are much more difficult to obtain. Fortunately, mathematical relationships do exist between the two. One such relationship is shown in Equation 2.128:

$$\gamma^S = 1.25\gamma_{\text{mpt}}^L + \frac{\partial \gamma^L}{\partial T}\left(T - T_{\text{mpt}}\right) \qquad (2.128)$$

Furthermore, the molar volume of the solid can be estimated from the molar volume of the liquid using Equation 2.129:

$$\bar{V}^S = \left(\frac{\bar{V}^L}{1 + \beta}\right) \qquad (2.129)$$

where β is a unitless number that depends on the element in question. Using Equations 2.127, 2.128, and 2.129, we can write a complete expression for the total Gibbs energy (Equation 2.130):

$$\Delta G_{\text{trans}}^{\text{total}} = \Delta G_{\text{trans}}^{\text{bulk}} + \frac{2}{r}\left[\gamma^L \bar{V}^L - 1.25\gamma_{\text{mpt}}^L + \frac{\partial \gamma^L}{\partial T}\left(T - T_{\text{mpt}}\right)\frac{\bar{V}^L}{1 + \beta}\right] \qquad (2.130)$$

At the melting temperature, the Gibbs energy $\Delta G_{\text{trans}}^{\text{total}} = 0$ for the pure nanoparticle of diameter r. The following example illustrates how Equation 2.130 can be used to predict the melting temperature of a nanoparticle.

Example 2.11   Predicting Melting Temperatures from the Gibbs Energy

Use the following known information to predict the melting temperature of Au nanoparticles of diameter 100 nm. The data was obtained from Niemelae et al., (1986), Dinsdale (1991), Ioda and Guthrie (1988), and Wittenberg and DeWitt (1972).

$$\Delta \bar{G}^{bulk} = 12552.0 - 9.385866T \ \text{(in J/mol)}$$

$$\gamma^L = 1.169 - 0.00025(T - 1336.15) \ \text{(in N/m)}$$

$$\bar{V}^L = 11.3 \times 10^{-6}[1.0 + 0.000069(T - 1336.15)]$$

$$\frac{\partial \gamma^L}{\partial T} = -0.00025, \ \gamma^L_{mpt} = 1.169 \ \text{N/m}, \ T_{mpt} = 1336.15K, \ \beta = 0.055$$

Solution   Substituting the above information into Equation 2.130 and setting $\Delta G^{total}_{trans} = 0$ gives

$$0 = 12552.0 - 9.385866T + \frac{2}{20 \times 10^{-9}}$$

$$\begin{bmatrix} 1.169 - 0.00025(T - 1336.15) \times 11.3 \times 10^{-6}[1.0 + 0.000069(T - 1336.15)] \\ -1.25(1.169) + (-0.00025)(T - 1336.15)\dfrac{11.3 \times 10^{-6}[1.0 + 0.000069(T - 1336.15)]}{1 + 0.055} \end{bmatrix}$$

The above equation can be simplified or one can use a program like Maple or Mathematica to solve directly for T.

The simplified equation is

$$0 = 9.238 \times 10^{-10}(T + 1.076 \times 10^5)(T - 1298)(T - 94677)$$

Solving this equation gives $T = 1300$ K.

End of chapter questions

1. In examining the graph in Figure 2.5, estimate the work done in stretching the DNA strand from $r = 1$ to $r = 1.6$.

2. Amorphous silicon dioxide (SiO_2) has a bulk density of 2.2 g/cm^3 and a formula mass of 60.08 g/mol. Calculate the number of atoms per cubic meter in SiO_2 and the minimum size of a SiO_2 nanosystem where classical thermodynamics is valid, assuming that we accept a relative thermal fluctuation of 1%. How does this size change if the density of the nanomaterial is only half the bulk density?

3. Consider a disclike nanostructure with diameter l and thickness h. The volume of such a disc is given by $\pi l^2 h/4$, and the surface area is $\pi l^2/2 + \pi l h$. Let d be the atomic diameter.

    a. Show that the ratio N/n is equal to $(4/3)d[1/h + 2/l]$.

    b. For nanowires, $h \gg l$. In this case show that $N/n = (8/3)d/l$.

    c. For nanofilms, $l \gg h$. In this case show that $N/n = (4/3)d/h$.

    d. Can nanofilms be regarded as extreme cases of nanodiscs?

4. The bulk phase melting temperatures of indium is 429.8 K. The atomic radius of In is 167 pm.

Determine the melting point for (a) an In nanowire of length 20 nm, and (b) an In nanofilm of thickness 15 nm.

5. Using a graphing program of your choice, generate a theoretical plot showing how the melting temperature of an indium nanowire changes from 2 nm to 20 nm. Use the information given in Problem 5.

6. For the reaction describing the synthesis of silica nanoparticles, determine the equilibrium constant at 298 K and at 350 K. Assume that the standard enthalpy of the reaction is constant in this temperature range. Refer to the data in Example 2.9 to answer this question.

7. Use the following known information to predict the melting temperature of Si nanoparticles of diameter 150 nm. The data was obtained from references 4–7.

$$\Delta \bar{G}^{bulk} = 50696.36 - 930.099439T + 2.0931$$
$$\times 10^{-21} T^7 \quad (\text{in J/mol})$$

$$\gamma^L = 0.865 - 0.00013(T - 1687.15) \ (\text{in N/m})$$

$$\bar{V}^L = 11.1 \times 10^{-6}[1.0 + 0.00014(T - 1687.15)]$$

$$\frac{\partial \gamma^L}{\partial T} = -0.00013, \ \gamma^L_{mpt} = 0.865 \, \text{N/m}, \ T_{mpt}$$
$$= 1687.15 \text{K}, \ \beta = -0.095$$

Cited references

Cluzel, P. et al. DNA: An Extensible Molecule. *Science* 1996, 271(5250): 792–794.

Dinsdale, A. T. SGTE Data for Pure Elements. *CALPHAD*, 1991, 15: 317–425.

Ioda, T. and Guthrie, R. I. L. *The Physical Properties of Liquid Metals*. 1988, Clarendon Press, Oxford, pp. 71, 132.

Niemelae, J., Effenberg, G., Hack, K. and Spencer, P. J. A Thermodynamic Evaluation of the Copper—Bismuth and Copper—Lead systems. *CALPHAD* 1986, 10: 77–89.

Qi, W. H. Size Effect on Melting Temperature of Nanosolids. *Physica B* 2005, 368: 46–50.

Smith, S. B., Cui, Y. and Bustamante, C. Overstretching B-DNA: The Elastic Response of Individual Double-Stranded and SingleStranded DNA Molecules. *Science*. 1996, 271(5250): 795–799.

Wittenberg, L. J. and DeWitt, R. Volume Contraction During Melting; Emphasis on Lanthanide and Actinide Metals. *J. Chem. Phys*. 1972, 56: 4526–4533.

Wautelet, M., Shirinyan, A. S. Thermodynamics: Nano vs. Macro. *Pure Appl. Chem.* 2009, 81: 1921–1930.

References and recommended reading

Hill, T. L. *Thermodynamics of Small Systems*, Parts I & II. 2013, Dover Publications. This book provides the best treatment of nanothermodynamics. It is highly recommended for the student who wants a thorough understanding of the field.

McQuarrie, D. A. and Simon, J. D. *Physical Chemistry: A Molecular Approach*. 1997, University Science Books. This book provides an excellent introduction to molecular and classical thermodynamics.

CHAPTER 3

Kinetics and Transport in Nanoscience

OVERVIEW

In the previous chapter, we learned about thermodynamic properties and the relative stability of reactants and products and how to quantify the spontaneity of a chemical reaction based on thermodynamic state functions such as Gibbs energy. However, thermodynamics does not provide insight on how long a reaction takes to reach equilibrium. Even though a reaction may have a negative Gibbs energy change, it may proceed very slowly. For example, bulk iron will spontaneously rust (oxidize) when exposed to air, but the reaction may take many years, while iron nanoparticles may oxidize very quickly. This chapter is concerned with how quickly chemical reactions occur and what we can learn about *how* reactions occur from the rates at which they occur. **Chemical kinetics**, the study of the rates of reactions and the implication of these rates, relies on the experimental observation of reactions and the effects of changing conditions on their rates. This chapter also makes an important connection between rates of chemical reactions, **transition states**, and the thermodynamic properties for processes at equilibrium. The final section of the chapter considers the effects of particle diffusion on reaction rates, which is important for nanoscale systems.

3.1 RATES OF CHEMICAL REACTIONS

3.1.1 The Rate of Reaction

During the course of a chemical reaction, reactants are consumed and products are formed. The reactant concentrations decrease and the product concentrations increase until equilibrium is reached. The **rate of reaction** can be quantified by measuring how the concentrations of the reacting species change as a function of time. Consider the formation of silver nanoparticles by the thermal decomposition of silver oxalate at 140°C:

$$Ag_2C_2O_4(s) \rightarrow 2Ag(s) + 2CO_2(g) \tag{3.1}$$

At the beginning of the reaction, the rate of disappearance of $Ag_2C_2O_4$ will be large, and this rate will decrease as the concentration of the reactant decreases, as seen in Figure 3.1. The rate of reaction is the slope of the reactant concentration at any point in time. The rate at $t = 0$ is the **initial rate**. The larger the initial concentration of $Ag_2C_2O_4$, the larger the initial rate will be. Once the rate of change of the concentration reaches zero, the reaction is complete. In general, we can define the **instantaneous rate** of the reaction, $v(t)$, as

$$v(t) = -\frac{d[Ag_2C_2O_4]}{dt} \tag{3.2}$$

The negative sign in front of the derivative in Equation 3.2 tells us that the concentration of $Ag_2C_2O_4$ is *decreasing* with time because $Ag_2C_2O_4$ is a reactant. The units of reaction rate are mol dm^{-3} s^{-1} and its numerical value changes with time. The rate of increase in product is proportional to

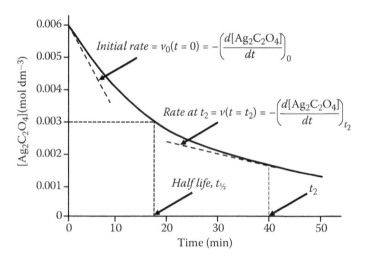

Figure 3.1 A plot of [Ag$_2$C$_2$O$_4$] versus time. The product concentration decreases exponentially. The instantaneous rate at any point in time is given by the derivative of concentration with respect to time at t. Instantaneous rates are shown at two different time points: $t = 0$ (the initial rate) and $t = t_2$. The half-life for the reaction is also indicated.

the rate of loss of reactant. Considering the stoichiometric coefficients in Equation 3.1, we can write

$$v(t) = -\frac{d[Ag_2C_2O_4]}{dt} = \frac{1}{2}\frac{d[Ag]}{dt} = \frac{1}{2}\frac{d[CO_2]}{dt} \tag{3.3}$$

For the general reaction

$$aA + bB \rightarrow cC + dD, \tag{3.4}$$

Equation 3.5 gives the relationship between the various rates.

$$v(t) = -\frac{1}{a}\frac{d[A]}{dt} = -\frac{1}{b}\frac{d[B]}{dt} = \frac{1}{c}\frac{d[C]}{dt} = \frac{1}{d}\frac{d[D]}{dt}. \tag{3.5}$$

If the reaction is reversible and has not reached equilibrium, then the net instantaneous rate is simply the forward rate minus the reverse rate. At equilibrium, these two rates will be equal and the concentrations of all species are constant. We will discuss the rates of reversible reactions in Section 3.1.3.

3.1.2 Rate laws and reaction orders

Experimentally, the rate of decomposition of silver oxalate (Equation 3.1) has been determined to be directly proportional to the concentration of $Ag_2C_2O_4$. Doubling $[Ag_2C_2O_4]$ doubles the rate. For this particular reaction, Equation 3.6 describes the relationship between rate and concentration.

$$v(t) = k[Ag_2C_2O_4] \tag{3.6}$$

Equation 3.6 is an example of a **rate law**. It mathematically describes the relationship between the concentration of $Ag_2C_2O_4$ and the rate at which the decomposition occurs. The proportionality constant, k, is called the **rate constant**. The magnitude of the rate constant provides a quantitative measure of how quickly the reaction progresses and forms silver nanoparticles. A plot of the reaction rate versus the concentration of $Ag_2C_2O_4$ will be linear, with a slope equal to the rate constant. Not all reactions exhibit this linear relationship. For example, the rate law describing the decomposition of ethane, C_2H_6, to two $\bullet CH_3$ radicals is

$$v(t) = k[C_2H_6]^2 \tag{3.7}$$

In this example, doubling the reactant concentration quadruples the reaction rate. A plot of initial rate versus $[C_2H_6]^2$ will be linear with a slope equal to the rate constant.

So far, we have considered the simple rate law of the form

$$v(t) = k[A]^n \tag{3.8}$$

When $n = 1$, as in the case of the $Ag_2C_2O_4$ decomposition reaction, we say that the rate is *first order* with respect to the concentration of the reactant, $Ag_2C_2O_4$. When $n = 2$, as in the case of the C_2H_6 decomposition reaction, we say that the rate is *second order* with respect to the reactant. A third-order reaction would have $n = 3$, and so on. If the order with respect to A is zero, then the rate is equal to the rate constant, since $[A]^0 = 1$. In other words, the concentration of A does not affect the reaction rate. For the general reaction described by Equation 3.4, the rate law can be defined as

$$v(t) = k[A]^n[B]^m \tag{3.9}$$

In this equation, n is the order with respect to A and m is the order with respect to B. The overall order of the reaction is $n + m$. Table 3.1 shows examples of reactions with various orders, including *fractional orders*, along with corresponding units of the rate constant. Note that the units of the rate constant are obtained by dividing the units of rate (mol dm^{-3} s^{-1}) by the product of the units corresponding to all concentrations in the rate law (e.g., $[A]^n \times [B]^m$).

Table 3.1

Some Examples of Zero-, First-, and Second-Order Reactions and Their Corresponding Rate Laws and Rate Constant Units

Reaction Order	Examples	Rate Laws	Units of Rate Constants
Zero	$2NH_3(g) \xrightarrow{Pt(s)} N_2(g) + 3H_2(g)$	$v(t) = k$	mol dm^{-3} s^{-1}
	$H_2(g) + Cl_2(g) \xrightarrow{hv} 2HCl(g)$	$v(t) = k$	
First	$N_2O_5(g) \rightarrow 2NO_2(g) + {}^1/_2O_2(g)$	$v(t) = k[N_2O_5]$	s^{-1}
	$CH_3NC \rightarrow CH_3CN$	$v(t) = k[CH_3NC]$	
	${}^{238}_{92}U \rightarrow {}^{234}_{90}Th + {}^4_2He$	$v(t) = k\left[{}^{238}_{92}U\right]$	
Second	$2HI(g) \rightarrow H_2(g) + I_2(g)$	$v(t) = k[HI]^2$	mol^{-1} dm^3 s^{-1}
	$NO(g) + O_3(g) \rightarrow NO_2(g) + O_2(g)$	$v(t) = k[NO][O_3]$	
Third	$H_2PO_2^-(aq) + OH^-(aq) \rightarrow HPO_3^{2-}(aq) + H_2(g)$	$v(t) = k[H_2PO_2^-][OH^-]^2$	mol^{-2} dm^6 s^{-1}
Fractional	$H_2(g) + Br_2(g) \rightarrow 2HBr(g)$	$v(t) = k[H_2][Br_2]^{1/2}$	mol^{-1} dm$^{3/2}$ s^{-1}

There is no simple way of predicting the order of a reaction based on the identities of the reactions and products; it cannot be arrived at from the stoichiometric equation. Only through experiments examining the effect of changing concentration of the reactants on the rate of that reaction can we determine the rate law. Example 3.1 illustrates the method of initial rates, which is a method for determining the order and rate constant for a reaction based on observations of the initial rate of a reaction under several different starting conditions.

Example 3.1   Method of Initial Rates

The reaction between nitrogen monoxide and chlorine at $-15^{\circ}C$ is given by the following equation:

$$2NO(g) + Cl_2(g) \rightarrow 2NOCl$$

The table below gives the results of three experiments involving this reaction. Use the data to obtain the rate law and the value of the rate constant.

Experiment	$[NO]_0$ (mol dm^{-3})	$[Cl_2]_0$ (mol dm^{-3})	Initial Rate v_0 (mol dm^{-3} min^{-1})
1	0.10	0.10	0.18
2	0.10	0.20	0.37
3	0.20	0.20	1.46

Solution   Begin by writing the rate law $v_0(t) = k[NO]_0^n[Cl_2]_0^m$, where all measurements are at their initial values. Using the values for experiments 1 and 2, where $[NO]_0$ is fixed, we can write the rate laws

$$0.18 = k[0.10]^n[0.10]^m \qquad \text{Experiment 1}$$

$$0.37 = k[0.10]^n[0.20]^m \qquad \text{Experiment 2}$$

Dividing the two equations gives

$$\frac{0.18}{0.37} = \left(\frac{0.10}{0.20}\right)^m$$

$$0.49 = 0.50^m$$

Taking logs of both sides gives

$$\log(0.49) = m\log(0.50)$$

Solving for m we get 1.03, where we can make the reasonable assumption that $m = 1$. Repeating these steps for experiments 2 and 3, where $[Cl_2]_0$ is fixed, gives

$$0.36 = k[0.10]^n[0.20]^m \qquad \text{Experiment 2}$$

$$1.46 = k[0.20]^n[0.20]^m \qquad \text{Experiment 3}$$

Dividing the two equations gives

$$\frac{0.36}{1.46} = \left(\frac{0.10}{0.20}\right)^n$$

$$0.25 = 0.50^n$$

$$\log(0.25) = n\log(0.50)$$

Solving for n we get 2.0.

Thus, the experimentally determined rate law is $v(t) = k[NO]^2[Cl_2]$.

We see that the orders with respect to NO and Cl_2 are 2 and 1, respectively. The reaction is overall third order. We can use the data from any one of the experiments to determine the value of the third-order rate constant. Considering experiment 1,

$$k = \frac{v(t)}{[NO]^2[Cl_2]} = \frac{0.36\,\text{mol}\,\text{dm}^{-3}\,\text{min}^{-1}}{\left(0.10\,\text{mol}\,\text{dm}^{-3}\right)^2\left(0.20\,\text{mol}\,\text{dm}^{-3}\right)}$$

$$= 180\,\text{dm}^6\text{mol}^{-2}\,\text{min}^{-1}$$

Zero-order reactions are rare in solution, primarily limited to photochemical reactions, but more common in heterogeneous systems that contain an interface between multiple phases of matter. To understand a zero-order reaction, let's consider an example of a heterogeneous reaction. The reverse Haber reaction, which is the reverse reaction for the standard industrial reaction for making ammonia, is a zero-order reaction. Gaseous NH_3 first adsorbs to a solid Pt **catalyst** before being converted into products. Once the reaction begins, the surface of the metal catalyst is quickly saturated by NH_3 molecules regardless of the pressure of NH_3. The concentration of NH_3 molecules on the surface remains constant even though the products are formed. This is why the reaction rate is unaffected by $[NH_3]$ (i.e., it is zero-order with respect to the reactant). Many reactions occurring on surfaces show zero-order behavior in one or more reactants. In later chapters we will see that surface chemistry plays a crucial role in determining properties and reactivity in nanosystems.

First-order reactions are common and exhibited by many processes such as nanoparticle decomposition and isomerization/rearrangement of molecular moieties in self-assembled films. Furthermore, first-order reactions are not limited to a single one-reactant process. For instance, an $A + B$ reaction may be first-order in A and zero-order in B (Table 3.1).

Reactions with orders larger than two are uncommon. Second-order reactions, however, are common and often take place between two different reactants. Nevertheless, there are plenty of examples of second-order reactions in which only a single reactant is present. One important example is the self-assembly of double-stranded DNA from the individual single strands (Figure 3.2).

Sometimes measuring a second-order reaction rate with different reactants A and B can be problematic. The concentrations of the two reactants must be monitored simultaneously. In order to overcome this problem, we can use a **pseudo-first-order approximation**. For example, if $[B]$ remains constant as the reaction proceeds, we can simplify the rate law as

$$v(t) = k[A][B] = k_{\text{obs}}[A] \tag{3.10}$$

The reaction can be considered **pseudo-first-order** because it depends on the concentration of only one reactant, in this case $[A]$. The *observed* pseudo-first-order rate constant is $k_{\text{obs}} = k[B]_0$.

One way to deliberately approach a pseudo-first-order reaction is to use a large excess of one of the reactants (e.g., $[B] \gg [A]$). As a result, only a very small amount of the reactant $[B]$ is consumed as the reaction progresses, and its concentration can be assumed to stay constant. By collecting k_{obs} values for a number of reactions with different excess concentrations of $[B]$, a plot of k_{obs} versus $[B]$ gives k as the slope. Some aqueous phase reactions involving H_2O as a reactant can be pseudo-first-order due to the large excess of H_2O present. For example, the hydrolysis of an ester, such

Figure 3.2 The second-order formation of double-stranded DNA from the individual single strands.

as ethyl acetate, follows pseudo-first-order kinetics when conducted in aqueous solution (Equation 3.11).

$$CH_3COOCH_3 + H_2O \rightarrow CH_3COOH + CH_3OH \qquad (3.11)$$

3.1.3 A note on reversible reactions

All spontaneous reactions eventually reach a state of equilibrium. Consider the following reversible reaction:

$$A + B \underset{k_r}{\overset{k_f}{\rightleftharpoons}} P \qquad (3.12)$$

k_f and k_r represent the rate constants for the forward and reverse reactions, respectively, and the corresponding rate laws governing the two processes are given by Equations 3.13 and 3.14:

$$v_f(t) = k_f[A][B] \qquad \text{for} \qquad A + B \overset{k_f}{\longrightarrow} P \qquad (3.13)$$

$$v_r(t) = k_r[P] \qquad \text{for} \qquad P \overset{k_r}{\longrightarrow} A + B \qquad (3.14)$$

The forward and the reverse rates will be equal to each other ($v_f(t) = v_r(t)$) when the reaction reaches equilibrium, and when this happens the apparent concentrations [A], [B], and [P] will not change with time. Thus

$$k_f[A]_{eq}[B]_{eq} = k_r[P]_{eq} \qquad (3.15)$$

We have used the subscripts eq in Equation 3.15 to emphasize equilibrium concentrations. Figure 3.3 shows changes in concentrations of species A, B, and product P as a function of time. Reactant concentrations decrease as product concentration increases. At time t_{eq}, the reaction reaches equilibrium and all concentrations become constant over time. Rearranging Equation 3.15 gives,

$$\frac{k_f}{k_r} = \frac{[P]_{eq}}{[A]_{eq}[B]_{eq}} = K \qquad (3.16)$$

Thus, according to Equation 3.16 we see that for a reaction at equilibrium, the ratio of the forward and reverse rate constants is numerically equal to the thermodynamic equilibrium constant, K. We will make use of this relationship from time to-time, especially when we discuss adsorption isotherms in Chapter 7.

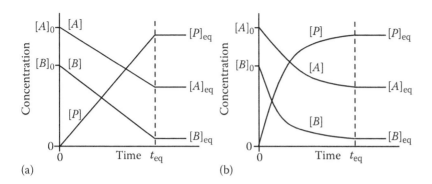

Figure 3.3 Changes in concentrations of species A, B, and product P as a function of time. Reactant concentration decreases as product concentration increases. At time t_{eq}, the reaction reaches equilibrium and all concentrations become constant over time. (a) Linear behavior. (b) Changes in concentration that appear to change exponentially.

3.1.4 Integrated rate laws

The explicit relationship between the concentration of a reactant and time can be obtained by integrating rate laws with respect to time. These relationships, known as integrated rate laws, can be used to determine the concentration of a reactant at any given time or to determine the rate constant from a series of concentrations at different times. As a concrete example, let's consider the following first-order reaction:

$$A \xrightarrow{k} products \tag{3.17}$$

Equation 3.18 describes the rate law corresponding to this first-order process:

$$-\frac{d[A]}{dt} = k[A] \tag{3.18}$$

This equation can be rearranged such that concentration terms are on the left-hand side and time is on the right-hand side (Equation 3.19):

$$\frac{1}{[A]} d[A] = -kdt \tag{3.19}$$

Integrating the above equation from the initial concentration $[A]_0$ at $t = 0$ to some final concentration $[A]$ at time t, or mathematically,

$$\int_{[A]_0}^{[A]} \frac{1}{[A]} d[A] = -k \int_0^t dt \tag{3.20}$$

yields

$$\ln\left(\frac{[A]}{[A]_0}\right) = -kt \qquad (3.21)$$

Taking exponentials of both sides gives

$$\frac{[A]}{[A]_0} = e^{-kt} \qquad (3.22)$$

which can be rearranged to the final form

$$[A] = [A]_0 e^{-kt} \qquad (3.23)$$

Equation 3.23 tells us that the concentration of the reactant A decreases exponentially with time for a first-order reaction (Figure 3.4b). Since we can rearrange Equation 3.21 to the form

$$\ln[A] = \ln[A]_0 - kt \qquad (3.24)$$

a plot of $\ln[A]$ versus t will be linear with a slope equal to the rate constant (Figure 3.4c).

One useful aspect of analyzing data for first-order processes is that we only require the ratios $[A]/[A]_0$ to obtain the rate constant using any of Equations 3.21 to 3.24. Therefore, we can use any technique that

Figure 3.4 Graphical representations of integrated rate equations. (a) Zero-order plot showing how reactant A and product P concentrations change with time, and (b) first-order plot showing how reactant A and product P concentrations change with time. (c) and (d) Linear plots for the first- and second-order processes, respectively.

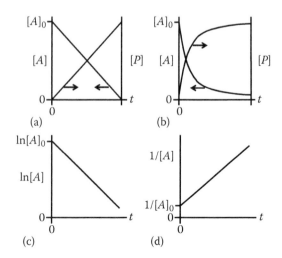

measures a physical quantity (such as optical absorbance) that is proportional to the concentration $[A]$ in that ratio.

An important characteristic of a first-order process is the **half-life** value
($t_{1/2}$). This is the time taken for the reactant concentration $[A]$ to reach
exactly one half of its initial value—the half-life of the $Ag_2C_2O_4$ decomposition reaction is illustrated in Figure 3.1. Substituting the condition
$[A] = [A]_0/2$ into Equation 3.21 yields

$$\ln\left(\frac{[A]_0/2}{[A]_0}\right) = -kt_{1/2} \tag{3.25}$$

which can be simplified and rearranged to Equation 3.26:

$$t_{1/2} = \frac{\ln 2}{k}. \tag{3.26}$$

The half-life is an intuitive measure of how fast a reaction proceeds.
However, the half-life has practical use for comparing reactions for first-
order processes because it is independent of the initial concentration $[A]_0$.
Table 3.2 summarizes rate laws and the corresponding integrated rate
equations for reactions of various orders.

Table 3.2

Rate Expressions for the General Reaction $A + B \rightarrow P$: Summary of Various Rate
Laws and Their Corresponding Integrated Rate Equations and Half-Lives

Reaction Order	Rate Law	Integrated Rate Equation	Half Life	Linear Plot
Zero	$-\dfrac{d[A]}{dt} = k$	$[A] = [A]_0 - kt$	$t_{1/2} = \dfrac{[A]_0}{2k}$	$[A]$ vs. t
First	$-\dfrac{d[A]}{dt} = k[A]$	$[A] = [A]_0\, e^{-kt}$	$t_{1/2} = \dfrac{\ln 2}{k}$	$\ln[A]$ vs. t
Second	$-\dfrac{d[A]}{dt} = k[A]^2$	$\dfrac{1}{[A]} = kt + \dfrac{1}{[A]_0}$	$t_{1/2} = \dfrac{1}{k[A]_0}$	$1/[A]$ vs. t
Second*	$-\dfrac{d[A]}{dt} = k[A][B]$	$\dfrac{1}{[A]_0 - [B]_0} \ln\dfrac{[B]_0[A]}{[A]_0[B]} = kt$	NA	$\ln\dfrac{[B]_0[A]}{[A]_0[B]}$ vs. t

* For $[A]_0 \neq [B]_0$.

3.2 THEORETICAL MODELS FOR REACTION RATES

3.2.1 Temperature dependence of the rate constant

The rate of decomposition of an ensemble of nanoparticles increases with temperature. This intuitive observation holds for any chemical reaction and is based on the premise that temperature represents average kinetic energy, and thus, molecules move faster at higher temperatures and are more likely to collide with each other, providing more opportunities to react. The empirical Arrhenius equation tells us that rate constant depends exponentially on temperature (Equation 3.27):

$$k = Ae^{-E_a/RT} \tag{3.27}$$

The constants A and E_a in the above equation are the preexponential factor and the activation energy, respectively. R is the molar gas constant in units J mol^{-1} K^{-1}. We will discuss the significance of the constants A and E_a in the following two sections. For now, we'll see how these constants can be determined for a particular reaction.

Taking the natural logarithm of Equation 3.27 yields Equation 3.28:

$$\ln k = \ln A - \frac{E_a}{RT} \tag{3.28}$$

Plotting $\ln k$ versus $1/T$ will yield a straight line with a slope $-E_a/R$ and an intercept $\ln A$. If we only have k values at two temperatures, say T_1 and T_2, then we can set up two equations corresponding to these two conditions (Equations 3.29 and 3.30):

$$\ln k(T_1) = \ln A - \frac{E_a}{RT_1} \tag{3.29}$$

$$\ln k(T_2) = \ln A - \frac{E_a}{RT_2} \tag{3.30}$$

Subtracting Equations 3.30 from 3.29 yields Equation 3.31:

$$\ln \frac{k(T_1)}{k(T_2)} = -\frac{E_a}{RT_1} + \frac{E_a}{RT_2} = \frac{E_a}{R}\left(\frac{1}{T_2} - \frac{1}{T_1}\right) \tag{3.31}$$

Example 3.2   The Arrhenius Equation

The rate constant for the gas phase reaction $2HI(g) \longrightarrow H_2(g) + I_2(g)$ was measured at two temperatures as shown:

$$k = 1.30 \times 10^{-6} \, dm^3 mol^{-1} s^{-1} \qquad 580 \, K$$
$$k = 2.50 \times 10^{-3} \, dm^3 mol^{-1} s^{-1} \qquad 715 \, K$$

Use this information to determine the preexponential factor and the activation energy for this reaction.

Solution   Using Equation 3.31,

$$\ln \frac{1.30 \times 10^{-6}}{2.50 \times 10^{-3}} = \frac{E_a}{R} \left(\frac{1}{716} - \frac{1}{580} \right)$$

$$-7.56 = \frac{E_a}{8.3145 \, JK^{-1} mol^{-1}} \left(-3.27 \times 10^{-4} K \right)$$

Solving for E_a gives $25.43 \, kJmol^{-1}$.

To obtain A, we can use Equation 3.27 for either one of the two sets of data. Using the rate constant value at 580 K,

$$A = \frac{k(T_1)}{e^{-E_a/RT_1}} = \frac{1.30 \times 10^{-6} \, dm^3 mol^{-1} s^{-1}}{e^{-25430 \, Jmol^{-1}/(8.3145 \, JK^{-1} mol^{-1})(580 \, K)}} = 2.54 \, dm^3 mol^{-1} s^{-1}$$

3.2.2 Collision theory

We can use our understanding of kinetic-molecular theory to provide a deeper understanding of kinetics. From general chemistry, this theory provides the molecular speeds, collision density, and collision cross section of a sample of gas particles in a container. In a typical gas phase reaction, the collision density is on the order of 10^{30} collisions per liter per second. The reaction rate corresponding to this collision density would be remarkably high (about $10^6 \, mol \, dm^{-3} \, s^{-1}$) if every collision resulted in the formation of a product molecule. Most reactions would be over within a matter of seconds! However, many reactions are known to proceed much slower, implying that only a fraction of these collisions actually lead to product formation. In other words, only a fraction of the reactant molecules have sufficient kinetic energy to break and/or form bonds during intermolecular collisions. The minimum energy required for a reaction to take place is the activation energy (E_a), the same activation energy described by the Arrhenius equation of the last section. Figure 3.5 shows the distribution of molecular kinetic energy, with the activation energy indicated. We see that at higher temperatures a greater fraction of molecules have kinetic energy in excess of E_a, leading to a chemical reaction.

In addition to E_a, the relative orientation of the particles at the exact time of collision is important. Orientation is not important for reactants between spherically symmetrical particles such as atoms (and sometimes spherical nanoparticles). However, for more complicated reactant molecules, their relative positions in space will affect the reaction rate. In particular, orientation factors may be important in surface reactions in which the reactive moiety of a surface bound molecule in a thin nanofilm must point "up" toward the bulk phase. Figure 3.6 illustrates this by describing the binding of an antibody to a nanofilm composed of antigen molecules. Only the correct relative orientation of the reactants will lead to binding. Any other relative orientation between the reactant molecules will not form products.

Transition state theory provides a more robust interpretation of the activation energy. This theory is based on the premise that reactant particles come together to form a very short-lived high-energy complex (the transition state) that very quickly leads to product formation. Figure 3.7 shows the relative energies of the reactants and products as a function of reaction coordinate or reaction progress. In transition state theory, the reactants $A + B$ pass a transition state, described as an activated complex, before being converted into product P. ΔE represents the energetic difference between reactant and product molecules.

Figure 3.5 Distribution of molecular kinetic energies for two temperatures T_1 and T_2, where $T_2 > T_1$. The fraction having kinetic energy in excess of the activation energy is indicated. At higher temperatures, there is a greater fraction of molecules possessing kinetic energy in excess of E_a.

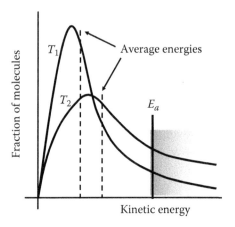

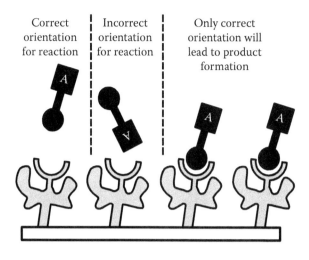

Correct orientation for reaction

Incorrect orientation for reaction

Only correct orientation will lead to product formation

Figure 3.6   Favorable approaches of an antigen and antibody molecules for a reaction to occur. Any other relative orientation between the reactant's molecules will not form products.

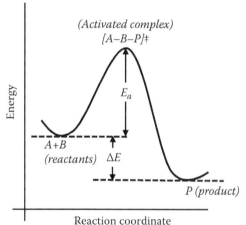

Figure 3.7 Relative energies of the reactants and products as a function of reaction coordinate or reaction progress. In transition state theory, the reactants $A + B$ pass a transition state, described as an activated complex, before being converted into product. ΔE represents the energetic difference between reactant and product molecules.

3.2.3 Catalysis

A catalyst is a species that when added to a reaction mixture speeds up the rate of the reaction. Catalysis has great relevance to reactions such as nanoparticle synthesis, quantum dot functionalization, and to reactions on surfaces. A catalyst increases the rate of a reaction by lowering its activation energy (Figure 3.8). We can distinguish two types of catalysis: homogeneous and heterogeneous. In homogeneous catalysis, the catalyst and reactants are always in the same phase. For example, the hydrolysis of ethyl acetate, encountered in Section 3.1.2, is catalyzed by sulfuric acid. Another example of homogeneous catalysis is the natural destruction of

Figure 3.8 The effect of a catalyst on the activation energy of a reaction. The line with lower activation energy represents the catalyzed reaction. Note that the catalyst does not change the relative energies of the reactants and products. It only lowers the energy of the activated complex. Therefore, although the catalyst speeds up the reaction, the Gibbs energy of reaction is unchanged.

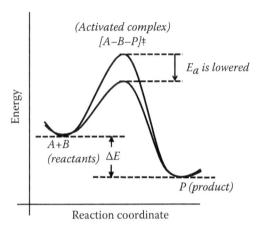

ozone in the atmosphere by species such as halogen radicals and nitric oxide, in which all species, including the catalyst, are in the gas phase.

In heterogeneous catalysis, the catalyst is in a different phase compared to the reactant. For example, in the Fischer–Tropsch process for converting carbon monoxide and hydrogen into liquid hydrocarbons, the reactions occur on surfaces of metals such as iron or cobalt. As another example, catalytic converters in vehicles use catalysts such as platinum, palladium, and rhodium to oxidize CO and hydrocarbons to CO_2 and to reduce nitrogen oxides to oxygen and nitrogen. There is considerable interest in replacing the expensive metals in these catalysts with high-surface-area nanomaterials such as carbon nanotubes.

How do catalysts manage to lower E_a? To answer this question we must appreciate that most reactions occur in a series of steps called **elementary steps**. These steps describe the **mechanism** of the reaction. Furthermore, these elementary steps add up to give a stoichiometric reaction. The overall rate of reaction is always governed the slowest elementary step in the mechanism. Such a step in the mechanism is known as the **rate-limiting** (or **rate-determining**) step. A catalyst changes the reaction mechanism, often providing an alternative pathway with overall lower activation energy (Figure 3.8). This may be due to some additional step that the catalyst offers that circumvents the rate-determining step.

Finally, it is worth noting that the number of reacting particles in an elementary step is described by the **molecularity** of the reaction. If two molecules encounter each other in an elementary step, the molecularity is 2 and the step is referred to as **bimolecular**. Bimolecular collisions resulting in

Table 3.3

Various Elementary Processes and the Corresponding Molecularities

Reacting Species in the Elementary Step	Molecularity	Type of Encounter
$A \longrightarrow P$	1	Unimolecular
$2A \longrightarrow P$	2	Bimolecular
$A + B \longrightarrow P$	2	Bimolecular
$A + B + C \longrightarrow P$	3	Termolecular
$2A + B \longrightarrow P$	3	Termolecular

the formation of an **intermediate** are common in surface reactions. An example would be the attachment of a small molecule onto a surface bound enzymatic nanofilm. Other molecularities are shown in Table 3.3.

3.3 MODELING SIMPLE MECHANISMS

In Section 3.1.3 we stated that forward and reverse rates are equal to each other when a reaction is at equilibrium. The same is true for elementary steps in a mechanism. Consider the general reaction $A \longrightarrow P$, via an intermediate I. Let's say the mechanism is

$$A \underset{k_{-1}}{\overset{k_1}{\rightleftharpoons}} I \overset{k_2}{\longrightarrow} P \qquad (3.32)$$

We see that the first step in the mechanism is reversible. The presence of the second step will interfere with the first as it approaches equilibrium. For instance if $k_2 \gg k_1$, then I is quickly converted to P and the first-step equilibrium may never be established. However, if the first-step equilibrium is achieved right away due to a slower second-step (i.e., $k_1 = k_{-1} \gg k_2$), then I is only slowly converted to P, but the first step instantaneously adjusts to a state of equilibrium.

Now consider a slightly different mechanism described by the nonequilibrium process:

$$A \overset{k_1}{\longrightarrow} I \overset{k_2}{\longrightarrow} P \qquad (3.33)$$

Let's assume that $k_2 \gg k_1$. Therefore, as soon as any I is formed, it is rapidly converted to P. We can conclude that if $k_2 \gg k_1$, then $[I]$ builds up to a small and constant value. This is the basis of the **steady-state approximation**. Mathematically, this approximation can be stated as

$$\frac{d[I]}{dt} = 0 \qquad (3.34)$$

Looking back at Equation 3.33, we can apply the steady-state approximation to I and write

$$\frac{d[I]}{dt} = k_1[A] - k_2[I] = 0 \qquad (3.35)$$

Solving the above equation for $[I]$ yields

$$[I] = \frac{k_1[A]}{k_2} \qquad (3.36)$$

and since the rate of formation of product is given by

$$\frac{d[P]}{dt} = k_2[I] \qquad (3.37)$$

we have

$$\frac{d[P]}{dt} = \frac{k_1 k_2}{k_2}[A] = k_{obs}[A] \qquad (3.38)$$

where

$$k_{obs} = \frac{k_1}{k_2} \qquad (3.39)$$

Thus we see that under steady-state conditions, the reaction (Equation 3.33) is first-order in A and has an observed rate constant given by Equation 3.39. In Section 3.5, we will revisit steady-state kinetics when describing diffusion-limited reactions involving nanoparticles.

Example 3.3   The Steady-State Approximation

Apply the steady-state approximation to the mechanism $A + B \underset{k_{-1}}{\overset{k_1}{\rightleftharpoons}}$ $I \overset{k_2}{\longrightarrow} P$ and obtain an expression for k_{obs}.

Solution   First, obtain an expression for $d[I]/dt$ and set the result to zero. Thus

$$\frac{d[I]}{dt} = k_1[A][B] - k_{-1}[I] - k_2[I] = 0$$

Rearranging gives

$$k_1[A][B] = k_{-1}[I] + k_2[I] = [I](k_{-1} + k_2)$$

Solving for $[I]$ gives

$$[I] = \frac{k_1}{k_{-1} + k_2}[A][B]$$

Since the rate of formation of product is

$$\frac{d[P]}{dt} = \frac{k_1 k_2}{k_{-1} + k_2}[A][B]$$

and so

$$k_{obs} = \frac{k_1 k_2}{k_1 + k_2}$$

3.4 BIMOLECULAR BINDING KINETICS

3.4.1 Kinetics of reversible binding

Let's consider the binding of a molecule, such as a protein, onto a solid surface. We can use Equation 3.40 to describe this process, where A refers to the protein molecule, S refers to the surface, and k_{on} and k_{off} refer to the rate constants in the forward and reverse directions, respectively:

$$A + S \underset{k_{off}}{\overset{k_{on}}{\rightleftharpoons}} AS \qquad (3.40)$$

We can do this experiment by continuously flowing an aqueous solution of the protein over the solid surface in the manner illustrated in Figure 3.9. In this way we always keep bulk phase concentration of the protein, $[A]$, constant. The rate of formation of surface bound protein AS is given by

$$\frac{d[AS]}{dt} = -\frac{d[S]}{dt} = k_{on}[A][S] = k'_{on}[S] \qquad (3.41)$$

Notice that we have defined $k'_{on} = k_{on}[A]$ in the above expression. This is a good approximation since $[A]$ is constant. Thus, k'_{on} is a pseudo-first-

Figure 3.9 Flow cell in which species *A* is continuously passed over a fixed substrate *S*. *A* may be a ligand or a protein molecule and *S* may represent either a binding site or an immobilized receptor molecule.

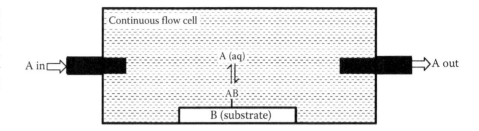

order rate constant. The adsorption process is reversible, so we need to consider the rate of dissociation of *AS* back to free *A* and exposed surface *S*. The overall rate of *AS* formation is the forward rate minus the reverse rate; that is

$$\frac{d[AS]}{dt} = k'_{on}[S] - k_{off}[AS] \tag{3.42}$$

The concentration of the available binding sites on the surface, $[S]$, will decrease exponentially over time (a first-order process) as *AB* is formed. Thus

$$[S] = [S]_0\, e^{-k'_{on}t} \tag{3.43}$$

At any time, the amount of *AB* is given by the following mass balance equation:

$$[AS] = [S]_0 - [S] \tag{3.44}$$

or

$$[S] = [S]_0 - [AS] \tag{3.45}$$

therefore

$$[S]_0 - [AS] = [S]_0\, e^{-k'_{on}t} \tag{3.46}$$

or

$$-[AS] = [S]_0 e^{-k'_{on}t} - [S]_0 \tag{3.47}$$

therefore

$$[AS] = [S]_0\left(1 - e^{-k'_{on}t}\right) \tag{3.48}$$

To determine a value for k'_{on}, we would need to plot experimental values of $[AS]$ versus time and then do a nonlinear fit to the data using the above equation. In order to do this kind of analysis, we need to measure the accumulation of A onto the surface in real time. There are many modern instruments capable of measuring the mass or concentration of a species on a surface as AS forms, as a function of time. We will discuss methods that measure the formation of nanofilms on surfaces in later chapters. If $[A]$ is known (say, the concentration of the protein in the solution flowing over the surface), then k_{on} can be determined, since $k'_{on} = k_{on}[A]$.

To determine k_{off}, we need to examine the rate of dissociation of AS (Equation 3.49):

$$\frac{d[AS]}{dt} = -k_{off}[AS] \tag{3.49}$$

The integrated form of this equation is

$$[AS] = [AS]_0 \, e^{-k_{off}t} \tag{3.50}$$

which can be linearized to

$$\ln[AS] = \ln[AS]_0 - k_{off}t \tag{3.51}$$

Thus, a plot of $\ln[AS]$ versus time will give a slope equal to $-k_{off}$. In practice, a solution of A is flowed over the surface until equilibrium is reached (i.e., $[AS]$ no longer changes with time). At this point, the flow is changed to the pure solvent (with no A) and A desorbs from the surface, decreasing AS as a function of time. It is the data collected from this last step that is used in Equation 3.51.

Finally, it is worth noting that the equilibrium constant is related to both k_{on} and k_{off} according to Equation 3.52:

$$K = \frac{[AS]}{[A][S]} = \frac{k_{on}}{k_{off}} \tag{3.52}$$

Thus, from our values of k_{on} and k_{off}, we can determine the thermodynamic equilibrium constant, K, from real-time kinetic data. The equilibrium constant provides a measure of how strongly A and S are bound in the form AS. In Problem 3.5, we use this analysis to determine the strength of binding between the drug labetalol to an important blood protein. In this problem, the latter is immobilized as a thin nanofilm on a surface and represents B in the analysis.

3.4.2 The Scatchard and Hill equations: Cooperativity of binding

In nanoscience, we often encounter reactions in which some number n of small molecules (or ligands, L) reversibly bind to distinct sites of a large nanoassembly (or macromolecule, M) to form a complex (ML_n). An important example is the binding of oxygen molecules to hemoglobin. Equation 3.53 describes the process:

$$nL + M \rightleftharpoons ML_n \tag{3.53}$$

The equilibrium constant for this process is

$$K = \frac{[ML_n]}{[L]^n[M]} \tag{3.54}$$

Since at any point in time the total concentration of M is $[M] + [ML_n]$ (mass balance), we can describe the fraction of bound molecules by Equation 3.55:

$$f = \frac{[ML_n]}{[M] + [ML_n]} \tag{3.55}$$

The above equation can be rearranged as

$$f = \frac{1}{\dfrac{[M]}{[ML_n]} + 1} = \frac{1}{\dfrac{1}{K[L]^n} + 1} = \frac{1}{\dfrac{1}{K[L]^n} + \dfrac{K[L]^n}{K[L]^n}} = \frac{1}{\dfrac{1 + K[L]^n}{K[L]^n}} = \frac{K[L]^n}{1 + K[L]^n} \tag{3.56}$$

Figure 3.10 shows how f varies with $[L]$ for a K value of 0.1 Lmol^{-1} with $n = 1$. Equation 3.56 can be linearized as described by the following steps:

$$f(1 + K[L]^n) = K[L]^n \tag{3.57}$$

$$f + fK[L]^n = K[L]^n \tag{3.58}$$

$$f = K[L]^n - fK[L]^n = K[L]^n(1 - f) \tag{3.59}$$

$$\frac{f}{[L]^n} = K(1 - f) \tag{3.60}$$

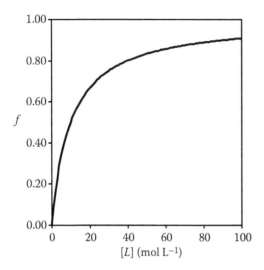

Figure 3.10 Fraction of ligands (f) bound to macromolecule M as a function of ligand concentration [L] for a K value of 0.1 Lmol^{-1}. The plot is shown with $n = 1$.

If all of the sites on the macromolecule are identical, independent, and noninteracting, then $n = 1$. In this case we can rearrange Equation 3.60 into Equation 3.61:

$$\frac{f}{(1-f)} = K[L] \tag{3.61}$$

The above equation is known as the **Scatchard equation**. However, many systems have binding sites that are affected by the presence of other bound ligands, in which case $n \neq 1$, and

$$\frac{f}{(1-f)} = K[L]^n \tag{3.62}$$

Taking logs of both sides yields

$$\log \frac{f}{(1-f)} = n \log[L] + \log K \tag{3.63}$$

The above expression is known as the **Hill equation**. A plot of the left-hand side of the above equation versus log [L] gives a straight line with slope equal to n and an intercept equal to log K. The value n is known as the Hill coefficient and its value tells us how the bound molecules are interacting with each other. A value $n > 1$ is interpreted as **cooperative** binding. This means that once L is bound to M, it promotes the binding of

Figure 3.11 Hill plot for a long-chain anionic surfactant binding to a polycationic nanofilm. There are two distinct linear regions with slopes of 1 and 3. Reproduced with permission from Johal, M. and Chiarelli, P. Polymer-Surfactant Complexation in Polyelectrolyte Multilayer Assemblies. *Soft Matter*, 2007, 3: 34–46.

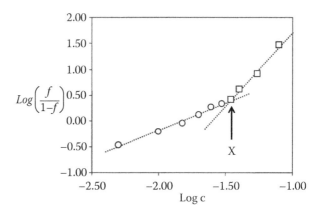

other L molecules to M. If $n < 1$ the binding is **anticooperative**—the bound ligand prevents the binding of other ligands. If $n = 1$, the binding is **noncooperative**—the bound ligand does not affect the binding of other ligands. The theoretical maximum value of n is the total number of binding sites on M. Figure 3.11 shows the Hill plot describing the binding of a small molecule to a polymer nanofilm. One can clearly see that the binding goes from noncooperative to cooperative at some critical concentration shown by the X. We can interpret this observation as a loosely packed layer of L on M at low concentrations, where the molecules are too far from each other to interact. At sufficiently high concentration we have a transition to a more densely packed layer, where the bound molecules are interacting strongly with their neighbors.

3.5 SOLUTION KINETICS AND DIFFUSION CONTROL

The rate-determining step in a mechanism governs the overall rate of reaction. Therefore, in a series of steps in which the reacting species approached each other slowly, the rate of the reaction would depend on the speed at which the reactants approached each other. This is common in solution phase reactions, where the *diffusion* of reactants is slower due to the greater number of collisions between reactant molecules and solvent molecules versus collisions between reactant molecules. Such reactions are known as **diffusion-controlled** (or diffusion-limited) reactions. The reaction rate of a diffusion-controlled reaction is the rate of transport of the reactants through the solution. In diffusion-controlled reactions, the formation of products from the activated complex is much faster than the diffusion of reactants. Thus, the rate is governed by the **collision frequency** between reacting species. Heterogeneous reactions are also often diffusion-controlled, as reactant molecules must reach the interface to participate in the reaction and are therefore common for reactions in which solid nanomaterials (like particles) react with other species in solution.

3.5.1 Some basic physics of diffusion of nanomaterials in solution

This section introduces some of the basic equations describing the diffusion of nanoparticles in solution. Diffusion is an entropically driven process; particles in solution will be more likely to move from regions of high concentration to regions of low concentration since there are a greater number of ways of moving a particle out of the high-concentration region than into it. Consider the movement of particles along a two-dimensional plate of area A. We will define flux (Φ) as the mass moving across this area per unit time. Fick's first law of diffusion tells us that the amount of material diffusing across an area A is directly proportional to the concentration gradient (dC/dt) and the area (Equation 3.64):

$$\Phi = -DA\frac{dC}{dt} \tag{3.64}$$

The proportionality constant D is known as the **diffusion coefficient**. The negative sign in the above equation tells us that diffusion takes place in the direction of decreasing concentration. If we were examining particles diffusing through a spherical shell of radius R per unit time (J), then Equation 3.64 becomes

$$J = 4\pi R^2|\Phi| = 4\pi R^2 D\frac{dC}{dt} \tag{3.65}$$

Diffusion is a result of the Brownian motion that describes the random collision of particles in a given medium. The average displacement, $\langle x \rangle$, of a particle from its original position after time t is given by Einstein's equation (Equation 3.66):

$$\langle x \rangle = \sqrt{2Dt} \tag{3.66}$$

A spherical particle's movement in a solution will depend on viscosity (η) and the radius of the particle (r). Large values of η and r will result in the particle having greater friction through the solution. The frictional coefficient is defined by Stokes' law and is given by Equation 3.67:

$$f = 6\pi\eta r \tag{3.67}$$

The diffusion coefficient is temperature-dependent and is related to the frictional coefficient of the particle by Einstein's law of diffusion (Equation 3.68):

$$D = \frac{k_B T}{f} \tag{3.68}$$

where k_B is the Boltzmann constant. Therefore, for spherical particles, the diffusion coefficient is inversely proportional to the solution viscosity according to Equation 3.69:

$$D = \frac{k_B T}{6\pi\eta r} \tag{3.69}$$

Since $k_B = R/N_A$, we can write Equation 3.69 as

$$D = \frac{RT}{6\pi N_A \eta r} \tag{3.70}$$

Substituting Equation 3.70 into Equation 3.66 gives

$$\langle x \rangle = \sqrt{\frac{RTt}{3\pi\eta r N_A}} \tag{3.71}$$

Example 3.4 The Diffusion Coefficient and Displacement of a Nanoparticle

Calculate the diffusion coefficient of 100-nm nanoparticles in water at 20°C. What is the average displacement of these particles after 1 hour? The viscosity of water at 20°C is 1.002×10^{-3} Ns/m^2.

Solution   Using Equation 3.70, we have

$$D = \frac{RT}{N_A 6\pi\mu r}$$

$$= \frac{8.314\,\text{Jmol}^{-1}\text{K}^{-1} \times 293\,\text{K}}{\left(6.023 \times 10^{23}\,\text{mol}^{-1}\right)(6)(3.14)\left(1.002 \times 10^{-3}\,\text{Nsm}^{-2}\right)\left(100 \times 10^{-9}\,\text{m}\right)}$$

$$= 2.14 \times 10^{-12}\,\text{m}^2\text{s}^{-1}$$

For displacement we use Equation 3.71 and set $t = 3600$ s (1 hour):

$$\langle X \rangle = \sqrt{\frac{RTt}{N_A 3\pi\mu r}}$$

$$= \sqrt{\frac{8.314\,\mathrm{Jmol^{-1}K^{-1}} \times 293\,\mathrm{K} \times 3600\,\mathrm{s}}{\left(6.023 \times 10^{23}\,\mathrm{mol^{-1}}\right)(3)(3.14)\left(1.002 \times 10^{-3}\,\mathrm{Nsm^{-2}}\right)\left(100 \times 10^{-9}\,\mathrm{m}\right)}}$$

$$= 1.24 \times 10^{-4}\,\mathrm{m}$$

3.5.2 Kinetics of diffusion control

We now relate some of the equations from the last section to the rate constant of a diffusion-controlled reaction. Consider the following process.

$$A + B \xrightarrow{k_{obs}} P \tag{3.72}$$

We can break the above reaction down into an initial diffusion of A and B to form the association complex AB. This process can be described by a diffusion-controlled bimolecular rate constant k_1. We note that a competing reverse reaction can also occur in which AB undergoes unimolecular dissociation back to A and B. Let's describe this reverse reaction by the rate constant k_{-1}. This initial reversible step is followed by the rapid unimolecular reaction in which AB converts to P, described by a rate constant k_2. The two steps are summarized below.

$$A + B \underset{k_{-1}}{\overset{k_1}{\rightleftharpoons}} AB \xrightarrow{k_2} P \tag{3.73}$$

We have encountered this type of mechanism in Section 3.3. In this particular case, k_1 describes the slow diffusion step. We have seen that under steady-state conditions, the observed rate constant is given by (see Example 3.3).

$$k_{obs} = \frac{k_1 k_2}{k_2 + k_{-1}} \tag{3.74}$$

For a diffusion-controlled reaction, the AB complex is rapidly converted to P as soon as it is formed. Therefore, $k_2 \gg k_{-1}$, and according to Equation 3.74 $k_{obs} = k_1$ (the diffusion-controlled rate constant). Under these conditions, the rate of reaction is limited by how fast A and B diffuse toward each other to form AB.

In a diffusion-controlled reaction, A and B must approach each other in order to form AB. Our goal is to determine the rate at which B diffuses toward A. To do this, let's first define r_C as the critical distance between A and B necessary for the reaction to occur. As an approximation, we can take r_C to be the sum of the radii of particles A and B (i.e., $r_C = r_A + r_B$). Consider a stationary A particle surrounded by B particles diffusing toward it. The concentration of B at $r = r_C$ is $[B]_{r_C} = 0$. This concentration increases to the bulk concentration at $r = r_\infty$, or $[B]_{r_\infty} = [B]_b$. According to Equation 3.65,

$$J = 4\pi r^2 D \frac{d[B]_r}{dr} \tag{3.75}$$

Let's assume that the total number of particles moving across any spherical shell around A per unit time is the same. In other words, we assume that J is independent of r. This assumption would be valid under steady-state conditions where the concentration gradient is independent of time. Equation 3.75 can be rearranged to

$$d[B]_r = \frac{J}{4\pi r^2 D} dr \tag{3.76}$$

and integrated,

$$\int_{[B]_r}^{[B]_b} d[B]_r = \int_r^\infty \frac{J}{4\pi r^2 D} dr \tag{3.77}$$

$$[B]_b - [B]_r = \frac{J}{4\pi D} \int_r^\infty \frac{1}{r^2} dr \tag{3.78}$$

$$[B]_b - [B]_r = \frac{J}{4\pi r D} \tag{3.79}$$

and so

$$[B]_r = [B]_b - \frac{J}{4\pi r D} \tag{3.80}$$

We can solve for J by realizing that at $r = r_C$, $[B]_r = 0$,

$$0 = [B]_b - \frac{J}{4\pi r_C D} \tag{3.81}$$

or

$$J = 4\pi r_C D[B]_b \tag{3.82}$$

Since both A and B are diffusing toward each other, we need to account for the diffusion coefficients of both species, represented by D_A and D_B, respectively. Thus, Equation 3.82 can be modified to Equation 3.83:

$$J = 4\pi r_C (D_A + D_B)[A][B] \tag{3.83}$$

It should be understood that $[A]$ and $[B]$ in the above equation are the bulk phase concentrations of A and B (i.e., $[A]_b$ and $[B]_b$), respectively.

For our diffusion-controlled reaction, the rate of product formation is determined by the rate at which A and B approach each other. Thus

$$\frac{d[P]}{dt} = k_1[A][B] \tag{3.84}$$

Comparing the above equation with Equation 3.83, we see that the rate constant for the diffusion-controlled reaction is

$$k_1 = 4\pi r_C (D_A + D_B) \tag{3.85}$$

The units of k_1 are $m^3 s^{-1}$. It is sometimes more desirable to express the units in terms of concentration per unit time. The rate constant in units of $mol L^{-1} s^{-1}$ is given by Equation 3.86 (see also Example 3.5):

$$k_1 = 4\pi r_C (D_A + D_B) N_A \times 10^3 \tag{3.86}$$

Example 3.5 Estimating the Diffusion-Controlled Rate Constant

Calculate k_1 by assuming that r_C is twice the radius of a spherical nanoparticle involved in a diffusion-controlled bimolecular collision in water at 20°C. The viscosity of water at 20°C is 1.002×10^{-3} Ns/m^2.

Solution Let r_{np} be the radius of the nanoparticle. Using Equation 3.69, we have

$$D_A = D_B = \frac{k_B T}{6\pi\mu r_{np}}$$

so

$$D_A + D_B = \frac{k_B T}{3\pi \mu r_{np}}$$

Also, $r_C = 2 \times r_{np}$. Substituting these two into Equation 3.85 gives

$$k_1 = 4\pi (2 r_{np}) \left(\frac{k_B T}{3\pi \mu r_{np}} \right) N_A \times 10^3 = \frac{8 k_B T}{3\mu}$$

$$= \frac{8 (1.380658 \times 10^{-23}\, \text{JK}^{-1})(293\, \text{K})}{3 \left(1.002 \times 10^{-3}\, \text{Ns/m}^2 \right)} = 1.077 \times 10^{-17}\, \text{m}^3\text{s}^{-1}$$

We can convert the value in m^3s^{-1} to $\text{molL}^{-1}\text{s}^{-1}$ by multiplying the above number by $10^3 \times N_A$. We get

$$k_1 = \left(1.077 \times 10^{-17}\, \text{m}^3\text{s}^{-1} \right) \left(10^3\, \text{Lm}^{-3} \right) \left(6.023 \times 10^{23}\, \text{mol}^{-1} \right)$$

$$= 6.48 \times 10^9\, \text{molL}^{-1}\text{s}^{-1}$$

End of chapter questions

1. A certain *self-assembly* process leads to the formation of a nanomaterial P as described by the following reaction:

$$A + B + 2C \xrightarrow{k_{obs}} P$$

The experimental rate law is found to be

$$v(t) = -\frac{d[A]}{dt} = k_{obs}[A][B]^2$$

An experiment is carried out where $[A]_0 = 1.5 \times 10^{-2}\, \text{molL}^{-1}$, $[B]_0 = 3.0\, \text{molL}^{-1}$, and $[C]_0 = 2.0\, \text{molL}^{-1}$. The reaction is initiated and after 10s $[A] = 3.5 \times 10^{-3}\, \text{molL}^{-1}$.

(a). Using your knowledge of pseudo-first-order kinetics, determine the rate constant k_{obs} for the reaction and the half-life corresponding to this experiment.

(b). Determine $[A]$ and $[C]$ after 20 s.

2. Many reactions involving the formation of nanomaterials are bimolecular. For the bimolecular reaction $A + B \xrightarrow{k} P$, the rate law is given by

$$\frac{d[A]}{dt} = \frac{d[B]}{dt} = -k[A][B]$$

By defining the variable $x = [A]_0 - [A] = [B]_0 - [B]$, show that

$$-\frac{dx}{dt} = -k([A]_0 - x)([B]_0 - x)$$

Integrate the above equation to obtain

$$kt = \frac{1}{[B]_0 - [A]_0} \ln \frac{[B][A]_0}{[B]_0[A]}$$

3. Apply the steady-state approximation to the mechanism below and obtain an expression for k_{obs}.

$$A + B + C \underset{k_{-1}}{\overset{k_1}{\rightleftharpoons}} I \underset{k_{-2}}{\overset{k_2}{\rightleftharpoons}} P$$

4. The reversible binding data of the drug labetalol to a nanofilm of the protein bovine serum albumin (BSA) is shown in the table below. The first two columns show the mass bound on the BSA surface as a function of time, and the last two columns show the loss of the drug from the BSA surface upon rinsing.

Binding of Drug to BSA (Adsorption)		Loss of Drug from BSA (Desorption)	
Time (s)	Mass of Drug on Surface (ng)	Time (s)	Mass of Drug on Surface (ng)
0	0	0	253.11
5	5.31	5	249.57
15	8.85	10	242.49
20	15.93	15	242.49
25	26.55	20	233.64
30	47.79	25	208.86
35	77.88	30	182.31
40	107.97	35	145.14
45	143.37	40	109.74
50	177.00	45	79.65
55	205.32	50	51.33
60	224.79	55	33.63
65	238.95	60	26.55
70	244.26	65	24.78
75	246.03	70	21.24
80	249.57	75	17.70
85	253.11	80	15.93

Estimate the binding constant of the drug to BSA.

5. Ethidium bromide is a common fluorescent stain used for detecting and visualizing DNA for microscopy and electrophoresis-based assays. The total dye concentration $[L]_0$ and fraction of dye bound is tabulated below as a function of total DNA base pair concentration $[M]_0$.

Dye Concentration $[L]_0$ (M)	Base Pair Concentration $[M]_0$ (M)	f (Fraction of Dye Bound)
1.01×10^{-4}	0.00×10^{0}	0.00
1.01×10^{-4}	1.80×10^{-5}	0.07
9.98×10^{-5}	3.55×10^{-5}	0.14
9.91×10^{-5}	5.27×10^{-5}	0.20
9.83×10^{-5}	6.94×10^{-5}	0.26
9.75×10^{-5}	8.58×10^{-5}	0.31
9.68×10^{-5}	1.02×10^{-4}	0.38
9.61×10^{-5}	1.17×10^{-4}	0.42
9.53×10^{-5}	1.33×10^{-4}	0.47
9.46×10^{-5}	1.48×10^{-4}	0.51
9.39×10^{-5}	1.62×10^{-4}	0.54
9.32×10^{-5}	1.77×10^{-4}	0.58
9.26×10^{-5}	1.91×10^{-4}	0.61
9.12×10^{-5}	2.18×10^{-4}	0.65
9.06×10^{-5}	2.31×10^{-4}	0.68
8.99×10^{-5}	2.44×10^{-4}	0.70

Using the identical-site binding model, $K_A = \frac{[ML]}{[L][M]}$, where $[L]$ is the concentration of the free dye and $[M]$ is the concentration of binding sites on the DNA, perform a Scatchard analysis to determine the affinity constant K_A and number of binding sites per DNA base pair, n.

6. Calculate the diffusion coefficient of 1-, 10-, 100-, and 1000-nm nanoparticles in water at 20°C. What is the displacement of these

particles after 1 hour? Provide a physical interpretation for the trends in D and x with increasing size. The viscosity of water at 20°C is 1.002×10^{-3} Ns/m^2.

7. Prove that the dimensions of the diffusion coefficient are m^{3}s^{-1} using Equation 3.69. Take the units of viscosity as being Ns/m^2.

8. If the process described in Equation 3.73 was not diffusion-controlled, but rather reaction-controlled, show the $k_{obs} = K \times k_2$, where K is

the equilibrium constant describing the formation of the AB complex.

9. Consider a reaction between two spherical nanoparticles in aqueous solution at 50°C. The average diameter of one nanoparticle is 50 nm and the average diameter of the other is 150 nm. Assuming the reaction is diffusion-limited, calculate the diffusion-controlled rate constant in molL^{-1}s^{-1}. The viscosity of water at 20°C is 0.547×10^{-3} Ns/m^2.

References and recommended reading

Berry, S. R., Rice, S. A., Ross, J. *Physical and Chemical Kinetics*, Oxford University Press, Oxford, UK, 2002.

Laidler, K. J. *Chemical Kinetics*, 3rd edition. HarperCollins, New York, 1987.

Tinoco, I.; Sauer, K.; Wang, J. C.; Puglisi, J. D., *Physical Chemistry: Principles and Applications in Biological Sciences*. 4th ed.; Prentice-Hall: Upper Saddle River, NJ, 2002.

CHAPTER 4

Quantum Effects at the Nanoscale

During our study of thermodynamics and kinetics in the last two chapters, we described the properties of bulk and nanoscale systems as state or path functions that could have any physically meaningful value (e.g., any nonnegative value for total energy or entropy, or any real value for changes in energy). Such a description is based in **classical physics**, a group of theories that had been developed by the end of the nineteenth century. Most of what you learned in general physics was classical physics, and to this day, classical physics provides an effective system for characterizing many bulk materials and processes as well as systems containing large numbers of particles. However, classical physics was unable to explain a number of the key aspects of the interaction between light and matter, leading to the discovery in the early twentieth century by Planck and Einstein that energy is *quantized*, meaning that the energy of a system has a number of discrete possible values. This led to the development of the field of **quantum mechanics**.

For a system that contains a large number of particles and/or has sufficient energy to access many states, the possible energies for the system appear continuous, and classical physics remains a good approximation. This is why, for example, we can represent the Gibbs energy of formation for silver nanoparticles from solution using classical physics—many particles are present. However, at the nanoscale, classical physics is no longer able to accurately represent individual particles, and we must account for quantum effects as well.

This chapter describes a number of models to understand the implications of quantized energy levels and wavefunctions that describe the behavior of small particles, such as electrons, in nanoscale regions of space. We will discover that these models allow us to make powerful predictions about the properties of simple nanosystems. As we will discover in later chapters, the widespread application of quantization has important implications in how energy is stored and distributed in nanosystems. It also provides the tools by which we can use light to probe particles and molecular systems. We begin by providing an introduction to quantum theory and the fundamental equation of quantum mechanics (the Schrödinger equation). The application of this equation will allow us solve for energy in particles confined to nanoscale regions of space.

4.1 QUANTUM CONFINEMENT IN NANOMATERIALS

We begin by describing the significance of placing geometrical constraints on materials to create nanosystems of different dimensionalities. We will then explain these consequences using simple quantum mechanical models in the proceeding sections. **Quantum confinement** describes the changes that occur in atomic structure due to very small length scales in particles, usually on the order of nanometers. The changes occur because electrons are trapped in regions in which they interact with the boundaries of the system. Quantum-confined structures can be classified into three different types as determined by the number of dimensions restricting the motion of electrons. Macroscopic bulk materials are not confined at all, with three free dimensions. However, a thin two-dimensional film has two free dimensions (say, on the yz plane) and motion is entirely confined in the x-direction. These two-dimensional planes are sometimes referred to as quantum well superlattices. A quantum wire has only one free dimension. An electron on this wire, for example, may be free to move along the x-direction. The electron is not allowed to move along the other two axes. Restricting motion in all three dimensions gives us a confined structure similar to an individual atom, where the electron is essentially confined to a "dot." Quantum dots fall into this category. One consequence of quantum confinement is that the optical and electronic properties of these materials differ substantially from the bulk phase. This is because the discrete electronic energy levels of an electron in a confined material are size-dependent. We now describe in detail the origin of this quantum size effect.

4.2 BASIC INTRODUCTION TO QUANTUM MECHANICS

4.2.1 Electromagnetic radiation

From general physics you learned that electromagnetic radiation (light) is composed of a varying magnetic and electric field (Figure 4.1). Here, we only consider the electric field component because it interacts to a significant degree with far more classes of matter than the magnetic component. The wavelength (λ) and frequency (ν) of light are related according to Equation 4.1, where c is the speed of light in a vacuum (2.98×10^8 ms^{-1}):

$$c = \nu\lambda \tag{4.1}$$

In quantum theory, light can be considered discrete particles of energy called **photons**. The energy (E) of the photon is directly related to its frequency (or wavelength) according to Equation 4.2:

$$E_{\text{photon}} = h\nu = \frac{hc}{\lambda} \tag{4.2}$$

The constant h is known as the Planck's constant (6.626×10^{-34} J·s) and is probably the most important constant in quantum mechanics. In some situations, Planck's constant is written as $\hbar = h/2\pi$. Also, if a system absorbs or emits electromagnetic radiation, the change in energy of the system is directly related to the wavelength of absorption or emission according to Equation 4.3:

$$\Delta E = \frac{hc}{\lambda} \tag{4.3}$$

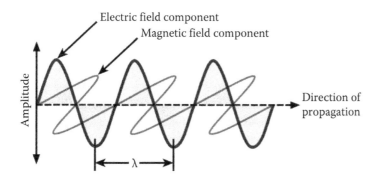

Figure 4.1 Light is composed of an oscillating electric field whose plane of oscillation is perpendicular to an oscillating magnetic field. The wavelength (λ) of the oscillating electric field is indicated.

We will discuss light–matter interactions further in Chapters 5 and 6, especially in the context of spectroscopic methods used to characterize nanomaterials.

4.2.2 Matter waves and the uncertainty principle

There are two ideas central to quantum theory worth discussing. The first is rooted in observations that very small moving particles like electrons have wavelike properties. In other words, a particle moving with a certain momentum (a particle-like property) will undergo wave phenomena such as diffraction and interference. In fact, if the moving particle is small enough it will have a measureable wavelength. This wave-particle duality is described by the de Broglie relationship (Equation 4.4):

$$\lambda = \frac{h}{p} \tag{4.4}$$

In the above equation, p is the particle's momentum, which is given by the product of the particle's mass (m) and velocity (v). Thus, a moving particle with momentum p will have a corresponding de Broglie wavelength, λ.

Example 4.1   The de Broglie Wavelength of a Nanoparticle

Calculate the de Broglie wavelength of a 10-ng particle moving at a speed of 2.0×10^5 ms^{-1}.

Solution   The particle's momentum is $m \times v = (1.0 \times 10^{-11}$ kg) $\times$ $(2.0 \times 10^5$ ms$^{-1}) = 2.0 \times 10^{-6}$ kg ms^{-1}. The corresponding wavelength is

$$\lambda = \frac{h}{p} = \frac{6.626 \times 10^{-34}\,\text{Js}}{2.0 \times 10^{-6}\,\text{kg ms}^{-1}} = 3.313 \times 10^{-28}\,\text{m}$$

This wavelength is too small to be measured (the diameter of a proton is on the order of 10^{-15} m). Thus, a fast moving nanoparticle will not exhibit appreciable wavelike behavior.

The second central idea in quantum theory is the **Heisenberg uncertainty principle.** This principal states that it is not possible to determine both the position and momentum of a small particle with infinite precision. Let's consider a stationary electron. We know that its momentum is zero but in order to locate its position we need to observe it. One way of observing it is to use a stream of photons to locate its position. However, when the photons strike the electron they impart a momentum to the electron, causing it move and consequently creating uncertainty in its position. The uncertainty principle is thus a statement on the inherent uncertainty in a quantum system created by the act of observation. The uncertainty principle also states that product of the uncertainty in position (Δx) and the uncertainty in momentum (Δp) is on the order of at least Planck's constant (Equation 4.5):

$$\Delta x\, \Delta p \geq \frac{h}{2\pi} \text{ or } \Delta x\, \Delta p \geq \hbar \qquad (4.5)$$

Example 4.2   Uncertainty within a Line of Nanoscale Dimension

Calculate the uncertainty in speed of an electron trapped within a 10-nm region on a line. The mass of an electron is 9.109×10^{-31} kg.

Solution   According to Equation 4.5, the uncertainty in momentum is

$$\Delta p \approx \frac{h}{\Delta x} = \frac{6.626 \times 10^{-34}\text{ Js}}{2\pi\left(1.0 \times 10^{-10}\text{m}\right)} = 1.05 \times 10^{-24}\text{ kg ms}^{-1}$$

Thus, the uncertainty in speed is

$$\Delta v = \frac{\Delta p}{m} = \frac{1.05 \times 10^{-24}\text{ kg ms}^{-1}}{9.109 \times 10^{-31}\text{kg}} = 1.15 \times 10^{-4}\text{ ms}^{-1}$$

In the above example, the uncertainty in speed is very small. This is because the 10-nm region is much larger than the size of the electron. If we trapped the electron in a region corresponding to the diameter of a hydrogen atom (120 pm), the uncertainty in its speed will be much greater ($9.65 \times 10^{5}\text{ ms}^{-1}$).

4.2.3 Bound systems and quantization

In this section we introduce the concept of quantization of energy. Classical physics tells us that a moving particle, in principle, may have any value of kinetic energy. There's no reason to think that certain values of kinetic energy are not permitted. Classically energy is a continuously varying function. Quantum theory, however, tells us that small particles, like electrons, behave differently. When confined, only certain values of energy are allowed. This is known as **quantization of energy**. As a simple system, let's examine the ground state of the hydrogen atom. You know from introductory courses in chemistry and physics that the electron normally resides in the $1s$ orbital. It can be excited to the $2s$ orbital where it possesses more energy. In fact, the energy of an electron in hydrogen is quantized and depends on a quantum number n (Equation 4.6):

$$E = -\frac{R_H}{n^2} \tag{4.6}$$

Equation 4.6 provides the energy levels available to an electron in the hydrogen atom. The negative sign in the above equation is used to describe an attractive potential energy interaction between the proton and the electron. A smaller negative value of energy corresponds to the electron having more energy and less attraction for the central proton. R_H is a constant known as the Rydberg energy (about 2.180×10^{-18} J) and the quantum number n takes on values 1, 2, 3, 4, and so on, depending on which energy level the electron is in.

Figure 4.2 shows an energy level diagram for the hydrogen atom, derived from Equation 4.6. We see that the energy levels become closer together. At $n = \infty$, the levels converge. This corresponds to ionization, where the electron is no longer bound to the nucleus. Once the electron is ionized (unbound), it can have any value of energy. It behaves classically. This is a general result in quantum theory. Quantization of energy is a natural consequence of bound systems, such as an electron confined to a small region of space defined by some potential energy of interaction. When there is nothing keeping the particle bound (i.e., no potential energy "boundary"), it behaves classically and its energy is no longer quantized.

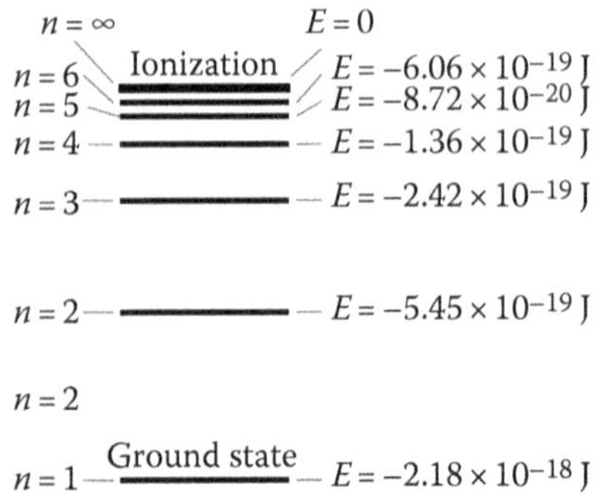

Figure 4.2 Discrete quantized energy levels of an electron orbiting a central proton. The energy levels get closer together as n increases, and eventually converge at $n = \infty$.

$n = \infty$ $E = 0$
$n = 6$ Ionization $E = -6.06 \times 10^{-19}$ J
$n = 5$ $E = -8.72 \times 10^{-20}$ J
$n = 4$ $E = -1.36 \times 10^{-19}$ J
$n = 3$ $E = -2.42 \times 10^{-19}$ J
$n = 2$ $E = -5.45 \times 10^{-19}$ J
$n = 2$
$n = 1$ Ground state $E = -2.18 \times 10^{-18}$ J

At this point it is worth saying something about **spatial quantization**. This is the allowed three-dimensional region given to a bound particle. An example is the $1s$ orbital in hydrogen. We know this to be a spherically symmetrical orbital that accommodates the ground state electron in hydrogen. A $2p$ orbital consists of dumbbell-shaped region occupied by the electron. It should be noted that spatial quantization provides only a probability density of finding a particle within the space defined by certain **boundary conditions**.

4.2.4 The wavefunction

In Sections 4.2.1 and 4.2.2, we established that light and matter can be represented as waves. Quantum confinement is observed when a dimension of a material is of the same magnitude as the de Broglie wavelength of the particles (typically electrons) moving within the material. Confined particles can be represented by an important mathematical function called the **wavefunction**. The wavefunction, usually given the symbol ψ, is a function that describes the wave properties of the bound particle. While a particle's wavefunction cannot be directly observed, the square of the wavefunction has an observable, probabilistic interpretation. For example let's consider an electron confined to a line of length a. This is a simple one-dimensional model where our line can be described along an x-axis, where the length goes from $x = 0$ to $x = a$. Since our model restricts the electron to being on the line, the electron has 100% probability of being found in this region. This is described mathematically by Equation 4.7:

$$\int_0^a \psi(x)^2 dx = 1 \qquad (4.7)$$

A wavefunction obeying Equation 4.7 is said to be a **normalized** wavefunction. In fact, since ψ^2 gives us probability, any well-behaved wavefunction should be normalizable, single-valued, and continuous. Once we have a well-behaved and normalized wavefunction, we can use it to determine the probability of finding our electron between any region on the line of length a. For example, we may want to find the probability of locating our electron between length $x = a/4$ to $x = 3a/4$. To do this, we have to solve the integral defined by Equation 4.8:

$$\int_{a/4}^{3a/4} \psi(x)^2 dx = \text{Probability between } \frac{a}{4} \text{ and } \frac{3a}{4} \qquad (4.8)$$

We discuss this model in more detail in Section 4.3, and show how equations like Equation 4.8 are solved.

4.2.5 The Schrödinger equation

We now present the fundamental equation of quantum mechanics that allows us to describe the energy of confined particles, such as electrons. The **Schrödinger equation**, also known as the Schrödinger wave equation, takes into account the wavelike nature of small particles and is completely consistent with the uncertainty principle. To use the equation, we need to know the particle's mass and define its region of enclosure (known as boundary conditions). For simplicity, we can define the region in one dimension (the x-direction). Solving the equation should allow us to determine the energy of the particle in the defined region. We will do the mathematical derivations in the next section, but for now we will simply present the Schrödinger equation as a simple axiom in quantum theory (Equation 4.9):

$$\hat{H}\psi(x) = E\psi(x) \tag{4.9}$$

The above equation is deceptively simple, so let's discuss each term separately. The term $\hat{H}$ is an **operator** (also known as the **Hamiltonian** operator). Operators describes some kind of mathematical operation, such as "take the square root," or "multiply by four," or "take the second derivative." The Hamiltonian operator is an energy operator, meaning that when it's used in the Schrödinger equation, it will yield the energy of the system. The Hamiltonian operator consists of a kinetic energy term ($\hat{K}$) and a potential energy term ($\hat{V}$). For example, if we had a three-dimensional system described by Cartesian coordinates, the Hamiltonian operator would take the form shown in Equation 4.10, where the kinetic and potential energy terms of the operator are clearly indicated:

$$\hat{H} = \hat{K} + \hat{V} = -\frac{\hbar^2}{2m}\underbrace{\left[\frac{\partial^2}{\partial x^2} + \frac{\partial^2}{\partial y^2} + \frac{\partial^2}{\partial z^2}\right]}_{\text{Kinetic Energy}} + \underbrace{V(x,y,z)}_{\text{Potential Energy}} \tag{4.10}$$

The term ψ is the wavefunction. As discussed in the last section, it's an equation that describes the wave properties of the particle in question and is subject to certain constraints. Thus, Equation 4.9 is an operator equation, which means that the operator $\hat{H}$ acts on the wavefunction ψ to yield the energy of the system E and regenerates the wavefunction ψ.

In operator algebra, wavefunctions are a type of **eigenfunction** and the energies obtained by operating on the wavefunction with the Hamiltonian are the **eigenvalues** of the wavefunction. The example below shows how eigenvalues are obtained from operators and eigenfunctions.

Example 4.3   An Eigenvalue Problem

Consider the operator $\hat{H} = \dfrac{d^2}{dx^2}$ and the eigenfunctions $\psi(x) = B\,e^{-ax}$, where a and B are both constants. Determine the corresponding eigenvalue.

Solution   According to Equation 4.9,

$$\hat{H}\psi(x) = \frac{d^2}{dx^2}\left[B\,e^{-ax}\right] = B\frac{d^2}{dx^2}e^{-ax} = a^2 B\,e^{-ax} = a^2\,\psi(x)$$

The eigenvalue is a^2.

We now apply the Schrödinger equation to more relevant systems in nanoscience. In particular, we focus on the consequences of electrons confined to nanoscale regions of space, as illustrated in Figure 4.3a. For comparison, Figure 4.3b shows a simple molecule undergoing rotational motion and vibrational motion, which will be discussed in Section 4.5. Our goal is to determine the wavefunctions that describe the behavior of an electron confined inside an imaginary "box," a region of space between potential barriers. Electrons become trapped within a box when they do not have enough energy to overcome some sort of potential barrier.

To determine a wavefunction, we must follow a series of steps. First, we must create a model of the system. From the model, we must determine the easiest coordinate system to use. For example, depending on the

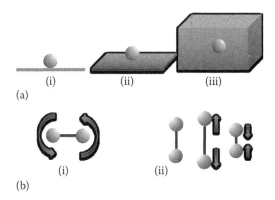

Figure 4.3 (a) A particle confined to (i) a one-dimensional line, (ii) a two-dimensional plane, and (iii) a three-dimensional box. (b) Two-body system undergoing (i) rotational motion and (ii) vibrational motion.

symmetry of the system, it may be more convenient to use Cartesian (e.g., particle on a sheet), spherical (e.g., hydrogen atom), or radial (e.g., particle on a ring) coordinates. We must then define the boundaries of the system and define our Hamiltonian operator based on the model. Finally, we must determine the wavefunctions that fit the constraints of the Hamiltonian operator and the boundary conditions.

Once a wavefunction has been found, the final step is to interpret the wavefunction to provide insight into how our system behaves. We will begin by considering the simple case of an electron confined to a one-dimensional region (a line). We will provide a descriptive rationale for what the wavefunction in this model should look like. Usually, determining wavefunctions requires a more complex mathematical treatment of the model, but this is beyond the scope of this book. Instead, the general form will be provided and this, with the appropriate Hamiltonian operator, will be used in the Schrödinger equation to solve for the eigenvalues, the set of quantized energy states of the system. The mathematical form of the wavefunction will also allow us to describe the position of the electron in terms of probability. This one-dimensional model will then be generalized to describe two (square) and three-dimensional (cube) regions by the simple extension of the one-dimensional wavefunction.

4.3 CONFINEMENT OF ELECTRONS IN BOXES

4.3.1 The one-dimensional model

Let's think about the simplest model describing an electron with mass m_e, trapped on a line segment, such as a nanowire. To constrain the electron on this segment, we impose "walls" of infinite potential at the beginning and the end such that the electron is always between the walls as shown in Figure 4.4. Because we are working with a one-dimensional system, it makes sense to use the x-dimension of a Cartesian coordinate system. The "walls" of our one-dimensional box exist at $x = 0$ and $x = L$. Mathematically

Figure 4.4 A particle of mass m confined to a line of length L. The potential energy is infinite below $x < 0$ and above $x > L$, and zero between $x = 0$ and $x = L$. The particle has only kinetic energy.

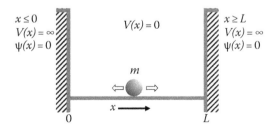

we describe this through boundary conditions by explicitly stating that the wavefunction has a value of zero at and below $x = 0$ and at and above $x = L$. Since the wavefunction is zero at any point beyond the line segment, the probability of finding the electron outside of this region is zero. We write this boundary condition mathematically as

$$\psi(x) = 0 \text{ when } 0 \leq x \text{ and } L \geq x \qquad (4.11)$$

For simplicity we do not allow gravity or electric fields to interact with our electron meaning that our potential operator, $\hat{V}$, is zero within the box and infinite outside. Since there is no potential energy term within the box, the Hamiltonian operator, according to Equation 4.10, becomes simply the kinetic operator, $\hat{K}$, as shown in Equation 4.12:

$$\hat{H} = \hat{K} = -\frac{h^2}{8\pi^2 m}\frac{d^2}{dx^2} \qquad (4.12)$$

As mentioned in the last section, the mathematical form of wavefunctions must obey certain rules: they must be well-behaved, obey the appropriate boundary conditions, and provide solutions to the Schrödinger equation. In this particle-on-a-line model, the rules are met by a sinusoidal wavefunction. This function describes the electron as a wave particle and always yields finite amplitude between $x = 0$ and $x = L$. We can view these wavefunctions as those that can fit an integer or half-integer number of de Broglie wavelengths between $x = 0$ and $x = L$. According to the boundary condition, the amplitude of this wave must be zero at $x = 0$ and $x = L$. Equation 4.13 is such a wavefunction and describes the behavior of a particle confined to a line of length L.

$$\psi_n(x) = A_n \sin\frac{n\pi x}{L} \qquad (4.13)$$

In Equation 4.13, A_n is a constant that normalizes the wavefunction to guarantee that the probability of finding the electron between the bounds is exactly 1. n can take on values of 1, 2, 3, 4, and so on, and is therefore the quantum number describing each wavefunction. As n increases, so does the energy of the electron. An important point to make is that *any* positive integer value of n provides a meaningful wavefunction so that there is not simply *one* solution to the Schrödinger equation obtained using the Hamiltonian in Equation 4.12. Instead, an infinite number of equations exist where each equation represents an individual energy state. The energy states and their wavefunctions are illustrated in Figure 4.5.

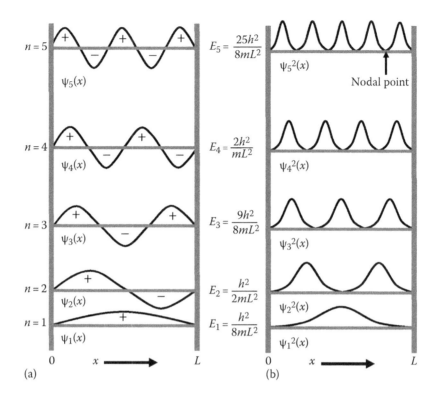

Figure 4.5 The wavefunctions (a) and the wavefunctions squared (b) for a particle confined to a line of length L. These are plotted as a function of n. Also shown are the energy values. The nodal points represent regions of zero probability.

Before we can interpret the wavefunction, we need to take care of a small piece of bookkeeping. Specifically, we need to know the value of the **normalization constant**—the constant that guarantees that the wavefunction will remain within the bounds we set. In the example below we illustrate this process.

Example 4.4   Normalizing the Wavefunction

The wavefunction described by Equation 4.13 is not normalized. Use Equation 4.7 to normalize the wavefunction and determine the normalization constant A_n.

Solution   A normalized wavefunction must obey

$$\int_0^L \psi(x)\psi(x)dx = 1$$

$$\int_0^L \left(A_n \sin \frac{n\pi x}{L} \right) \left(A_n \sin \frac{n\pi x}{L} \right) dx = 1$$

$$A_n^2 \int_0^L \sin^2 \frac{n\pi x}{L} dx = 1$$

Using the fact that the integral $\int \sin^2 ax\, dx = \dfrac{x}{2} - \dfrac{\sin(2ax)}{4a}$

we get $A_n^2 \left| \dfrac{x}{2} - \dfrac{\sin(2n\pi x/L)}{4n\pi} \right|_0^L = 1$

substituting in the integration limits yields

$$A_n^2 \left(\frac{L}{2} - \frac{\sin(2n\pi)}{4n\pi} - \frac{0}{2} - \frac{\sin(0)}{4n\pi} \right) = 1$$

which simplifies to $A_n^2 \left(\dfrac{L}{2} \right) = 1$, because $\sin(2n\pi) = 0$

Thus, the normalization constant is $A_n = \sqrt{\dfrac{2}{L}}$

As mentioned in Section 4.1.4, normalization of the wavefunction ensures that the electron has a probability of 1 of being found within the region governed by the boundary conditions. The above example showed us that the normalized wavefunction for a particle confined along a line of length L is

$$\psi_n(x) = \sqrt{\frac{2}{L}} \sin \frac{n\pi x}{L} \tag{4.14}$$

By applying the Hamiltonian operator (Equation 4.12) on this wavefunction, we obtain the quantized energy values of the electron in this region (Equation 4.18):

$$\hat{H}\psi(x) = E\psi(x) \tag{4.15}$$

$$\hat{H}\psi(x) = -\frac{h^2}{8\pi^2 m} \frac{d^2}{dx^2} \sqrt{\frac{2}{L}} \sin \frac{n\pi x}{L} \tag{4.16}$$

$$-\frac{h^2}{8\pi^2 m} \sqrt{\frac{2}{L}} \frac{d^2}{dx^2} \sin \frac{n\pi x}{L} = \frac{h^2}{8\pi^2 m} \frac{n^2 \pi^2}{L^2} \sqrt{\frac{2}{L}} \sin \frac{n\pi x}{L} \tag{4.17}$$

$$\therefore E_n = \frac{n^2 h^2}{8mL^2} \tag{4.18}$$

Equation 4.18 shows that the electron can only have discrete energy values that are dictated by the value of n. For example, the ground state energy of the electron will be given when $n = 1$. This energy according to Equation 4.18 is $h^2/8mL^2$. Figure 4.5 summarizes the allowed energies as a function of n. The spacing between the energy levels increase as n increases. Also sketched in Figure 4.5 are the wavefunctions and the wavefunctions squared. Interpretation of the wavefunction squared allows us to examine the probability profile of the electron along the x-direction.

In this one-dimensional electron on a line segment model, the electron does not exist as a discrete particle moving along the line. Rather, it resembles a standing wave whose exact form depends on the value of n. The standing waves can be viewed as a cloud of electron density with regions of high and low electron probability. Regions where the wavefunction equals zero correspond to a zero electron probability and are known as **nodes**. Because we interpret the wavefunction in terms of probabilities, we can determine the likelihood of finding the electron in a given region of space. We illustrate this in Example 4.5, where we see that the probability of finding the electron in one half of the interval $(0,L)$ is 50%. In fact, this is true regardless of the value of n, as illustrated in Figures 4.5 and 4.6.

Example 4.5   Determining Probabilities from Normalized Wavefunctions

Calculate the probability of finding the electron between $x = 0$ and $x = L/2$ (i.e., between one half of the line). By selecting any one of

Figure 4.6 The wavefunctions squared for a particle confined to a line of length L, for the $n = 1$ and 2 states. Also shown is the probability of finding the electron between 1 and $L/2$ (one half of the line) as indicated by the shaded region. The shaded area in both cases is 50% of the total area under the ψ^2 curve.

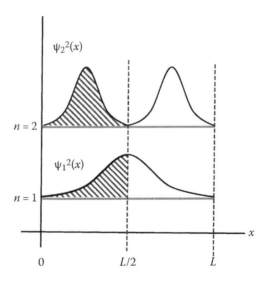

the wavefunctions, verify this probability by considering the area under the plot of $\psi(x)^2$ versus x between $x = 0$ and $x = L/2$.

Solution   This probability is given by the square of the normalized wavefunction integrated between 0 and $L/2$.

$$\int_0^{L/2} \psi*(x)\psi(x) = \frac{2}{L} \int_0^{L/2} \sin^2 \frac{n\pi x}{L} dx$$

Using the trigonometric identity $\sin^2 x = \frac{1}{2}(1 - \cos 2x)$ we can write the above expression as

$$\frac{2}{L} \int_0^{L/2} \frac{1}{2}\left(1 - \cos 2\frac{n\pi x}{L}\right) dx = \frac{2}{L} \int_0^{L/2} \left(\frac{1}{2} - \frac{1}{2}\cos 2\frac{n\pi x}{L}\right) dx$$

Integrating gives

$$\frac{2}{L} \left| \frac{1}{2}x - \frac{1}{2}\frac{L}{2n\pi} \sin 2\frac{n\pi x}{L} \right|_0^{L/2}$$

Subtracting the integration limits and realizing that $\sin(n\pi x)$ is zero gives

$$\frac{2}{L} \left(\frac{L}{4} - \frac{L}{4n\pi} \sin n\pi - 0 - \frac{L}{4n\pi} \sin(0) \right) = \frac{2}{L} \times \frac{L}{4} = \frac{1}{2}$$

Figure 4.6 shows a plot of $\psi(x)^2$ versus x between $x = 0$ and $x = L$ for two representative states ($n = 1$ and $n = 2$). The total area under these curves represents the probability of finding the electron between $x = 0$ and $x = L$. For normalized wavefunctions this is equal to one. Also indicated by the shaded region in the figure is the area under the curve between $x = 0$ and $x = L/2$. In both cases, we see that this is exactly 50% (or 0.5). This result is consistent with the calculation in Example 4.5, which shows that the probability of finding the electron between $x = 0$ and $x = L/2$ is always 50%, regardless of the value of n.

There are a few important features in Figure 4.5 that are in stark contrast to the classical picture. The energy is quantized (it increases with n) as opposed to a continuously varying function. Also, the probability of finding the electron at $x = 0$ and $x = L$ is zero for all values of n. For instance, in the quantum mechanical model the most likely position of the electron is centered at $L/2$ for the $n = 1$ energy level. For the $n = 2$ level we see a single node centered at $L/2$, which tells us that the electron cannot ever be at the center of our one-dimensional box. The number of nodes increases as n increases; in fact the number of nodes is given by n-1. This is in contrast to a classical picture in which no nodes exist and the probability of finding the electron at any point along x is always the same.

Example 4.6   Approaching the Classical Limit

By referring to Figure 4.3, explain how the probability distribution changes as n increases and the implications of this change on the behavior of the electron trapped in the one-dimensional box.

Solution   At low quantum numbers the probability distribution for each energy state is uneven in different regions from $x = 0$ to L. As the quantum number increases, the distribution gets more and more even such that at $n = \infty$ the distribution is even across the entirety of the box. This is the expected classical result and occurs because as n approaches ∞, the energy E of the system approaches ∞. Thus at very large values of n, the likelihood that the electron is bound decreases and so the effect of quantization diminishes.

Since this particle-on-a-line model yields quantized values of energy, it presents the opportunity to investigate energy transitions—an electronic transition in the case of an electron. For example, an electron described by this model may absorb a photon and undergo a transition from the ground state, $n = 1$, to the $n = 2$ level (Figure 4.7). This represents an absorption process, but an emission process may also occur if an excited electron falls back to a lower energy level. The latter process will emit a

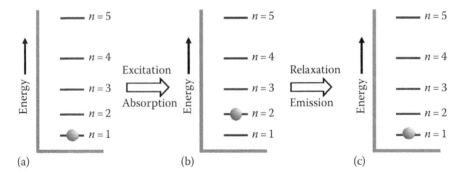

Figure 4.7   A particle on a line in two different energy levels. The transition from (a) to (b) represents an energy absorption process in which the particle is excited from the $n = 1$ state to the $n = 2$ state. The transition from (b) to (c) represents an energy emission process in which the particle relaxes from the $n = 2$ state to the $n = 1$ state.

photon whose energy matches the difference between the energy levels corresponding to the transition. Since we know the dependence of energy on the quantum state from Equation 4.18, we can take the difference in energy between two states to determine the energy of the transition. To describe this transition, we can label the initial quantum number n_i and the final n_f, regardless of whether we have an absorption or an emission process. The energy of the transition (ΔE) is then given by Equation 4.19 and can be directly related to wavelength (λ) of absorption (or emission):

$$\Delta E = E_{n_f} - E_{n_i} = \frac{n_f^2 h^2}{8mL^2} - \frac{n_i^2 h^2}{8mL^2} = \frac{h^2}{8mL^2}\left(n_f^2 - n_i^2\right) = \frac{hc}{\lambda} \qquad (4.19)$$

It is important to note that the value of ΔE in the above expression will be positive for absorption and negative for emission. It is also important to note that due to the **Pauli exclusion principle**, only two electrons (one spin-up, one spin-down) can occupy each state. The electron on a line model has some value in predicting the absorption properties of electrons confined to a line on the order of nanometers.

Example 4.7 Freely Moving Electrons on a Nanowire

Consider six electrons moving along a nanowire of length 2 nm. Determine the wavelength of light required to excite an electron from the ground state to its first excited state in this material.

Solution We have six electrons confined to the nanowire (one free dimension). We need to construct an energy level diagram describing these six electrons. Only two electrons occupy each energy level, and they do so with opposite spins. Figure 4.8 shows

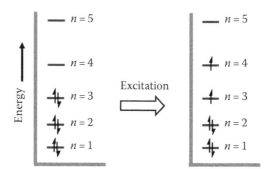

Figure 4.8 Six electrons arranged in the various energy levels on a nanowire. Each energy level accommodates two electrons with opposite spin. An electron from the highest occupied state ($n = 3$) can be excited to the lowest unoccupied state ($n = 4$) by absorbing a photon of the appropriate energy. This excitation process represents the lowest energy transition.

the transition in which the electron is excited from the $n = 3$ to the $n = 4$ level. This represents the lowest energy transition from the ground state to the first excited state. According to Equation 4.19, the energy of the transition is given by

$$\Delta E = \frac{h^2}{8\,mL^2}\left(n_f^2 - n_i^2\right)$$

$$= \frac{\left(6.626 \times 10^{-34}\ J \cdot s\right)^2}{8\left(9.10939 \times 10^{-31}\ kg\right)\left(2.00 \times 10^{-9}\ m\right)^2}(16 - 9)$$

$$= 1.054 \times 10^{-19}\ J$$

And since

$$\Delta E = \frac{hc}{\lambda}, \lambda = \frac{hc}{\Delta E} = \frac{\left(6.626 \times 10^{-34}\ J \cdot s\right)\left(2.9979 \times 10^{-8}\ ^{m}/_{s}\right)}{1.054 \times 10^{-19}\ J}\left(\frac{10^9\ nm}{1\ m}\right)$$

$$= 1982\ nm$$

Now that we have developed an expression for the energy of the electron, we can explore the effect of size on energy level spacing. To determine a general expression for the energy spacing, we can define a particular state by its n value and then describe its neighboring level by $n + 1$. The energy spacing is then

$$\Delta E = E_{n+1} - E_n = \frac{h^2(n + 1)^2}{8mL^2} - \frac{h^2 n^2}{8mL^2} = \frac{h^2}{8mL^2}(2n + 1) \qquad (4.20)$$

Equation 4.20 tells us that energy spacing decreases as the length L increases. In fact, for bulk macroscopic dimensions, L may be so large that the spacings are so small that the levels converge to a continuum. At the other extreme, L may be so small that spacing are too large for transitions to occur. Figure 4.9 illustrates how length scale affects energy level spacing. In the atomic scale to nanoscale, we see clear discrete energy

Figure 4.9 The density of energy states for different length scales.

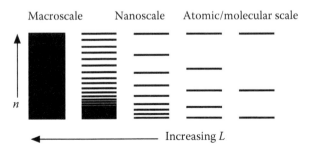

levels compared to the macroscale, and this, as we've seen, has consequences on the electronic and optical properties of the material.

4.3.2 The two- and three-dimensional models and the concept of degeneracy

The particle on a line model can be extended to describe the particle moving in both a two-dimensional rectangular region (a quantum well) and a three-dimensional cubical region (a quantum cube). Let's first discuss the two-dimensional model. Figure 4.10a shows a rectangle of length a and b along the coordinates x and y, respectively. In order to solve the Schrödinger equation for this system, we require the two-dimensional kinetic energy operator described by Equation 4.21:

$$\hat{H} = -\frac{h^2}{8\pi^2 m}\left(\frac{\partial^2}{\partial x^2} + \frac{\partial^2}{\partial y^2}\right) \tag{4.21}$$

The wavefunction we need now depends on the two variables x and y. One way of describing such a wavefunction is to write it as a product of two functions, one that depends only on x and one that depends only on y. Splitting up the wavefunction in this way is known as a **separation of variables** and is commonly used to describe wavefunctions that depend on more than one variable. In this case we can write

$$\psi(x,y) = X(x)Y(y) \tag{4.22}$$

The two-dimensional wavefunction can be reduced to the product of two one-dimensional wavefunctions, each of which is described by Equation 4.14. Thus

$$\psi(x,y) = \sqrt{\frac{2}{a}}\sin\frac{n_x\pi x}{a}\sqrt{\frac{2}{b}}\sin\frac{n_y\pi y}{b} = \sqrt{\frac{4}{ab}}\sin\frac{n_x\pi x}{a}\sin\frac{n_y\pi y}{b} \tag{4.23}$$

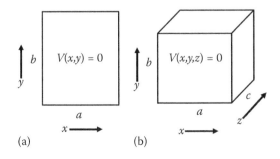

(a)   (b)

Figure 4.10 The two-dimensional (a) and three-dimensional (b) regions within which the particle resides. The wavefunction and potential energy outside of the region is zero and infinite, respectively.

where the quantum numbers n_x and n_y independently assume values of 1, 2, 3, 4, and so on. The boundary conditions for this model are similar to the one-dimensional case in that the wavefunction is zero at the edges and outside of the rectangle. The energy levels corresponding to this wavefunction can be written as a sum of the individual one-dimensional energies given by Equation 4.18. Thus

$$E = \frac{n_x^2 h^2}{8ma^2} + \frac{n_y^2 h^2}{8mb^2} \qquad (4.24)$$

Factoring out the constants gives

$$E = \frac{h^2}{8m} \left(\frac{n_x^2}{a^2} + \frac{n_y^2}{b^2} \right) \qquad (4.25)$$

The above equation can be used to model the energy of electrons freely moving in a 2D quantum well of nanoscale dimension. According to Equation 4.25, the ground state energy of an electron in the well is given when $n_x = 1$ and $n_y = 1$. The next, higher-energy level occurs when one of the n quantum numbers is 2, but exactly which one depends on the values of a and b. If a is the larger of the two, then $n_x = 2$ (and $n_y = 1$) because it will yield the smallest value of E (but still higher than the ground state value). If the length $a = b$ then the electron is trapped in a square of length a. This leads to Equation 4.26, which describes the energy of an electron in a perfect square.

$$E = \frac{h^2}{8m} \left(\frac{n_x^2}{a^2} + \frac{n_y^2}{a^2} \right) = \frac{h^2}{8mL^2} \left(n_x^2 + n_y^2 \right) \qquad (4.26)$$

In the above equation, we have replaced a with L, the length of the square. By changing the rectangle into a square, we have made our region more symmetrical. This symmetry has a very important consequence on the energy levels of the electron. When $n_x = 1$ and $n_y = 1$, the energy is simply $E = 2h^2/8mL^2$. However, when we consider the next energy level we find that there are two combinations of n that lead to the same energy value, these being $n_x = 1$, $n_y = 2$, and $n_x = 2$, $n_y = 1$. These two combinations of n lead to a quantum state in which we have two levels of the same energy ($E = 5h^2/8mL^2$) (Figure 4.11). We refer to this state as being twofold **degenerate**, or doubly degenerate, or having a degeneracy of two. We use the term **degeneracy** to describe a quantum state in which we have a number of levels with the same energy; we previously encountered

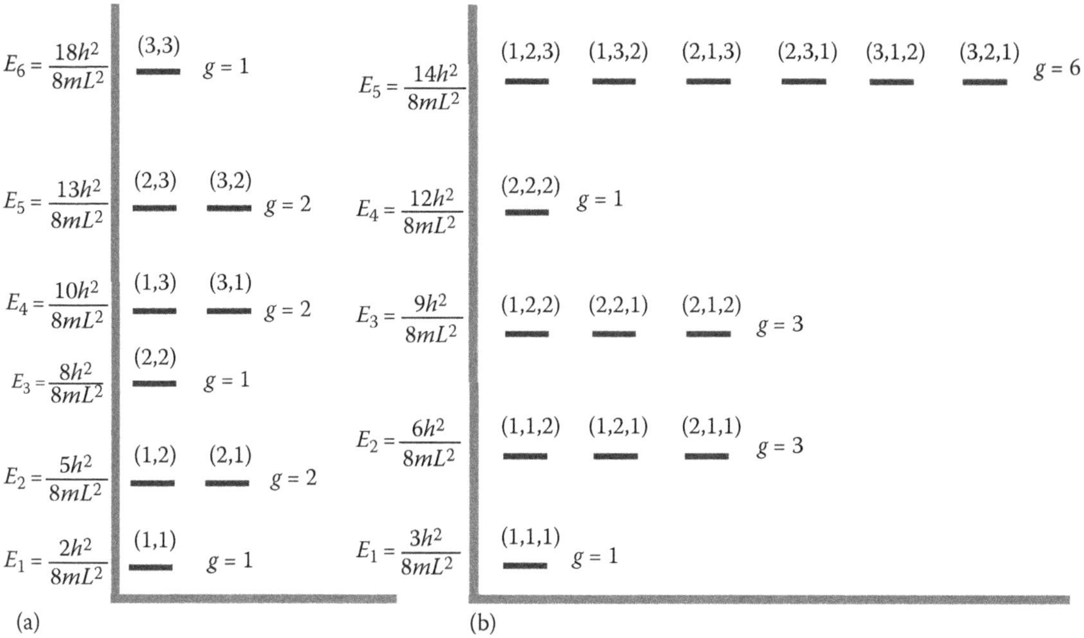

(a)

$E_6 = \dfrac{18h^2}{8mL^2}$ (3,3) $g = 1$

$E_5 = \dfrac{13h^2}{8mL^2}$ (2,3) (3,2) $g = 2$

$E_4 = \dfrac{10h^2}{8mL^2}$ (1,3) (3,1) $g = 2$

$E_3 = \dfrac{8h^2}{8mL^2}$ (2,2) $g = 1$

$E_2 = \dfrac{5h^2}{8mL^2}$ (1,2) (2,1) $g = 2$

$E_1 = \dfrac{2h^2}{8mL^2}$ (1,1) $g = 1$

(b)

$E_5 = \dfrac{14h^2}{8mL^2}$ (1,2,3) (1,3,2) (2,1,3) (2,3,1) (3,1,2) (3,2,1) $g = 6$

$E_4 = \dfrac{12h^2}{8mL^2}$ (2,2,2) $g = 1$

$E_3 = \dfrac{9h^2}{8mL^2}$ (1,2,2) (2,2,1) (2,1,2) $g = 3$

$E_2 = \dfrac{6h^2}{8mL^2}$ (1,1,2) (1,2,1) (2,1,1) $g = 3$

$E_1 = \dfrac{3h^2}{8mL^2}$ (1,1,1) $g = 1$

Figure 4.11 The degeneracy of the energy levels of a particle confined to a perfect square (a) and a perfect cube (b). The letter g states the degeneracy and the numbers in parentheses describes the quantum number combinations.

this concept in the discussion of the entropy of systems. Degeneracy in this model is a natural outcome of the symmetry of the perfect square. If we remove this symmetry (i.e., go back to the rectangle) we effectively *lift* the degeneracy because $a \neq b$, and all energy levels have a unique quantized value.

Recall that the probability profile of the one-dimensional model contained points of zero probability (nodes). Such a probability profile can be generated for a particle in a two-dimensional region of space. Figure 4.12a illustrates this when $n_x = 2$ and $n_y = 2$. One can see regions of zero probability defined by nodal *lines*.

We now consider an extension of the two-dimensional model to a box of length a, height b, and width c, which lie along the x, y, and z coordinates, respectively (Figure 4.10b). By following the same argument we used in going from one-dimension to two-dimensions, we can write the operator (Equation 4.27), the normalized wavefunction (Equation 4.28), and the energy levels (Equation 4.29) for the electron in a box:

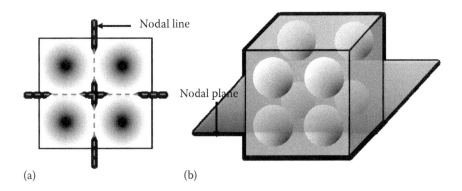

Figure 4.12   (a) The probability density profile for a particle in a two-dimensional region of space. The state shown corresponds to $n_x = 2$ and $n_y = 2$. Regions of zero probability are defined by nodal *lines*. (b) The three-dimensional case corresponding to $n_x = 2$, $n_y = 2$, and $n_z = 2$. Regions of zero probability defined by nodal *planes*.

$$\hat{H} = -\frac{h^2}{8\pi^2 m}\left(\frac{\partial^2}{\partial x^2} + \frac{\partial^2}{\partial y^2} + \frac{\partial^2}{\partial z^2}\right) \tag{4.27}$$

$$\psi(x, y, z) = \sqrt{\frac{8}{abc}}\,\sin\frac{n_x \pi x}{a}\,\sin\frac{n_y \pi y}{b}\,\sin\frac{n_z \pi z}{c} \tag{4.28}$$

$$E = \frac{h^2}{8m}\left(\frac{n_x^2}{a^2} + \frac{n_y^2}{b^2} + \frac{n_z^2}{c^2}\right) \tag{4.29}$$

Equation 4.29 gives the **nondegenerate** quantized energy value for the electron. When $n_x = n_y = n_z = 1$, the electron is in the ground state. The next level of higher energy occurs when one of the n quantum numbers is 2, but exactly which one depends on the values of a, b, and c. If a is the larger of the three, then $n_x = 2$ (and $n_y = n_z = 1$) because it will yield the smallest value of E (but still higher than the ground state value).

Degeneracy of the energy levels emerge when we increase the symmetry of our model so that it's a perfect cube of length L (i.e., $a = b = c$). The energy expression is now simply

$$E = \frac{h^2}{8mL^2}\left(n_x^2 + n_y^2 + n_z^2\right) \tag{4.30}$$

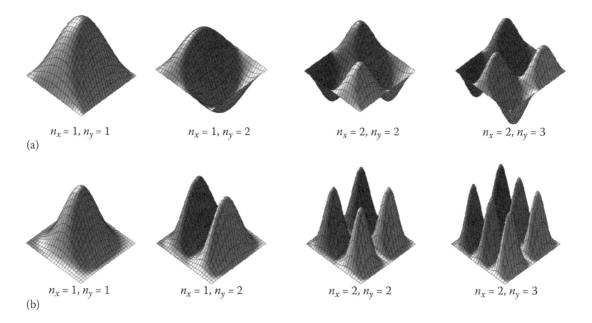

(a) $n_x = 1, n_y = 1$ $n_x = 1, n_y = 2$ $n_x = 2, n_y = 2$ $n_x = 2, n_y = 3$

(b) $n_x = 1, n_y = 1$ $n_x = 1, n_y = 2$ $n_x = 2, n_y = 2$ $n_x = 2, n_y = 3$

Figure 4.13 Computer generated three-dimensional plots representing (a) the wavefunction and (b) the square of the wavefunction, for a particle in a two-dimensional plane. The plots are shown for the first four quantum states. In (b), the amplitude of the curve indicates the probability profile of the particle.

The ground state is still nondegenerate, but when one of the n values is 2, we are in a state in which *three* levels have the same energy ($E = 6h^2/8mL^2$). Figure 4.11 shows how the degeneracy changes for the first few quantum states for a particle trapped in a perfect square and in a cube of length L. In addition to knowing the energy, a probability profile can be generated for a particle in a three-dimensional region of space. Figure 4.12b illustrates this for the state having quantum numbers $n_x = 2$, $n_y = 2$, and $n_z = 2$. We can see *planes* of zero probability—these are **nodal planes**. Figures 4.13 and 4.14 show computer-generated three-dimensional plots representing the probability profile for an electron in a square and cube, respectively.

Equation 4.30 has some important applications. For example, it can be used to calculate the optical properties of an electron trapped in a cubical region of nanoscale dimension. Problem 4.9 goes through this exercise for a solvated electron.

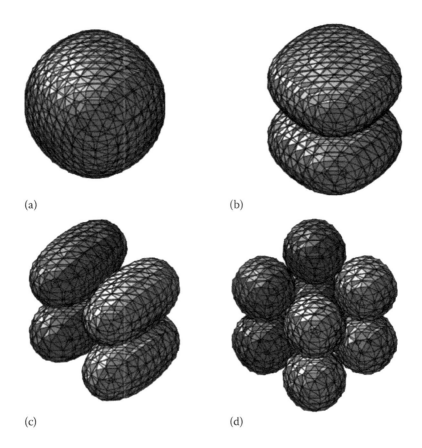

(a)   (b)

(c)   (d)

Figure 4.14   Computer-generated three-dimensional plots representing the probability profile (square of the wavefunction) for a particle in a three-dimensional cube. Shown are (a) $n_x = 1$, $n_y = 1$, and $n_z = 1$; (b) $n_x = 1$, $n_y = 1$, and $n_z = 2$; (c) $n_x = 1$, $n_y = 2$, and $n_z = 2$; (d) $n_x = 2$, $n_y = 2$, and $n_z = 2$.

4.4 NANOSCALE CONFINEMENT ON RINGS AND SPHERES

Our final two simple models for quantum confinement consider electrons in which the confining region can be defined by a ring or sphere, such that the Hamiltonian and wavefunction depend on angular coordinates. Such models are useful for systems such as aromatic molecules and spherical nanoparticles.

4.4.1 The particle on a ring model

Consider an electron confined to a ring of radius r (Figure 4.15) whose position can be described by the angle ϕ. Once again we consider a kinetic energy operator, but this time we select one that is appropriate for particles in rotational motion. The operator is given by

$$\hat{H} = -\frac{h^2}{8\pi^2 I}\frac{d^2}{d\phi^2} \tag{4.31}$$

In this equation we have replaced the mass of the electron, m, by its rotational equivalent, I, the moment of inertia, which is simply $I = mr^2$. Equation 4.32 gives the correct wavefunction describing this system:

$$\psi(\phi) = Ae^{in\phi} \tag{4.32}$$

where A is a constant, $i = \sqrt{-1}$, and n can take on values 0, ±1, ±2, ±3, ±4, and so on. The boundary condition for this model dictates that the wavefunction repeats itself after the particle has made a complete rotation around the ring; in other words,

$$\psi(\phi) = \psi(\phi + 2\pi) \tag{4.33}$$

The value of the normalization constant A is determined in Problem 10. The normalized wavefunction is given by Equation 4.34:

$$\psi(\phi) = \sqrt{\frac{1}{2\pi}}e^{in\phi} \tag{4.34}$$

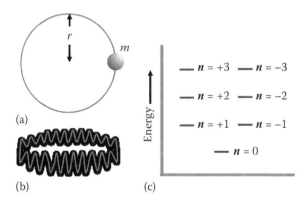

(a)

(b)

(c)

Figure 4.15 (a) A particle confined to a ring of radius r. (b) The wavefunctions for this model look like complete waves around the ring. (c) The energy-level diagram corresponding to the particle on a ring model.

Solving the Schrödinger equation for this system yields the energy levels described by Equation 4.35:

$$E = \frac{h^2}{8\pi^2 I} n^2 \tag{4.35}$$

According to the above equation, the ground state energy is zero because $n = 0$. The presence of n^2 in Equation 4.35 implies the levels above this are always twofold degenerate because $n = \pm1, \pm2, \pm3, \pm4$, and so on. An energy level diagram for this model is shown in Figure 4.15c.

Example 4.8   Estimating the Size of a Ring Nanoantenna

Imagine a single electron confined to a ring "nanoantenna." This electron absorbs light of wavelength 1000 μm (1 mm, in the microwave region of the electromagnetic spectrum), which corresponds to the lowest electronic energy transition (Figure 4.16). Estimate the radius of the ring.

Solution   We first change the absorption wavelength into the corresponding energy.

$$\Delta E = \frac{hc}{\lambda} = \frac{(6.626 \times 10^{-34}\ \text{Js})\,(2.998 \times 10^8\ \text{ms}^{-1})}{1 \times 10^{-3}\ \text{m}} = 1.99 \times 10^{-22}\ \text{J}$$

Assuming that the $n = 0$ to the $n = 1$ transition is associated with the absorption wavelength of 1 mm, we rearrange Equation 4.35 to determine the moment of inertia:

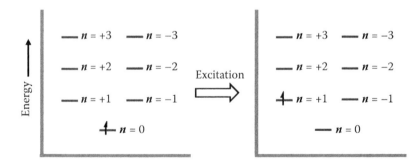

Figure 4.16   An electron in the ground state for the particle on a ring model. The electron from highest occupied state ($n = 0$) can be excited to the lowest unoccupied state ($n = \pm 1$) by absorbing a photon of the appropriate energy. This excitation process represents the lowest energy transition.

$$I = \frac{h^2\left(1^2 - 0^2\right)}{\Delta E\left(8\pi^2\right)} = \frac{\left(6.626 \times 10^{-34} \text{ Js}\right)^2}{1.99 \times 10^{-22} \text{ J} \times \left(8\pi^2\right)} = 2.80 \times 10^{-47} \text{ Js}^2$$

$$= 2.80 \times 10^{-47} \text{ kgm}^2$$

(Note: $1\text{Js}^2 = 1 \text{ kgm}^2$)

Finally, the radius of the ring can be estimated by realizing that $I = mr^2$:

$$r = \sqrt{\frac{I}{m_e}} = \sqrt{\frac{2.80 \times 10^{-47} \text{ kgm}^2}{9.109 \times 10^{-31} \text{ kg}}} = 5.54 \times 10^{-9} = 5.54 \text{ nm}$$

4.4.2 The particle in a sphere model

One quantum mechanical model that can be used to describe spherical nanocrystals or quantum dots is the spherical potential well. This is spherical region of a given radius with zero potential inside and infinite potential outside that radius. Solving for the energy states of a particle in this spherical potential well is beyond the scope of this book, but the solutions are worth noting. If we describe the particle using spherical polar coordinates, it position can be described by some distance r from the center and two angles (the azimuthal and the colatitude angles θ and ϕ, respectively). The wavefunction can then be written as a product of a radial function $R(r)$ and an angular function $Y(\theta,\phi)$ (Equation 1.1).

$$\psi(r, \theta, \phi) = R(r)Y(\theta, \phi) \tag{4.36}$$

The radial equation below is used to obtain the eigenvalues E for the system (Equation 1.2).

$$-\frac{\hbar^2}{2m}\left[\frac{d^2R(r)}{dr^2} + \frac{l(l+1)\hbar^2}{2mr^2} + V(r)\right]R(r) = ER(r) \tag{4.37}$$

l in the above equation is a quantum number taking values 0,1,2,3,..., and in the case of an infinite spherical potential well the term $V(r) = 0$. The solutions of Equation 1.2 are expressed in terms of another set of functions called the Bessel functions, but we will not discuss them here. In the special case of $l = 0$ (spherical symmetry) the corresponding Bessel function yields the eigenvalue given below.

$$E = \frac{h^2}{8mr^2} \tag{4.38}$$

The energy eigenvalue in Equation 4.38 represents the difference in the energy of the first excited state of a nanoparticle due to quantum confinement from the energy that would be observed in a bulk material, which allows control of the optical and electronic properties of quantum dots (semiconductor nanoparticles). We will discuss the semiconducting properties of quantum dots in Chapter 9.

4.5 QUANTIZATION OF VIBRATION AND ROTATION

4.5.1 Quantization of vibrational motion: The harmonic oscillator

We now consider the quantization of two-body systems. Consider two masses, m_1 and m_2, connected by a spring (Figure 4.17a). Let's assume that the spring obeys Hooke's law: the restoring force (F) is proportional to the displacement (x) from the equilibrium or "resting" position x_0 (Equation 4.39):

$$F = -k(x - x_0) \tag{4.39}$$

The proportionality constant, k, is known as the force constant and provides a measure of the stiffness of the spring. This system is called a

Figure 4.17 (a) A simple harmonic oscillator comprised of two masses (m_1 and m_2) attached to a spring that obeys Hooke's law. (b) The rigid rotator in which the two masses are attached to a rigid nonvibrating rod. Both of these simple systems can be used as models for describing the vibrations and rotations of a diatomic molecule.

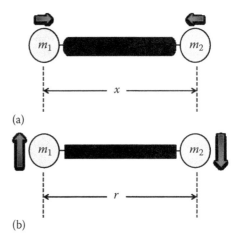

harmonic oscillator and vibrations occur with an amplitude $A = x-x_0$. Classical physics provides the vibrational energy as

$$E = \frac{1}{2}kA^2 \qquad (4.40)$$

We see that the energy is related to the amplitude squared. Now let's imagine this system to be very small, such that m_1 and m_2 are the sizes of atoms. At this scale, the system can be described using quantum mechanics. We can replace the spring by a bond that also obeys Hooke's law. In this rather simple model of a diatomic molecule, it turns out that energy is no longer a continuous function as described by Equation 4.40. Instead, the energy of the vibrating system is dictated by a quantum number. Solving the Schrödinger equation for this system yields the energy levels given by Equation 4.41:

$$E = h\upsilon\left(v + \tfrac{1}{2}\right) \qquad (4.41)$$

In this equation v is a quantum number, known as the **vibrational quantum number**. It can take values $v = 0, 1, 2, 3, 4$, and so on. Unlike our particle-in-a-box model, the lowest value of v is zero. The quantity υ in Equation 4.41 is known as the **fundamental vibration frequency**, which describes the number of vibrations per second. It depends on the force constant (which now is a measure of the strength of the bond between the two atoms) and the masses m_1 and m_2 (Equation 4.42):

$$\upsilon = \frac{1}{2\pi}\sqrt{\frac{k}{\mu}} \qquad (4.42)$$

The quantity μ is known as the **reduced mass** and is given by

$$\mu = \frac{m_1 m_2}{m_1 + m_2} \qquad (4.43)$$

Using a harmonic oscillator model, we have just described motion of a diatomic molecule whose frequency is given by Equation 4.42. It should be noted that the above two equations are also classical mechanics results. The vibrational frequency is directly proportional to the square root of the force constant and inversely proportional to the square root of the reduced mass. Thus, diatomic molecules with stiffer (larger k values) bonds have larger vibrational frequencies; those composed of heavier (larger μ values) atoms have smaller vibrational frequencies. Furthermore, the energy of vibration is quantized according to Equation 4.41.

Example 4.9 Units of the Fundamental Vibration Frequency

Show that Equation 4.42 has units of reciprocal seconds (s^{-1})

Solution We only need to consider the units of the term $\sqrt{\dfrac{k}{\mu}}$

Substituting the corresponding SI units into the above term gives

$$\sqrt{\frac{N/m}{kg}} = \sqrt{\frac{kgms^{-2}/m}{kg}} = \sqrt{\frac{ms^{-2}}{m}} = \sqrt{\frac{1}{s^2}} = \frac{1}{s} = s^{-1}$$

According to Equation 4.41, the lowest possible energy possessed by a harmonic oscillator occurs when $v = 0$. However, the energy of the ground state is not zero; it has a value of $E = 1/2hv$. This nonzero ground state energy is referred to as the the *zero-point vibrational energy*. Figure 4.18 shows an energy-level diagram, illustrating that the levels are nondegenerate and equally spaced. If the harmonic oscillator is in the $v = 1$ state it has more energy and vibrates with greater "spread" compared to the $v = 0$ ground state (Figure 4.19). Like the previous models we've discussed, the quantum mechanical harmonic oscillator can undergo energy transitions between the various vibrational energy levels by the absorption or emission of radiation. Vibrational spectroscopy will be discussed in Chapter 6.

Figure 4.18 The energy levels for a harmonic oscillator. Shown in the same diagram is the potential energy profile of the vibration. The distance r_e is the equilibrium separation and represents the lowest (most stable) potential energy.

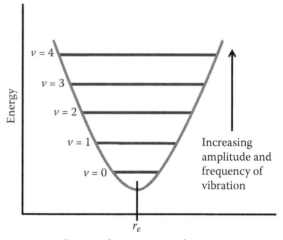

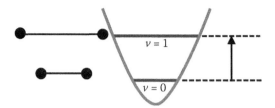

Figure 4.19   The fundamental vibrational transition in which a quantum mechanical oscillator goes from the $v = 0$ to the **v** = 1 level. The transition results in a vibrationally excited state in which the oscillator has greater vibrational energy (and amplitude). This system can be used to describe the vibrational excitation of a diatomic molecule.

Example 4.10   Vibrational Energy Transitions

Consider two atoms that vibrate across a bond with a fundamental frequency of 6.42×10^{13} s^{-1}. Determine the wavelength of the photon required to excite the system to its first vibrationally excited state.

Solution   According to Equation 4.41,

$$\Delta E = h\upsilon\left(v_f + \tfrac{1}{2}\right) - h\upsilon\left(v_i + \tfrac{1}{2}\right) = h\upsilon\left(1 + \tfrac{1}{2}\right) - h\upsilon\left(0 + \tfrac{1}{2}\right)$$

where v_i is the initial state ($v = 0$) and v_f is the final state ($v = 1$).

$$\Delta E = h\upsilon\left(\tfrac{3}{2}\right) - h\upsilon\left(\tfrac{1}{2}\right) = h\upsilon\left(\tfrac{3}{2} - \tfrac{1}{2}\right) = h\upsilon$$

Thus, ΔE is $(6.42 \times 10^{13}$ s$^{-1}) \times (6.626 \times 10^{-34}$ Js$) = 4.25 \times 10^{-20}$ J

$$\lambda = \frac{hc}{\Delta E} = \frac{\left(6.626 \times 10^{-34}\,\text{Js}\right)\left(2.998 \times 10^{8}\,\text{ms}^{-1}\right)}{4.25 \times 10^{-20}\,\text{J}} = 4.67 \times 10^{-6}\,\text{m}$$

This value is 4.67 μm and corresponds to the wavelength of an infrared photon.

4.5.2 Quantization of rotational motion: The rigid rotator

The final quantum mechanical model we will discuss is that of the rigid rotator. In this model we have a mass m_1 and m_2 connected by a rigid rod of length r analogous to that shown in Figure 4.17b. Although this system cannot vibrate, it can undergo rotational motion. When using classical physics to describe rotational motion, the rotational kinetic energy is given by

$$E = \frac{1}{2}I\omega^2 \tag{4.44}$$

In this equation, ω is the angular velocity of rotation and I is the moment of inertia, which is

$$I = \mu r^2 = \frac{m_1 m_2}{m_1 + m_2} r^2 \tag{4.45}$$

for a diatomic molecule. The quantum mechanical solution provides the quantized values of rotational energy (Equation 4.46):

$$E = BJ(J + 1) \tag{4.46}$$

The constant B is known as the rotational constant and its value depends on the moment of inertia I of the molecule (Equation 4.47):

$$B = \frac{h^2}{8\pi^2 I} \tag{4.47}$$

Just as the harmonic oscillator can be used to describe the vibrational states of a diatomic molecule, the rigid rotator can be used to describe the rotational energy states available to a diatomic molecule. As can be seen in Equation 4.47, a molecule with large masses m_1 and m_2 will have a relatively large moment of inertia (I), and thus a correspondingly small rotational constant.

Example 4.11   Units of the Rotational Constant

Use Equation 4.47 to show that the units of B are joules.

Solution   We only need to consider the units of the term $\frac{h^2}{I}$.

Substituting the corresponding SI units into the above term gives

$$\frac{J^2 s^2}{kg\ m^2} = \frac{N^2 m^2 s^2}{kg\ m^2} = \frac{N^2 s^2}{kg} = \frac{kg^2 m^2 (s^{-2})^2 s^2}{kg} = kgm^2 s^{-2} = Nm = J$$

According to Equation 4.47, the lowest possible energy possessed by a rigid rotator occurs when $J = 0$, yielding an energy of zero. Thus the ground state rotational energy of a rigid rotator is zero. In other words there is no rotational motion in the ground state. This result is in contrast to the models discussed so far, where there is finite energy in the ground state. It should be noted that each rotational energy state has a degeneracy of $2J + 1$.

Figure 4.20 Rotational energy levels of a rigid rotator. Note that the energy levels get progressively more spaced out at higher quantum numbers. The degeneracy of each level is also shown.

Figure 4.20 shows an energy level diagram for a quantum mechanical rigid rotator. The diagram shows the relative energy spacings of the various rotational states and the corresponding degeneracies. If the rigid rotator is in the fivefold degenerate $J = 2$ state it has more energy and rotates with greater "rotational spread" compared to the threefold degenerate $J = 1$ state. The quantum mechanical rigid rotator can undergo energy transitions between the various rotational energy levels by the absorption or emission of radiation.

Example 4.12   Rotational Energy Transitions

Consider a two-body system with a rotational constant equal to 2.10×10^{-22} J. Determine the wavelength of the photon required to excite the system to its first rotationally excited state.

Solution   According to Equation 4.46,

$$\Delta E = BJ_f(J_f + 1) - BJ_i(J_i + 1) = 2B$$

where J_i is the initial state ($J = 0$) and J_f is the final state ($J = 1$).

$$\Delta E = 2 \times \left(2.10 \times 10^{-22}\,\text{J}\right) = 4.20 \times 10^{-22}\,\text{J}$$

$$\lambda = \frac{hc}{\Delta E} = \frac{\left(6.626 \times 10^{-34}\,\text{Js}\right)\left(2.998 \times 10^{8}\,\text{ms}^{-1}\right)}{4.20 \times 10^{-22}\,\text{J}} = 4.73 \times 10^{-4}\,\text{m}$$

This value corresponds to the wavelength of a microwave photon.

4.6 QUANTUM MECHANICAL TUNNELING

4.6.1 Implications of finite energy barriers

A key feature of the particle in box models described in Section 4.3 is the presence of infinite potential barriers, which serve to preclude the presence of the particle outside of the region of interest. If the potentials barriers were not infinite, the solution to the Schrödinger equation tells us that the particle has a finite probability of residing beyond the line region, a phenomenon known as **quantum mechanical tunneling**. This happens even if the particle does not have enough energy to classically cross the barrier. However, the probability density in the regions beyond the line decays very rapidly. Figure 4.21 illustrates this situation for the one-dimensional model in which both sides of the line have a finite potential. The figure shows that the wavefunction leaks outside of line boundary beyond the limits defined by $x = 0$ and $x = L$. However, the wavefunction decays exponentially beyond the boundaries, with the rate of this decay depending on the exact finite potential term $V(x)$. Since the wavefunction squared can be interpreted as probability, this leaky wavefunction implies that the electron has a small but finite probability of being present beyond this region.

Some experimental methods actually take advantage of quantum mechanical tunneling. For example, we will see in Chapter 8 that some techniques use tunneling to detect the presence of another body or particle. The idea is

Figure 4.21 The particle on a one-dimensional model in which both sides of the line have (a) an infinite potential and (b) a finite potential. For the finite case, (b) and (c) shows that the wavefunction leaks outside of the region defined by $x = 0$ and $x = L$. The wavefunction decays exponentially beyond the boundaries, with the rate of this decay depending on the exact finite potential term $V(x)$.

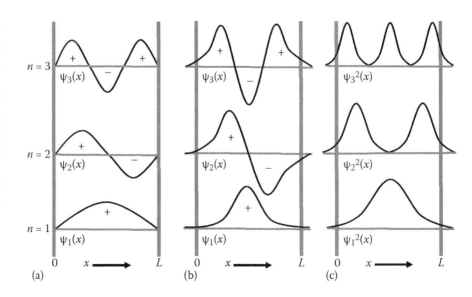

based on measuring a "tunneling" current that is produced when two bodies approach each other without ever touching. The electron essentially jumps from one body to the other. If there is a potential difference applied between the two bodies, then tunneling electrons will produce an electrical current. It is important to appreciate that this kind of electron tunneling occurs because two bodies are so close together that their electronic wavefunctions begin to overlap. Scanning tunneling microscopy (STM) is an example of how quantum mechanical tunneling can be used to probe the nanostructure of surfaces. The technique operates by monitoring the "tunneling" current that is produced when a sharp tip is brought extremely close to a surface that is able to conduct electricity.

4.6.2 Implications for the quantum mechanical harmonic oscillator

In Section 4.5.1 we discussed the energy levels of a harmonic oscillator. The general form of the wavefunctions for this model are rather complicated because they rely on complex functions known as Hermite polynomials. However, the first few wavefunctions have simple forms and the first four of these are given by Equations 4.48 to 4.51, corresponding to the $v = 0$ to $v = 3$ levels:

$$\psi_0(x) = \left(\frac{\alpha}{\pi}\right)^{1/4} e^{-\alpha x^2/2} \tag{4.48}$$

$$\psi_1(x) = \left(\frac{4\alpha^3}{\pi}\right)^{1/4} x e^{-\alpha x^2/2} \tag{4.49}$$

$$\psi_2(x) = \left(\frac{\alpha}{4\pi}\right)^{1/4} \left(2\alpha x^2 - 1\right) e^{-\alpha x^2/2} \tag{4.50}$$

$$\psi_3(x) = \left(\frac{\alpha^3}{9\pi}\right)^{1/4} \left(2\alpha x^3 - 3x\right) e^{-\alpha x^2/2} \tag{4.51}$$

where

$$\alpha = 2\pi \left(\frac{k\mu}{h^2}\right)^{1/2} \tag{4.52}$$

The squares of the above wavefunctions give the probability of finding masses m_1 and m_2 along the length of the spring (or bond) when the

Figure 4.22 The harmonic oscillator wavefunctions. The $\psi^2(x)$ curves for each quantum state show that the masses have a finite probability of being located beyond their classical limits.

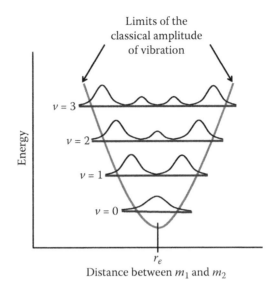

system is in that state. These are plotting in Figure 4.22 for $v = 0$ to $v = 3$. The solid parabola in Figure 4.22 defines the classical potential energy of the oscillator. It clearly shows how the energy changes as a function of x. However, the $\psi^2(x)$ curves for each quantum state show that the masses have a finite probability of being located beyond their classical limits. This is another example of quantum mechanical tunneling and is a direct result of using a finite harmonic oscillator potential energy term in the Hamiltonian operator used to solve the Schrödinger equation for this system.

4.7 SUMMARY

In this chapter we have seen that the precise energy value of an electron confined to a well-defined region is dictated by a quantum number. The particle's spatial (or probability) distribution can be described by the corresponding wavefunctions, again the exact distribution depending on a quantum number. Thus, these models exhibit quantization in both intrinsic energy and their spatial characteristics. We have also considered quantization of vibrational and rotational motion, where discrete energy levels once again determine the exact energy of the vibrating or rotating system. All of the models discussed in this chapter underscore the stark contrast between quantum mechanical outcomes and their classical equivalents. For example, the presence of nodes (as points, lines, or planes), the appearance of degeneracy, the existence of tunneling, and the probabilistic nature of the wavefunction are features of the quantum mechanics and have no classical analog. In subsequent chapters we discuss the implications of quantization on nanomaterial properties and the methods used to characterize them.

End of chapter questions

1. Consider an electron free to move along a line of length 200 nm. What is the ground state energy of the electron? What is the energy of the electron in the $n = 3$ state? What is the effect on the spacing between neighboring energy levels if the length of the line increased from 200 nm to 300 nm?

2. Calculate the energy of the $n = 2$ and $n = 6$ level of an electron constrained to move along a line of length 100 nm. Determine the energy corresponding to an electronic transition between these two levels. What wavelength of light is required to affect this transition? Draw an energy level diagram to illustrate this transition.

3. Consider the wavefunction $\psi_m = Ae^{im\varphi}$, where A is a constant and $i = \sqrt{-1}$. If this wavefunction is an eigenfunction of the operator $-i\dfrac{h}{2\pi}\dfrac{d}{d\varphi}$, determine the corresponding eigenvalue.

4. Consider the wavefunction $\psi(x) = \sqrt{\dfrac{2}{L}} \sin \dfrac{\pi x}{L}$. Explain why this wavefunction is *well-behaved*. Which energy state does this wavefunction correspond to? Either by plotting the function or by differentiation, find the value of x for which $\psi(x)$ is a maximum.

5. Consider a ring of diameter 4 nm containing four free electrons. Draw a diagram showing the electrons occupying the various energy levels when the system is in its ground state. What wavelength of light is required to excite the electron from the ground state to the first excited state?

6. A particle on a line is constrained between 0 and L. With the aid of a diagram (a plot of $\psi^2(x)$ from $x = 0$ to $x = L$), estimate the probability of finding the particle from $L/4$ to $3L/4$ for the $n = 1, 2, 3, 4,$ and 5 states. Do you see a pattern?

7. Calculate the probability of finding the particle in a 1D box of length L between the interval $\dfrac{1}{10}L$ to $\dfrac{1}{4}L$ for the $n = 3$ state. Sketch the probability ($\psi_3^2(x)$) from $x = 0$ to $x = L$) for the $n = 3$ state and shade in the region from $\dfrac{1}{10}L$ to $\dfrac{1}{4}L$. Show that the percentage of the area shaded is in agreement with the calculated probability.

8. Is there a relationship between the peak absorption wavelength (λ_{max}) and the length of a nanowire? If so, plot the relationship on a graph. Your answer must be quantitative and use the particle on a line model to calculate the relevant information.

9. When sodium dissolves in liquid ammonia, it reacts with the solvent to form Na^+ and solvated electrons, which are not associated with any particular atom or molecule but are stabilized by the solvent molecules surrounding them, just like a typical ion. Solvated electrons can be used as a powerful reducing agent or to give a material unusual electronic or optical properties. The solvated electron can be treated as a particle in a three-dimensional box. Assume that the box is cubic with an edge length of 1.35 nm and suppose that excitation

occurs in all directions simultaneously from the lowest state ($n = 1$, $n = 1$, $n = 1$) to the excited state ($n = 2$, $n = 2$, $n = 2$). What wavelength, in nm, of radiation would the electron absorb? Would you expect the solution to be colored?

10. Boundary conditions are important because they describe the confinement of our particle of interest. It is this confinement that naturally leads to the quantization of the energy levels of a system. By considering some of the models described in this chapter, describe the how quantization changes as we excite the system to extremely high quantum states.

11. The wavefunction described by Equation 4.32 is not normalized. We can use the condition $\int \psi\,(x)\ \psi^*\,(x)\ dx = 1$ to normalize the wavefunction and determine the normalization constant A.

(a) Using the fact that $A^2 \int_0^{2\pi} e^{in\phi}e^{-in\phi}d\phi = 1$ and the general relation $e^a \times e^b = e^{(a+b)}$, show

$$A^2 \int_0^{2\pi} d\phi = 1$$

(b) Integrate the above equation and show that the normalization constant is

$$A = \sqrt{\frac{1}{2\pi}}$$

References and recommended reading

Atkins, P. W. Quanta: A Handbook of Concepts, Second ed. 1991, Oxford University Press. This is more of a non-mathematical and highly visual approach to understanding ideas in quantum mechanics. The book is full of detailed illustrations that make grasping concepts easy.

Brus, L. E. A simple model for the ionization potential, electron affinity, and aqueous redox potentials of small semiconductor crystallites. *The Journal of Chemical Physics*. 1983, 79: 5566–5571. This paper introduces the application of the particle-in-a-sphere model to quantum dots. Additional explanation and a more detailed model can be found in Liu, H. Brozek, C. K., Sun, S., Lingerfelt, D. B., Gamelin, D. R., and Li, X., A Hybrid Quantum-Classical Model of Electrostatics in Multiply Charged Quantum Dots. *Journal of Physical Chemistry C*. 2017, 121: 26086–26095.

Kuhn, H., Försterling, H.-D. and Waldeck. D. H. *Principles of Physical Chemistry*, 2nd ed. 2009, John Wiley and Sons, Inc., New Jersey. This physical textbook takes a particle-in-a-box approach to understanding the essence of quantum mechanics and why atoms and molecules exist.

Lowe, J., and Peterson, K. *Quantum Chemistry*, 3rd ed. 2005, Academic Press. A great introduction to basic quantum mechanics that assumes little mathematical or physics knowledge. The book is written to enable students and researchers to comprehend the current literature.

CHAPTER 5

Intermolecular Interactions and Self-Assembly

CHAPTER OVERVIEW

Nanostructures assemble, often spontaneously, from simple molecular building blocks. It is therefore important to begin this chapter with a discussion of the forces between such molecules. The types of noncovalent intermolecular interactions (for example, ion–ion, ion–dipole, dipole–dipole, dipole-induced dipole, London forces, hydrogen bonds, and electrostatic forces) will ultimately determine the degree and type of intermolecular aggregation as well as the structure of the resulting aggregate. Such interactions are examined in both bulk media and on surfaces. This chapter concludes with some coverage on how the quantum mechanical models discussed in Chapter 4 can be used to predict some optical properties of nanomaterials. In particular, conjugation in simple organic molecules is used to make important connections among electronic structure, intermolecular interactions, and molecular self-assembly.

5.1 INTERMOLECULAR FORCES AND SELF-ASSEMBLY

This section introduces selected fundamental physical ideas relating to the assembly and properties of nanomaterials in order to provide a sufficient background for understanding subsequent chapters. Intermolecular interactions play a central role in surface chemistry and the process of self-assembly, both of which affect the structure and properties of nanomaterials. Such interactions also determine the properties of surfactants, influence adsorption phenomena, and even affect interactions between molecules and electromagnetic radiation.

Self-assembly is the process during which discrete structures such as molecules spontaneously and often reversibly organize themselves into nanomaterials. The organization of these molecular building blocks is driven by a combination of thermodynamic factors and kinetic factors, many of which can be understood through examining the underlying intermolecular interactions. The interactions may be covalent in nature, leading to strong bonds between the molecules and resulting in an irreversibly self-assembled nanostructure. Covalent interactions—heavily involved in much of what you learned in general chemistry—are mentioned throughout this chapter. Examples of specific self-assembled materials are further discussed in Chapters 9 and 10. However, many self-assembly processes involve weaker **noncovalent interactions**, which can have significant influence on the thermodynamics and kinetics of self-assembly due to being ubiquitous and varied in type and range of interaction distances. It is important to appreciate that self-assembly can be spontaneous and directed. We begin this chapter by reviewing some important noncovalent intermolecular interactions that govern the formation of self-assembled nanomaterials.

Various forces are responsible for intermolecular interactions. Most of the forces are electrostatic in origin, and we discuss them from a classical perspective, although it should be noted that a quantum mechanical approach to understanding intermolecular forces is perhaps more correct.

Any interaction between two molecules can be thought of as a sum of a variety of different forces. We will discuss many of these forces, including ion–ion (Coulomb) forces, ion–dipole forces, dipole–dipole forces, induced dipole forces, dispersion forces, and hydrogen bonds. Depending on the types of atoms or molecules interacting, one force or another may predominate.

Scientists often express intermolecular interactions not as forces, but as **intermolecular potentials** (or the potential energy of interaction). The potential energy (U) and force (F) between two interacting molecules are related by

$$F(r) = -\frac{dU(r)}{dr} \qquad (5.1)$$

where r is the distance between the two molecules. The intermolecular distance, r, may be defined differently for different types of interacting molecules. The negative sign on the derivative means that as the potential energy of the interacting molecules increases with increasing r, the force will tend to move it toward smaller r to decrease the potential energy.

The **van der Waals interaction** is the collective term used to describe attractive or repulsive forces, or noncovalent interactions, between molecules. Named after Dutch scientist Johannes Diderick van der Waals, this type of intermolecular interaction generally refers to molecules involving ion–ion, ion–dipole, dipole–dipole forces, and interactions involving induced dipoles (including **London dispersion forces**). van der Waals forces play a key role in biology, polymer science, surface science, nanotechnology, and material science. They govern self-assembly processes, protein–protein interactions, and crystallization processes. These interactions are also found in nature. For example, the ability of geckos to climb smooth surfaces (such as glass) is attributed to van der Waals interactions and likely involves a nanofilm of water trapped between the glass surface and the foot. In fact, research is currently being done in many nanoscience laboratories to mimic this behavior and allow people to scale walls or to create "gecko tape" that exploits this ability.

Geckos possess the ability to cling to nearly any surface, sometimes even by a single toe, because they have millions of branching hairs called setae on their toepads that present enough surface area for van der Waals interactions to have an influence at the macroscopic scale. Recent advances in nanotechnology have yielded reusable "adhesives" that are four times more "adhesive" than a gecko's foot. These adhesives are comprised of flexible polymers connected by silicon bases to carbon nanotubes, which are cylindrical graphene columns held together by van der Waals interactions. Structure, properties, and uses of carbon nanotubes will be further discussed in Chapter 9.

5.1.1 Ion–ion interactions

Ion–ion forces are perhaps the most well-known intermolecular forces and are among the strongest intermolecular forces of those we'll be discussing. Ion–ion forces arise between two ionic (charged) species, such as the force between Na^+ and Cl^- that holds together crystals of common table salt. The potential energy of interaction $U(r)$ between two charges q_1 and q_2 is often called the **Coulombic energy** and is given as

$$U(r) = \frac{q_1 q_2}{4\pi\varepsilon_0 r_{12}} \qquad (5.2)$$

where ε_0 is the permittivity of free space (8.854×10^{-12} m^{-3} kg^{-1} s^4 A^2) and r_{12} is the distance between the two ionic species. For atomic or molecular ions, q is often calculated as $q = ze$ where z is the formal charge on the ion and e is the charge on an electron, 1.60217×10^{-19} C.

We can express the Coulombic force using the relationship between potential energy and force (Equation 5.1) as

$$F(r) = -\frac{dU(r)}{dr} = \frac{q_1 q_2}{4\pi\varepsilon_0 r_{12}^2} \tag{5.3}$$

Equation 5.3 is obtained by differentiating Equation 5.2 and realizing that $d/dr(1/r)$ is $-1/r^2$. From Equation 5.3, we see that the Coulombic force between two ions changes as $1/r^2$. We also see that the force is negative when the two ions are attracted to each other (when q_1 and q_2 have opposite signs) and positive when they repel each other.

Example 5.1   The Coulombic Energy between Ions

The ionic radius of Na$^+$ is determined to be 95 pm and the ionic radius of Cl$^-$ is 181 pm. Calculate the Coulombic energy between two isolated ions of Na$^+$ and Cl$^-$ if they are in contact, as shown in Figure 5.1.

Solution If the two ions are "in contact," then the distance between their centers is the sum of their two ionic radii. So

$$r_{Na-Cl} = 95 \text{ pm} + 181 \text{ pm} = 2.76 \times 10^{-10} \text{ m}$$

Each ion has a formal charge of +1 or −1, so

$$q_{Na^+} = ze = (+1)\left(1.60217 \times 10^{-19} \text{ C}\right) = 1.60217 \times 10^{-19} \text{ C}$$

$$q_{Cl^-} = ze = (-1)\left(1.60217 \times 10^{-19} \text{ C}\right) = -1.60217 \times 10^{-19} \text{ C}$$

Figure   5.1 A   schematic depiction of Na$^+$ and Cl$^-$ "in contact" with each other.

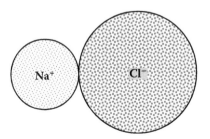

Then, using Equation 5.2,

$$U(r)_{Na-Cl} = \frac{q_{Na^+}q_{Cl^-}}{4\pi\varepsilon_0 r_{Na-Cl}} = \frac{(1.602 \times 10^{-19} \text{ C})(-1.602 \times 10^{-19} \text{ C})}{4\pi(8.854 \times 10^{-12} \text{ m}^{-3}\text{kg}^{-1}\text{s}^4\text{A}^2)(2.76 \times 10^{-10} \text{ m})}$$

$$= -8.36 \times 10^{-19} \text{ J}$$

where $1 \text{ J} = 1 \text{ kg m}^2 \text{ s}^{-2}$ and $1 \text{ C} = 1 \text{ A s}$.

One example of ion–ion interactions can be found in the formation of a charged polymer layer, or nanofilm, on a silicon surface. Silicon usually has an oxide layer (SiO_2) of about 1 nm in thickness on its surface. In a network of surface SiO_2 groups, each silicon atom has a tetrahedral molecular geometry, resulting in a layer of oxygen atoms, each covalently bonded to a single silicon atom at the surface. At very low pH, these oxygen atoms are protonated, but at neutral to high pH, the oxygen atoms are deprotonated and thus create a layer of negative ionic charge ($Si\text{-}O^-$) along the surface. Polyethylenimine (PEI), a polycation with several amine functional groups, can then be exposed to this negatively charged surface to create a positively charged PEI layer of relatively uniform thickness. A negatively charged ion or polymer can then be exposed to this surface-bound PEI layer to create a secondary layer, resulting in a new negatively charged surface that allows the process to be repeated. This process is called electrostatic self-assembly and will be used to develop several techniques in future chapters.

5.1.2 Ion–dipole interactions

Many molecules possess permanent dipoles and are classified as polar molecules. Polar molecules do not have a permanent charge, but because of the differing electronegativities of the atoms bound in the molecule, certain regions of the molecule may have a partial positive or a partial negative charge. In certain cases this partial charge can lead to a permanent dipole. For example, a water molecule has a permanent dipole due to its bent geometry. The oxygen atom has a much higher electronegativity than the hydrogen atoms and so it tends to draw more electrical charge to itself. As a result, the hydrogens have a partial positive charge and the oxygen has a partial negative charge. The net dipole moment passes through the oxygen atom and bisects the hydrogen atoms as shown in Figure 5.2. Note that there are two common (and opposite) dipole conventions in common use, one where the dipole moment is defined as pointing toward the partial positive charge (sometimes called

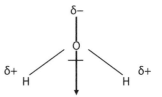

Figure 5.2 The electronegative oxygen atom in a water molecule pulls electron density away from the hydrogen atoms, leaving them with a partial positive charge δ+. Each O–H bond has a dipole moment that points toward the partial positive charge of the hydrogen, resulting in a net dipole moment that passes through the O and bisects the Hs.

the physicist's convention), and one where the dipole is defined as pointing toward the partial negative charge (sometimes called the chemist's convention). We will use the convention of the dipole pointing toward the partial positive charge.

Polar molecules such as H_2O are often characterized by their dipole moment μ, which is defined for a single bonds as

$$\mu = qL \tag{5.4}$$

where L is the distance separating the partial positive and partial negative charges of magnitude q, as shown in Figure 5.3. The total dipole moment of a molecule is the vector sum of the dipole moments of all of the bonds within the molecule. Dipole moments of some common molecules are shown in Table 5.1. As shown in the table, dipole moments are often given in units of Debye (D), where $1\ D = 3.336 \times 10^{-30}\ C\ m$.

When an ion draws near to a molecule with a dipole, there is an electrostatic interaction between the dipole and the ionic species. The potential energy of this interaction is given as

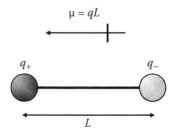

Figure 5.3 The dipole moment μ between two charges q_+ and q_- separated by a distance L is calculated as $\mu = qL$.

Table 5.1
Dielectric Constants of Common Solvents

Solvent	μ (D)	n	ε(0) at 20°C	Onsager ε (20°C)
Toluene	0.38	1.49	2.38	2.40
Diethyl ether	1.10	1.35	4.27	3.19
Dichloromethane	1.60	1.42	8.93	7.88
Acetone	2.88	1.36	21.01	18.28
Acetonitrile	3.93	1.34	36.64	44.95
Water	1.85	1.33	80.1	28.90*

Source: Data from Haynes, W.N., ed. *CRC Handbook of Chemistry and Physics*, 97th Ed., 2017, CRC Press.
* Onsager model does not account for hydrogen bonding.

$$U(r, \theta) = \frac{q_{ion}\mu \cos \theta}{4\pi\varepsilon_0 r_{12}^2} \tag{5.5}$$

where μ is the dipole moment, q_{ion} is the charge on the ion, r_{12} is the distance between the ion and the center of the dipole moment, and θ is the angle between L and r_{12}. A schematic of the geometries involved in a typical ion–dipole interaction is shown in Figure 5.4.

From Equation 5.5 we see that the potential energy of an ion–dipole interaction is angle-dependent, which makes sense intuitively. For example, consider the interaction between a cation (a positively charged ion) and a dipole. The cation attracts the negative region of the dipole but repels the positive region. If the negative region is oriented toward the cation and the positive region is pointing away from it ($\theta = \pi$), we expect the magnitude of the potential energy to be maximized. Likewise, if the

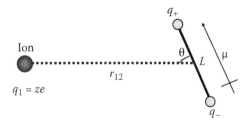

Figure 5.4 A schematic depiction of the variables involved in an ion–dipole interaction. L is the distance between the centers of the two partial charges of the dipole. r_{12} is the distance between the center of the ion and the midpoint of L. θ is the angle between L and r_{12}. q_1 is the charge on the ion, calculated as ze. μ is the dipole moment.

positive region is pointing toward the cation and the negative region is pointing away ($\theta = 0$), the magnitude of the potential energy is maximized (but it is repulsive rather than attractive). If, however, the dipole is oriented perpendicularly with respect to the cation ($\theta = \pi/2$), the potential energy is zero because the attractive interaction between the cation and the negative region is balanced out by the repulsive interaction with the positive region. Therefore, the angle dependence of the ion–dipole interaction is as expected.

It should be noted that some molecules (such as benzene or CO_2) may have partial charge separations but do not have a net dipole moment. For example, CO_2 has no net dipole moment despite two polar bonds—recall that the total dipole is the vector sum of bond dipoles, and for CO_2, the two bond dipoles are equal in magnitude and opposite in direction. However, in these types of molecules, higher-order **multipoles** such as quadrupoles or octupoles may exist and also results in an intermolecular force between the molecule and an ionic species. Different equations than those above must be used to calculate the interaction between molecules with electrical multipoles and ionic species; however, these are beyond the scope of this text.

5.1.3 Dipole–dipole interactions

Molecules with permanent dipoles may also interact with each other through electrostatic means. As might be expected, the strength of the interaction is also angle-dependent. This type of interaction is analogous to the magnetic attraction between two bar magnets—the attraction between the two magnets depends on the angle of rotation of each magnet relative to the other. A schematic representing the interaction between two molecular dipoles is depicted in Figure 5.5. The potential energy for such an interaction between two dipole moments μ_1 and μ_2 can be calculated as

$$U(r, \theta_1, \theta_2, \phi) = -\frac{\mu_1 \mu_2}{4\pi\varepsilon_0 r_{12}^3}(2\cos\theta_1\cos\theta_2 - \sin\theta_1\sin\theta_2\cos\phi) \qquad (5.6)$$

where θ_1, θ_2, and ϕ are defined in Figure 5.5.

It might be tempting to think that the attraction between two dipoles is always maximized when the two dipoles are in line "head to tail" with the partial positive charge of one dipole pointing directly toward the partial

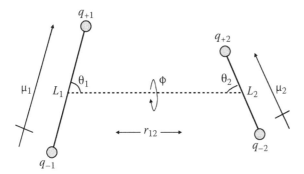

Figure 5.5 A schematic depiction of the variables involved in a dipole–dipole interaction. L is the distance between the centers of the two partial charges of either dipole 1 or 2. r_{12} is the distance between the midpoints of L_1 and L_2. θ_1 and θ_2 are the angles between L_1 and r_{12} or L_2 and r_{12}, respectively. ϕ is the angle of rotation between dipoles 1 and 2. q_+ and q_- are the partial positive or negative charges in each dipole. μ is the dipole moment.

negative charge of the other. However, depending on the lengths L of the interacting dipoles, the most attractive interaction may often be when the two dipoles are antiparallel to each other, with the positive region of one dipole directly adjacent to the negative region of the other. This anti-parallel orientation allows the molecules to draw closer together, reducing the value of r_{12} and thereby maximizing the attractive interaction energy.

In the purification technique of column chromatography, a solid material containing many polar bonds, such as silica or alumina, is placed in a vertical glass column and is referred to as the stationary phase. The liquid mobile phase of solution to be purified is then flowed through the column. Because dipole–dipole interactions between polar molecules in the mobile phase and the polar surfaces of the particles in the stationary phase slow the polar molecules' descent, compounds in the solution flow out the bottom of the column, or elute, in order of increasing polarity and can thus be separated. This same technique works for ions in solution as well because they are slowed by ion–dipole interactions. The polarity of the solvent dictates the rate of movement of compounds through the column. If a solvent is too polar, the stationary phase attracts the solvent rather than the solutes and no separation occurs. However, if a solvent is not polar enough, some more polar solutes may not make it all the way through the column.

5.1.4 Interactions involving induced dipoles

When an ion approaches a nonpolar molecule, the electrons of the nonpolar molecule experience the effect of the electric field produced by the ion. As a result, the electron cloud surrounding the nonpolar molecule becomes distorted. For example, when a cation approaches a nonpolar molecule, the electron cloud of the nonpolar molecule is pulled slightly toward the cation. The result of this distortion of the electron cloud is charge separation in the nonpolar molecule; this transient charge separation is called an **induced dipole**. A schematic of an ion-induced dipole interaction is shown in Figure 5.6. Polar molecules, as well as ions, are also capable of inducing dipoles in nonpolar molecules.

The extent to which the electron cloud of a molecule becomes distorted in the presence of an ion or polar molecule is termed its polarizability. Polarizability, α is mathematically defined in terms of the strength of the dipole induced in a molecule due to an electric field of strength E,

$$\mu_{induced} = \alpha E \tag{5.7}$$

Molecules with high polarizabilities have a larger induced dipole moment in the presence of an electric field than those with low polarizabilities. Typical polarizabilities of various atoms and molecules are shown in Table 5.2. Note that the units of polarizability given in the table are 10^{-24} cm^3 divided by $4\pi\varepsilon_0$.

The presence of an induced dipole moment in a nonpolar molecule means that a potential energy of interaction exists between the nonpolar molecule and the ion or polar molecule that is inducing the dipole. Using Coulomb's law and Equation 5.3, we can calculate the electric field produced by an ion as a function of the distance r from the center of the ion:

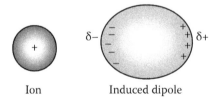

Ion   Induced dipole

Figure 5.6 As a cation approaches a polarizable atom or molecule, its electric field produces a distortion of the electron cloud surrounding the polarizable atom or molecule. The proximity of the cation to the molecule results induces a dipole moment in the molecule.

Table 5.2
Polarizabilities of Representative Atoms and Molecules

Molecule	α (10^{-24} cm^3/$4\pi\varepsilon_0$)
H_2	0.804
He (atom)	0.205
C (atom)	1.67
Methane (CH_4)	2.59
Benzene (C_6H_6)	10.0
Toluene (C_7H_8)	12.3
n-Hexane (C_6H_{14})	11.9
Ammonia (NH_3)	2.10
Acetonitrile (C_2H_3N)	4.40
Water (H_2O)	1.45
Heavy water (D_2O)	1.26
Diethyl ether ($C_4H_{10}O$)	10.2
Acetone (C_3H_6O)	6.4
Dichloromethane ($C_2H_2Cl_2$)	7.93
Si (atom)	5.53
Au (atom)	4.13

Source: Data from Haynes, W.N., ed. *CRC Handbook of Chemistry and Physics*, 97th Ed., 2017, CRC Press.

$$E(r) = \frac{F}{q} = \frac{q}{4\pi\varepsilon_0 r^2} \tag{5.8}$$

The interaction energy between an ion and an induced dipole can then be given as

$$U(r) = \frac{-q_1\alpha}{2(4\pi\varepsilon_0)^2 r_{12}^4} \tag{5.9}$$

Example 5.2 Perturbation of Electron Clouds Due to Ions

The center of a sodium ion (Na^+) is located 0.35 nm from the center of a gold atom. If the atomic radius of gold is 144 pm, by what percentage of its atomic radius would the electron cloud of the gold atom be shifted due to the presence of the sodium ion?

Solution   The electric field induced by Na^+ at a distance 0.35 nm from its center is calculated using Equation 5.8.

$$E(r) = \frac{q_{Na^+}}{4\pi\varepsilon_0 r^2} = \frac{(1.602 \times 10^{-19} \text{ C})}{4\pi(8.854 \times 10^{-12} \text{ m}^{-3} \text{ kg}^{-1}\text{s}^4\text{A}^2)(0.35 \times 10^{-9} \text{ m})^2}$$

$$= 1.18 \times 10^{10} \frac{\text{J}}{\text{Cm}}$$

Then, using Table 5.1, the dipole moment induced in the gold atom is

$$\mu = \alpha E = 4\pi\varepsilon_0 \left(5.8 \times 10^{-30} \text{ m}^3\right)\left(1.18 \times 10^{10} \frac{\text{J}}{\text{Cm}}\right)$$

$$= 7.61 \times 10^{-30} \text{ Cm}$$

Now, from Equation 5.4, we know that $\mu = qL$, and so per unit charge (e) we can calculate L:

$$L = \frac{\mu}{q} = \frac{7.61'10^{-30} \text{ Cm}}{1.602 \times 10^{-19} \text{ C}} = 4.75 \times 10^{-11} \text{ m} = 47.5 \text{ pm}$$

Then the electron cloud of Au is shifted by 47.5 pm/144 pm = 33% of its atomic radius.

In an analogous manner, the electric field strength at a given point in space produced by a polar molecule with dipole moment μ is a function of the orientation of the dipole moment with respect to that point in space and is calculated as

$$E(r, \theta) = \frac{\mu\left(3\cos^2\theta + 1\right)^{1/2}}{4\pi\varepsilon_0 r^3} \tag{5.10}$$

The potential energy of interaction between a polar molecule with permanent dipole μ_1 and an induced dipole is therefore

$$U(r, \theta) = \frac{-\mu_1^2\alpha\left(3\cos^2\theta + 1\right)}{2(4\pi\varepsilon_0)^2 r_{12}^6} \tag{5.11}$$

where θ is the angle between the dipole moment of the polar molecule and the line connecting the midpoint of the polar molecule with the center of the induced dipole.

Finally, we note that the interaction between an ion or a polar molecule and an induced dipole is always attractive. It is inherently so because the electric field produced by the ion or polar molecule always induces a dipole in the polarizable molecule that is oriented such that it is attracted toward the species inducing the dipole.

5.1.5 Dielectric screening

The interactions that we have discussed thus far are electrostatic in origin, involving charged species (ions) and either permanent or induced dipoles. We have discussed these interactions in terms of **pair potentials**, or interactions between two particles without regard to the presence of any other particles that could interact. However, in dense systems relevant to nanoscience, such as liquids, solids, and colloids, these interactions are affected by the presence of their neighbors. These complex interactions can be treated by three major approaches: (a) averaging these interactions into a simple parameter, (b) running computer simulations that contain many interacting particles, and/or (c) explicitly modeling these interactions using potentials involving three or more particles. We will focus on the first approach; the latter two approaches are beyond the scope of the present text.

When molecules are exposed to an electric field, E, whether that field is external (e.g., a beam of light or a voltage applied to a pair of electrodes) or internal (e.g., other molecules/ions in the system), the molecules reorient and polarize to minimize their energy of interaction with the field. Figure 5.7 shows how water molecules can reorient in response to the presence of either ions (left) or charged surfaces (right). The alignment of the permanent and induced dipoles of the other molecules in the system produces an electric field that counteracts the applied field. This field is known as the polarization field, P, and is defined as

$$P = (\varepsilon - 1)\varepsilon_0 E \qquad (5.12)$$

The **dielectric constant** (or relative permittivity), ε, is a dimensionless quantity that reflects the system's ability to polarize, and can be experimentally determined from the electrical capacitance of a material sandwiched between two parallel electrodes as

$$\varepsilon = \frac{Cd}{A} = \frac{Qd}{VA} \qquad (5.13)$$

where C is the capacitance, A is the area of the plates, Q is the total charge on each plate, V is the electrostatic potential (voltage) between the plates, and d is the distance between the plates. Figure 5.7 (right side) shows how when the dipoles of water align between two charged plates, they produce an electric field in the opposite direction of their dipole moments that reduces the total electric field between the plates by a factor of ε. The dielectric constant can be determined from either measuring the voltage

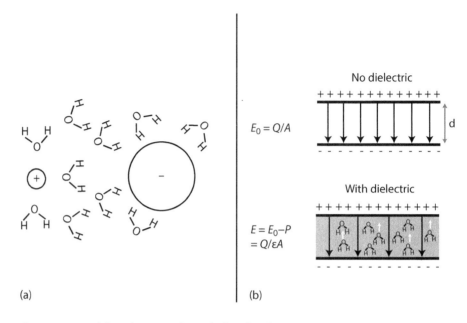

(a)

(b)

Figure 5.7 Dielectric screening of electric charges. (a) Water molecules orient around an anion and a cation to maximize energetically favorable ion–dipole interactions. The presence of multiple layers of solvent molecules screens the two ions from each other and reduces the strength of their interaction. (b) The presence of a dielectric such as water between two charged plates reduces the electric field between the plates.

between the plates at constant charge (voltage decreases with increasing ε), or measuring the charge on the plates at constant voltage (amount of stored charge increases with increasing ε).

The dielectric constant can also be estimated from theoretical models or computer simulations that incorporate both the interactions of the molecules with each other and with an external electric field; however, these are beyond the scope of this text.

One simple method for estimating the dielectric constant for a molecule with a known dipole moment (or vice versa) is the **Onsager model**, which assumes that molecules are spherical and are only acting with the average electric field of all other molecules in the system. Under these approximations and for a pure liquid,

$$\frac{(\varepsilon - 1)(2\varepsilon + n^2)}{\varepsilon_0 \varepsilon (n^2 + 2)} = \frac{N\mu^2}{9kT} \tag{5.14}$$

where n is the refractive index of liquid, N is the number density (concentration) of the liquid, μ is the dipole moment of an isolated molecule of the liquid, k is Boltzmann's constant, and T is the temperature in K. Table 5.1 compares the experimental dielectric constants of several common solvents with the prediction of the Onsager model. Note how more polar liquids have higher dielectric constants. Also note how the Onsager model performs poorly for water; it does not account for the strong hydrogen bonding interactions (Section 5.1.9) found in water and alcohols, and prediction of the dielectric constant for these systems is not trivial. Accurate models for intermolecular interactions of water are still an active area of research.

Of the physical properties used as parameters in the Onsager model, the refractive index, n, requires further explanation since we will encounter it in several other contexts in later chapters. It is a key optical property of a material related to the speed at which electromagnetic fields propagate through a material. While the speed of light *in vacuum* is constant in all inertial frames of reference according to special relativity, the speed of light within a material is dependent on the interaction between the light and the material. The refractive index of a medium is defined as the ratio of the speed of light in a vacuum (3.0×10^8 m/s) to the speed of light in the medium,

$$n = \frac{c_0}{c_{\text{material}}}\tag{5.15}$$

The refractive index of a material and the polarizability (α) of its constituent molecules are related by the **Clausius–Mossotti equation**,

$$\frac{n^2 - 1}{n^2 + 2} = \frac{N\alpha}{3\varepsilon_0}\tag{5.16}$$

The more polarizable the molecules in the material are, the stronger their interaction with the electric field of light passing through the material. In fact, at frequencies at which a material does not absorb light and assuming that the material does not undergo any significant magnetic response, the refractive index *is* the dielectric constant of the material at optical frequencies. The electric field of the light is oscillating too fast for permanent dipoles to rotate, such that only induced dipoles contribute to the polarization (P) of the material. While we have only explored high- and low-frequency limits, the dielectric constant of a material is a frequency-dependent property that depends on the motions of atoms, molecules, or electrons, for which a rigorous examination is beyond the scope of this text.

In the context of intermolecular interactions, the dielectric constant represents a simple scaling factor that reduces the strength of interactions. For example, in a dielectric environment, Equation 5.2 becomes

$$U(r) = \frac{q_1 q_2}{4\pi\varepsilon\varepsilon_0 r_{12}} \tag{5.17}$$

The screening of interactions by a dielectric can be very substantial in a solvent like water ($\varepsilon \sim 80$). In addition to specific contributions from ion-dipole interactions, this is one of the factors that allow salts to be more soluble in polar solvents.

5.1.6 Dispersion forces

Aside from the forces that are essentially electrostatic in nature described in the previous sections, a force also exists between all molecules, regardless of charge or polarity, which results from the quantum mechanical correlation between electrons in neighboring molecules. This force is called the dispersion or London force. Although dispersion forces are inherently quantum mechanical in nature and a rigorous description of their origin is beyond the scope of this book, we can gain an intuitive grasp of dispersion forces by considering in a somewhat classical manner their contribution to the interaction between two neutral, nonpolar molecules.

Even though a neutral, nonpolar molecule has no permanent dipole moment, at any given moment the distribution of its electrons may be asymmetrical, resulting in an instantaneous or momentary dipole moment. This instantaneous dipole moment creates an electric field that perturbs the electrons of neighboring molecules, producing induced dipole moments and resulting in attractive forces between the molecules.

In order to calculate the dispersion force between two molecules, a quantum mechanical calculation must be performed, the accuracy of which generally depends on the level of theory used. One of the earliest calculations was performed by London in the 1930s using quantum mechanical perturbation theory. His calculation produces reasonably accurate results, and although more precise calculations have been performed in more recent years, London's equation is less complex and therefore more suitable for our purposes.

According to London's results, the approximate potential energy of interaction due to dispersion between two molecules can be calculated in terms of their electronic polarizabilities, α, and their ionization potentials I. For two identical molecules (or atoms), the result is

$$U(r) = \frac{-3}{4(4\pi\varepsilon_0)^2} \frac{\alpha^2 I}{r_{12}^6} = \frac{-C_{\text{dispersion}}}{r_{12}^6} \qquad (5.18)$$

and for two different molecules

$$U(r) = \frac{-3}{2(4\pi\varepsilon_0)^2} \frac{\alpha_1\alpha_2}{r_{12}^6} \frac{I_1 I_2}{(I_1 + I_2)} \qquad (5.19)$$

As with the interaction energy for dipole-induced dipole interactions in Section 5.1.4 (Equation 5.11), we see that the potential energy of interaction for a dispersion interaction according to London's equations goes as $1/r^6$ and is always attractive between any two molecules.

The dispersion interaction plays an important role in the liquid and solid phases of many materials and is the main contributor to cohesion. However, it turns out that the strength of the dispersion interaction does not vary much between different types of molecules (i.e., the interaction between any two given molecules is of similar strength). Therefore, the electrostatic interactions described in the earlier sections, and not dispersion interactions, are generally responsible for such behaviors as phase separation and self-assembly in condensed phases, behaviors that are of utmost importance in the development and study of nanomaterials.

5.1.7 Overlap repulsion

In our discussion of the different types of intermolecular potentials in the previous sections, we ignored the fact that atoms and molecules occupy some finite space. For example, if we examine the equation for Coulombic force (Equation 5.3) by itself, we would be led to conclude that two oppositely charged ions are drawn toward each other with increasing force until they occupy the same point in space. Obviously, this does not occur with atoms and molecules in nature. To account for the finite size of atoms and molecules, we then include another contributor to the interaction potential energy between two atoms or molecules called overlap repulsion. Overlap repulsion is the interaction that accounts for two atoms or molecules being unable to occupy the same point in space, which is driven by both the electrostatic repulsion between their electron "clouds" and quantum mechanical effects for which the derivation is beyond the scope of this text.

What, then, is the size of an atom or a molecule? This is not a trivial question. From the results of quantum mechanics, we realize that the electron "clouds" of atoms and molecules do not have definite boundaries. Determining where an atom "ends" is therefore somewhat tricky. As a result, radii of atoms are often experimentally defined, and depending on the type of measurement made (and consequently the property measured), a different result might be obtained. For example, one way to measure the radii of atoms would be to assume that in solids they act as tiny, hard spheres that have packed closely together (see Figure 5.8). Using x-ray or neutron diffraction methods, one could then observe how closely the atoms pack together in a crystal and thereby deduce the atomic radius. The results of this method yield a type of atomic radius called the *hard sphere radius* or *van der Waals packing radius*. Other methods also exist, such as measuring the distance between two atoms in a covalent bond (rather than the distance in a crystal). This method yields the *covalent bond radius*. The atomic radius calculated depends on the method used. In certain cases the results obtained from these different methods might vary by as much as 30%. The type of measurement one chooses to use generally depends on the type of system being studied.

After using the most suitable method to determine an atomic radius, one can then calculate the overlap repulsion between two atoms. A variety of

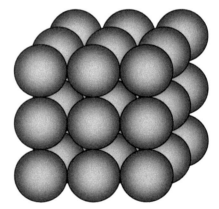

Figure 5.8   Atoms in crystal lattices can often be modeled as tiny, hard spheres in order to calculate their atomic radius. X-ray or neutron diffraction methods can then be used to experimentally determine the atomic radius.

models with increasing sophistication and complexity are used to calculate the repulsive potential energy between two atoms due to overlap. Perhaps the simplest model is to characterize the atom as a "hard sphere" with a definite boundary (i.e., the repulsive force between two atoms would be infinite at any distance smaller than the atomic radius). This hard sphere model between two atoms at a distance r from each other can be represented mathematically as

$$U(r) = \left(\frac{\sigma}{r}\right)^{\infty}$$
(5.20)

where σ is the atomic or molecular diameter (i.e., two times the atomic radius). As expected, when $r > \sigma$, then $V(r)$ is essentially zero and when $r < \sigma$, $V(r)$ is infinitely large. A graph of $V(r)$ versus r for the hard sphere model is shown in Figure 5.9a.

A more realistic model is the soft sphere model, which assumes that atoms are "compressible" to some degree and do not have completely rigid boundaries. One mathematical representation for the soft sphere model can be given as a power law as

$$U(r) = \left(\frac{\sigma}{r}\right)^{n}$$
(5.21)

where n is usually an integer between 9 and 16 and σ is defined as before. In this model, $V(r)$ quickly becomes quite small when r is much bigger

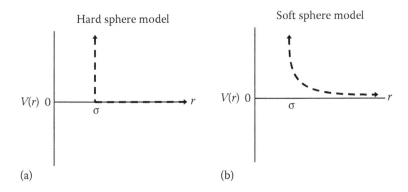

Figure 5.9   (a) The hard sphere model of overlap repulsion. r is the intermolecular distance and σ is the molecular diameter. (b) The soft sphere model of overlap repulsion. Note that r can assume some values slightly smaller than σ without $V(r)$ becoming infinitely large, as is the case with the hard sphere model.

than σ. Conversely, the overlap repulsion becomes large rather quickly when the distance between the two atomic centers is less than σ. Figure 5.9b depicts a graph of $V(r)$ versus r for the soft sphere power law model.

Thus far in our discussion of overlap repulsion, we have operated under the assumption that the atomic or molecular geometry is essentially spherical. While this assumption is relatively sound for atoms and some small molecules (CH_4, for instance, can be modeled as nearly spherical), most molecules possess other geometries. The concepts of overlap repulsion we have developed still apply to these species, but different calculations of their interaction energy are required to account for their differing geometries. However, such methods of calculation are beyond the scope of this text.

5.1.8 Total intermolecular potentials

The previous sections provided a fundamental description of the main forces involved in van der Waals intermolecular interactions. Ultimately, the total interaction potential energy between two molecules is the sum of all the different interactions that we have discussed (as well as a few more complex interactions).

In a very basic treatment of the interactions between two atoms or molecules, the total intermolecular potential is often modeled by the Lennard-Jones potential, which is the sum of a soft sphere repulsion term and an attractive term that goes as $1/r^6$ (analogous to the London dispersion attractive interaction). The Lennard-Jones potential is given as

$$U(r) = 4\varepsilon \left[\left(\frac{\sigma}{r} \right)^{12} - \left(\frac{\sigma}{r} \right)^{6} \right] \tag{5.22}$$

where $-\varepsilon$ is the minimum energy and σ is a constant parameter (not the molecular diameter). A graph of the Lennard-Jones potential is shown in Figure 5.9.

Although the Lennard-Jones potential is a relatively primitive model of the total intermolecular potential between two molecules, it provides us with a qualitatively useful picture of a common interaction between two molecules. Starting from the far right-hand side of Figure 5.10, we see that the potential energy decreases as the distance between the molecules grows smaller, until the energy reaches a minimum value of $-\varepsilon$. If r decreases beyond this minimum energy value, then the potential energy

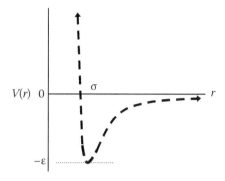

Figure 5.10 The Lennard-Jones total intermolecular potential curve. $-\varepsilon$ is the minimum energy. r is the intermolecular distance.

quickly increases (i.e., the force between the two molecules becomes strongly repulsive) due to the overlap repulsion term.

A more complete total intermolecular potential would be the sum of all interaction potential energies. Using only the interactions we have discussed so far, the complete total intermolecular potential would look like that shown in Equation 5.23. However, it must be realized that simple systems will not exhibit all of these kinds of interactions, so that some terms in Equation 5.23 will be zero.

$$
\begin{aligned}
U(r)_{\text{total}} = {} & U(r)_{\text{ion–ion}} + U(r,\theta)_{\text{dipole–dipole}} \\
& + U(r,\theta)_{\text{ion–dipole}} + U(r,\theta)_{\text{ion-induced dipole}} \\
& + U(r,\theta)_{\text{dipole-induced dipole}} + U(r)_{\text{dispersion}} + U(r)_{\text{overlap}}
\end{aligned}
\qquad (5.23)
$$

If the sum of attractive interaction terms is greater than the repulsive interaction terms, then the two molecules are drawn together until the repulsive interactions eventually overwhelm the attractive interactions (remember that the overlap repulsion quickly becomes prohibitively large at distances smaller than the atomic or molecular radii).

In conclusion and as a cautionary addendum, note that many of the models of intermolecular forces discussed in this and previous sections are mathematically convenient—simplified approximations to a more complex underlying potential. However, the interactions we have discussed are qualitatively very useful for the purposes of this text and provide the conceptual tools to understand the intermolecular forces at play in the realm of nanomaterials.

5.1.9 Hydrogen bonds

The hydrogen bond is a special type of intermolecular interaction of great importance in colloidal systems and nanomaterials in general. The hydrogen bond is essentially electrostatic in origin and so is a subset of the dipole interactions already discussed. However, it is of particular importance and strength, and so has acquired a special classification.

Hydrogen bonds occur between molecules that have hydrogen covalently bonded to a strongly electronegative atom such as N, O, or F. In such cases, the electron density surrounding the hydrogen atom is mostly drawn toward the more electronegative atom, leaving the hydrogen atom "exposed" with a strong partial positive charge through which it may form a strong dipole–dipole interaction with the electronegative element on an adjacent molecule. Perhaps more importantly, the reduced electron density around the hydrogen atom means that the neighboring molecule with which it is hydrogen bonding can draw much closer than it could otherwise (within ~1.5 to 2.0 Å). In other words, the overlap repulsion is minimized because the electron "cloud" surrounding the hydrogen atom has been reduced in size. Because the neighboring molecule can draw much closer, the magnitude of the attractive interaction energy is much greater than it would normally be (remember that in our equations for electrostatic interaction energies, r_{12} was in the denominator and usually raised to some power; so if r_{12} is smaller, then the magnitude of the interaction potential energy is much greater). Figure 5.11 depicts a hydrogen bond between two water molecules.

5.1.10 The hydrophobic effect

The van der Waals interactions discussed so far are responsible for many of the physical properties (e.g., solubility) of organic molecules. For example, in methanol (CH_3OH), the hydrocarbon portion of the molecule is relatively small and the polar hydroxyl group is largely responsible for the weak intermolecular van der Waals interactions. However, as the length of the hydrocarbon moiety increases [e.g., as in decanol, $CH_3(CH_2)_9OH$] the nonpolar hydrocarbon portion of the molecule dominates intermolecular interactions and defines its solubility. Hydrocarbon chains are essentially oil and have little or no tendency to interact with water. If the hydrocarbon chain is long enough the molecules may drop out of solution (precipitate) and interact with themselves instead of the water molecules. This brings us to a discussion of the hydrophobic effect,

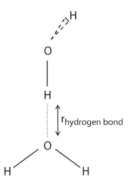

Figure 5.11 A hydrogen bond between two water molecules. The electronegative oxygen atom pulls much of the electron density surrounding the hydrogen atom to itself, giving the oxygen a large partial-negative charge and leaving the hydrogen atom with a partial-positive charge and very little electron density. The oxygen atom of a neighboring water molecule can therefore approach much closer to the hydrogen atom than would normally be possible for a dipole–dipole interaction.

which describes the tendency of nonpolar molecules to form aggregates of like molecules in water.

Experience shows us that at the macroscopic level oil and water do not mix, but rather form separate layers. We also know that water forms beads, or distinct droplets, on oil-like surfaces such as the surface of a leaf. Often when organic molecules containing large nonpolar hydrocarbon moieties are placed in water they spontaneously self-organize themselves into larger aggregates because this arrangement minimizes overall contact between the hydrocarbon part of the molecule and the water molecules of the solvent. The origin of this phase separation is primarily to maximize the water–water intermolecular interactions because the dipole–dipole interactions are stronger than the interactions between nonpolar molecules. In other words, the enthalpic force is in the water, not the nonpolar phase, meaning that this type of phase separation is enthalpically driven.

In order to complete our understanding of the hydrophobic effect, we need to say something about the contributions of entropy to intermolecular interactions. As discussed in Chapter 2, entropy is a thermodynamic state function related to the number of energy levels among which the energy of a system is spread. A collection of water molecules confined and "ordered" within a small region of space will have smaller entropy compared to the same molecules "disordered" and occupying a larger volume. Since entropy changes are positive for spontaneous processes, it is favorable for a water molecule to be "released" from a small space and enter a larger volume, especially if such a process leads to a loss of order.

Consider an organic molecule with a large hydrocarbon chain [e.g., $CH_3(CH_2)_9OH$] placed in water. It turns out that water molecules, in avoiding contact with the hydrocarbon chain, form a cage around the chain. This cage contains ordered water molecules essentially immobilized around the organic molecule. When two such "hydrated" molecules approach each other and make contact, this cage is disrupted and the confined water molecules are liberated into the bulk solution. This process is accompanied by an increase in entropy of the water molecules. In a sense the van der Waals attractions between the two hydrocarbon chains is driven in part by the large increase in entropy due to the disruption of the cage. If this process occurs among many organic molecules it will lead to aggregation, typically resulting in nanoscale entities dispersed within the aqueous solution.

The hydrophobic effect is paramount in many self-assembly processes, including some biological processes such as the formation of the cell membrane. Reference to the effect is made throughout the text, and in particular, Section 5.3 contains a discussion of the hydrophobic effect in the context of surfactant chemistry.

5.2 ELECTROSTATIC FORCES BETWEEN SURFACES: THE ELECTRICAL DOUBLE LAYER

Surface chemistry plays a vital role in the self-assembly of nanomaterials. The forces discussed so far (van der Waals interactions, hydrogen bonds, hydrophobic interactions, etc.) may exist between a planar surface and a molecule some distance away. The strength and nature of these interactions will determine the extent to which molecules adsorb to the surface and perhaps initiate the growth of a nanomaterial. Furthermore, surface forces play an important role in catalysis, where a surface-bound molecule may be immobilized on the surface in an optimal geometry for a reaction to ensue. This section focuses on electrostatic interactions at surfaces. A more thorough treatment of the subject can be found in Israelachvili's classic book *Intermolecular and Surface Forces*.

5.2.1 The electrical double layer

Our discussion of electrostatic forces would not be complete without a brief overview of the electrostatic double layer and its role in the interactions between surfaces in liquids at the nanoscale. The electrostatic double layer is the term given to the diffuse layer of counterions in a solution that are associated with a charged surface. As shown below, the electrostatic double layer plays an important role in determining the forces that operate between charged surfaces in liquids.

When a surface comes into contact with a liquid, it may become charged by adsorbing ions from solution or releasing ions into solution. Many surfaces, for instance, are pH labile, and at either high or low pH might become positively or negatively charged. For example, a surface containing primary amine groups becomes positively charged at pH < 10 as the amine groups acquire an extra proton. Another common example of a surface-charging mechanism is the binding of Ca^{2+} ions by the zwitterionic headgroups of many phospholipid bilayers, resulting in a positively charged surface.

In solution, we would expect the charges on a surface to be balanced by the appropriate counterions that have been released from the surface itself or drawn in from the surrounding solution. Indeed, this is the case and the result is the formation of two regions of counterion charges to neutralize the surface charge. The first region is a compact layer of counterions that is closely bound to the charged surface. This compact region of bound counterions is called the *Stern* or *Helmholtz* layer. It should be noted that the counterions in the Stern layer are not necessarily irreversibly bound to the surface and can often be exchanged with those in the surrounding solution. The second region is a more diffuse and extensive layer of counterions that is in rapid equilibrium with the surrounding solution. This region is referred to as the *electrical double layer* or the *diffuse electrical double layer* and is the focus of our discussion. These regions are shown schematically in Figure 5.12.

A common example of the electrical double layer can be found in milk, which is a solid–liquid colloid. As a mixture of primarily nonpolar butterfat droplets in water, milk particles would seem to be expected to aggregate and coagulate into butter due to hydrophobic interactions. However, trace amounts of the highly polar phosphoprotein casein at the water–milk interface result in an electrical double layer forming around each milk particle. This double layer creates enough repulsion to overcome the hydrophobic particle's tendency to aggregate. Inks, paints,

Figure 5.12 The Stern/Helmholtz layer and the diffuse electrical double layer. Ions within the Stern/Helmholtz layer are bound to the surface, although generally not rigidly.

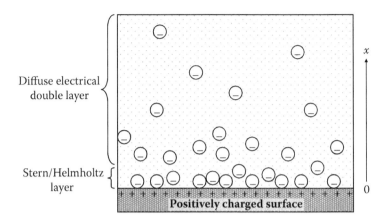

and blood provide further examples of heterogeneous liquid mixtures that are stabilized by electrical double layers.

The presence of the electrical double layer is a direct consequence of the tug-of-war between the energy and entropy of the charged-surface/bulk solution system. The electrostatic energy of the system is minimized when the charge separation is at a minimum—that is, when the counterions in solution become closely associated with the charged surface to the point of neutralization. The entropy, on the other hand, is maximized when the counterions are able to move freely through the entire volume of the bulk solution. Based on energy considerations alone, we would not expect an electrical double layer to exist—the surface charges would be completely neutralized by counterions closely bound to the surface. Entropic considerations, however, demand some sort of give-and-take. The resulting compromise between energy minimization and entropy maximization of the system produces the diffuse double layer with an equilibrium concentration of counterions that gradually decreases with distance away from the charged surface until it reaches a value equivalent to that of the bulk solution.

The actual distribution of the counterions at equilibrium can be calculated by the Poisson–Boltzmann equation

$$\frac{d^2\Psi}{dx^2} = -\left(\frac{e}{\varepsilon\varepsilon_0}\right)\sum_i z_i \rho_{i0} \exp(-z_i e\Psi(x)/kT) \qquad (5.24)$$

where z_{i-} is the valency of the ith electrolyte (i.e., +1 for Na$^+$), e is the standard unit of charge, ε is the static dielectric constant of the medium, k is Boltzmann's constant, and T is temperature. $\Psi(x)$ is the electrostatic potential at a distance x away from the surface. The zero of the potential can be arbitrarily assigned and ρ_{i0} is the number density of the ith electrolyte at that same distance x (often chosen to be the limit as x approaches ∞, which is equivalent to the bulk phase).

5.2.2 The Debye length

For most situations, the solutions to the Poisson–Boltzmann equation are rather complicated and should be obtained numerically by a computer. However, in the limit of a very small electrostatic potential such that $ze\Psi(x)/kT << 1$, then the Poisson–Boltzmann equation reduces to

$$\frac{d^2\Psi}{dx^2} = \left(\frac{e}{\varepsilon\varepsilon_0}\right)\sum_i z_i\rho_{i0}(z_ie\Psi(x)/kT) \tag{5.25}$$

$$\frac{d^2\Psi}{dx^2} = \kappa^2\Psi(x) \tag{5.26}$$

where

$$\kappa = \left(\frac{\sum_i \rho_{i0}z_i^2 e^2}{\varepsilon\varepsilon_0 kT}\right)^{1/2} \tag{5.27}$$

and has units of m^{-1}. In this case, ρ_{i0} is defined as the number density of the ith electrolyte in the bulk solution.

The second-order differential equation in Equation 5.26 is called the Debye–Hückel equation and has a well-known solution of

$$\Psi(x) = \Psi_0 e^{-\kappa x} \tag{5.28}$$

where Ψ_0 is the potential at the charged surface.

From Equation 5.28, we see that the characteristic decay length of the electrostatic potential for the Debye–Hückel model is $1/\kappa$. This length is often called the *Debye length* or the *Debye screening length* and can be used as a rough approximation for the "thickness" of the electrical double layer. If a charge is within the Debye length, it "feels" the effect of the charged surface, and if it is too far outside the Debye length, it effectively is screened from the charged surface by the intervening cloud of counterions.

From Equation 5.27 we also see that the Debye length is independent of the properties of the surface itself—that is to say for a given liquid at a certain temperature it depends only on the concentrations and valencies of ions in the bulk solution, not on the surface charge or surface potential. For example, for an aqueous 100-mM NaCl solution at 25°C, the Debye length is 0.96 nm, independent of the charge density or potential of the surface itself.

5.2.3 Interactions between charged surfaces in a liquid

In general, we can solve for the pressure due to the presence of ions at a position x between two charged surfaces as

$$P(x) = kT \sum_i \rho_i(x) \tag{5.29}$$

where $\rho_i(x)$ is the number density of the ith electrolyte at x (measured in molecules per cubic meter). Since the distribution of ions at any given point must be calculated using the Poisson–Boltzmann equation, then solving the Poisson–Boltzmann equation for the system must precede a calculation of the pressure between two surfaces. However, as mentioned previously, the Poisson–Boltzmann equation is rather complicated to solve for most systems of practical interest. It is beyond the scope of this text to discuss the solutions to this equation, but students should be aware of this general approach for calculating the pressure between two surfaces. We limit our discussion to qualitative descriptions of the forces operating between two interacting surfaces, descriptions resulting from the application of the approach described previously.

In the most elementary example, suppose we have a flat, neutral surface approaching a flat, charged surface in a parallel orientation, such as that shown in Figure 5.13. Before the approach of the neutral surface, the charged surface has associated with it a diffuse electrical double layer extending out into solution. As the neutral surface approaches, however, the counterions in the double layer must become confined to a smaller and smaller volume, resulting in a decrease in entropy of the system. The approach of the neutral surface likely causes some of the counterions to bind to the charged surface, resulting in a slight decrease in energy. However, this favorable decrease in energy is offset by the much larger decrease in the entropy of the system. For this reason, the interaction between a neutral and charged surface in a liquid must always be repulsive.

Now let's consider the interaction between two charged surfaces of like charge, as shown in Figure 5.14. As the surfaces approach each other, their electrical double layers begin to overlap, resulting in an effective decrease in the entropy of the system and making the interaction unfavorable. Hence, the interaction between two surfaces of like charge is

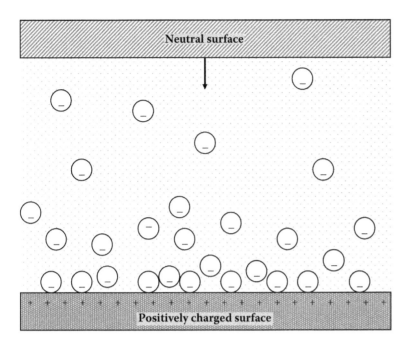

Figure 5.13 The interaction between a neutral and charged surface is repulsive.

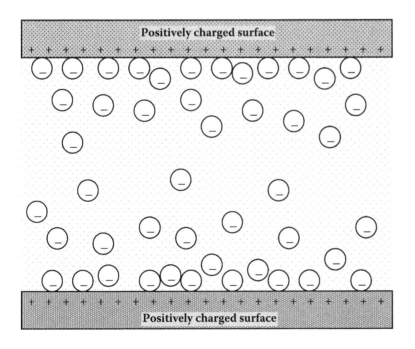

Figure 5.14 The interaction between two surfaces of like charge is repulsive.

always repulsive. By comparison, this repulsive interaction begins to occur when the two surfaces are at a greater distance than if one of the surfaces was neutral.

Finally, let's consider the case of the interaction between two surfaces of opposite charge. As we would expect, electrostatic attraction between the surfaces dominates at long range. One surface can, in effect, operate as the "counterion" for the apposing surface. As the surfaces approach, counterions are released into solution and expelled from the gap between the two surfaces, resulting in an increase in entropy of the system. If the charge densities of each surface are equivalent, the electrostatic attraction continues up until contact between the surfaces. If the charge densities are not equal, some counterions must remain in the gap. Gradually they become more concentrated as the surfaces approach each other, and at some point the repulsive force from these counterions balances the electrostatic attraction between the surfaces.

5.3 INTERMOLECULAR FORCES AND AGGREGATION

Supramolecular chemistry is dominated by the host of noncovalent interactions present in molecular subunits. A simple illustration can help us understand the interplay between the various interactions discussed so far, and how this interplay leads to molecular self-assembly into nanomaterials with a specific structure. Consider the set of generic molecules shown in Figure 5.15. For simplicity, only characteristics emphasizing interactions are shown, such as ionic moieties, hydrophobic regions, dipoles, and hydrogen bonding groups. The molecules represent molecular building blocks and the aggregation of these individual blocks will be affected by intermolecular interactions.

The organization of the building blocks into more complex structures will largely be driven by thermodynamics, whether energetics or entropy. The latter factor, for example, can be an entropic gain due to the hydrophobic effect. Minimization of energy will be achieved by minimizing unfavorable interactions such as bringing two like-charged moieties to the same vicinity. Possible ways the molecular building blocks may assemble are shown in Figure 5.15b. Self-assembly will lead to a three-dimensional aggregate, but if the assembly is occurring on a surface, then a two-dimensional aggregate will form. A planar support will also impose some restrictions on the exact structure. For example, in

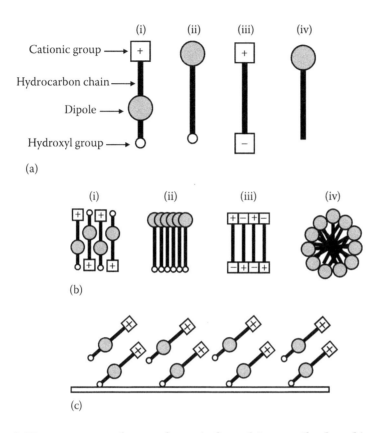

(a)

Cationic group ⟶
Hydrocarbon chain ⟶
Dipole ⟶
Hydroxyl group ⟶

(b)

(c)

Figure 5.15 (a) Representation of four molecular building blocks containing various interacting functionalities. (b) Possible aggregation patterns driven by (i) like-charge repulsion, (ii) dipole–dipole and H-bonding interactions, (iii) opposite-charge attraction, and (iv) strong hydrophobic interactions. Hydrophobic interactions probably play a role in all of these aggregates. (c) Strong substrate–molecule interactions cause the molecules to tilt in order to minimize like-charge interactions.

Figure 5.15c, a supported monolayer is forced into a tilted and/or staggered state because of strong substrate–molecule interactions and intermolecular headgroup repulsions.

A thermodynamically stable structure will be formed based on the conditions of the assembly process (temperature, the presence of a surface, pH, concentration, etc.). If the structure is formed in a solvent such as water, it is important to appreciate that the molecules comprising the aggregate may be in dynamic equilibrium with "free monomers" in solution. One consequence of this is that the aggregate size and shape may change with monomer concentration and other conditions such as pH, temperature, and salt concentration.

Up until now, we have ignored direct electron coupling between neighboring molecules in a self-assembled aggregate. Although induced dipole effects are electronic in origin, the molecules may be relatively far apart so that electrons are still localized on each molecule. When one is confronted with a dense aggregate comprised of molecules that are essentially touching one another, the electronic characteristics of the material may change. For example, molecular building blocks may result in a nanostructure in which the electrons are delocalized over the entire aggregate. This will change the material's optical and electronic properties, which will be determined by the shape and size of the aggregate. The following sections provide some background in basic electronic structure pertaining to electron delocalization and the effect of size.

5.4 SIMPLE MODELS DESCRIBING ELECTRONIC STRUCTURE

Electrons interact with radiation, and this interaction is responsible for the absorption and emission of radiation. Phenomena such as fluorescence, phosphorescence, and photoelectricity depend on how light interacts with molecules. This interaction can be exploited to gain information about molecular structure (the basis of spectroscopy). Spectroscopy and the nature of light–matter interactions are covered in Chapter 6. Here some pertinent elements of electronic structure are covered. It is assumed that the student has a general chemistry level grasp of Lewis structures, molecular orbital (MO) theory of simple molecules, and a quantum mechanical interpretation of light (photons) and electronic structure (energy levels). A basic understanding of these processes can be developed from the quantum mechanical models in Chapter 4. One important equation worth recalling is Planck's equation, which relates energy between two energy levels (ΔE) to the wavelength (λ) of light absorbed or emitted as a result of an electronic transition between these two energy levels (Equation 5.30):

$$\Delta E = h\nu = \frac{hc}{\lambda} \tag{5.30}$$

The wavelength and frequency (ν) of light is related by $\lambda = c/\nu$, where c is the speed of light in a vacuum ($2.998 \times 10^8 \text{ ms}^{-1}$) and h in Equation 5.24 is Planck's constant (6.626×10^{-34} Js).

5.4.1 Applications of the particle-in-a-box model

Chemical reactivity and physical phenomena such as the absorption of light is largely determined by the electronic structure in molecules. Since electronic energy levels are quantized, Equation 5.30 provides the wavelength (or color) of light absorbed or emitted due to transitions between such levels. For example, electron configurations of atoms and molecules provide an excellent explanation of such observations. However, as we discussed in Chapter 4, in many cases an electron (or electrons) is free to move within a certain region of space, such as a nanoscale aggregate or other nanomaterial (nanoparticles, nanowires, etc.). Metals and conjugated molecules are other examples where electrons are not restricted to the individual nuclei, but rather are delocalized over a larger region of space.

The particle-in-a-box model can be used to model delocalized electrons in conjugated systems. Recall that in the simplest, one-dimensional, case, the particle is confined between infinite potential barriers at $x = 0$ and $x = a$, with zero potential energy between the barriers. These values of energy are given by Equation 5.31, where a is the length of the line the particle is confined on, m is the mass of the electron, and n is a quantum number that can have any positive integer value (1,2,3...):

$$E = \frac{h^2 n^2}{8ma^2} \tag{5.31}$$

Example 5.3 Calculating the Energy of an Electron in a One-Dimensional Nanoscale Region

Consider an electron that is free to move along a line of length 200 nm. What is the ground state energy of the electron? What is the energy of the electron in the $n = 3$ state? What is the effect on the spacing between neighboring energy levels if the length of the line increased from 200 nm to 300 nm?

Solution   The ground state energy represents the lowest energy the electron can have. This is the case when $n = 1$. The mass of the electron is 9.109×10^{-31} kg and $h = 6.626 \times 10^{-34}$ Js. Using Equation 5.25,

$$E = \frac{h^2 n^2}{8ma^2} = \frac{\left(6.626 \times 10^{-34}\ \text{Js}\right)^2 (1)^2}{8\left(9.109 \times 10^{-31}\ \text{kg}\right)\left(200 \times 10^{-9}\ \text{m}\right)^2}$$

$$= \frac{4.390 \times 10^{-67}\ \text{J}^2\text{s}^2}{2.915 \times 10^{-43}\ \text{kgm}^2} = 1.506 \times 10^{-24}\ \text{J}$$

(Note: 1 J = 1 kgm^{2}s^{-2})

When $n = 3$, the energy is

$$E = \frac{h^2 n^2}{8ma^2} = \frac{\left(6.626 \times 10^{-34} \text{ Js}\right)^2 (3)^2}{8\left(9.109 \times 10^{-31} \text{ kg}\right)\left(200 \times 10^{-9} \text{ m}\right)^2}$$

$$= \frac{3.951 \times 10^{-66} \text{ J}^2\text{s}^2}{2.915 \times 10^{-43} \text{ kgm}^2} = 1.355 \times 10^{-23} \text{ J}$$

The difference between neighboring energy levels (ΔE) is given by

$$\Delta E = E_2 - E_1 = \frac{h^2}{8ma^2}\left(n_2^2 - n_1^2\right)$$

The subscripts 1 and 2 denote lower and upper energy levels, respectively. If a increases from 200 nm to 300 nm, ΔE will decrease by $2^2/3^2$ or about 45%.

The particle-in-a-box model is arguably the simplest quantum mechanical model describing the energy and probability density of an electron. Despite its simplicity it has been extremely valuable in describing the absorption properties of simple molecules in which the electron is freely moving along a line. Such molecules can be linear carbon chains containing alternating single and double bonds (conjugation) in which electrons are delocalized along the entire chain.

Conjugated molecules have more than one free electron moving along the chain. For example, hexatriene has six free electrons, as shown in Figure 5.16. Each carbon atom is sp^2 hybridized, and overlap of these orbitals forms the σ backbone of the carbon chain, including all the C–H bonds. The electrons due to σ-bonding are localized between the two atoms forming the bond. The particle-in-a-box model does not apply to these electrons. However, each carbon atom has an unhybridized p-orbital containing one electron. The p-orbitals on all the carbon atoms are able to overlap forming the π-framework (shown by the shaded regions in Figure 5.16a). The six π-electrons are delocalized along the chain and can be described using the particle-in-a-box model.

In order to deal with the six π-electrons in hexatriene, an energy-level diagram is constructed like that shown in Figure 5.16b. The energy levels are filled up with the appropriate number of electrons such that each level contains a maximum of two electrons. The Pauli exclusion principle tells us that only two electrons can enter a given energy level, and this pairing occurs with the electrons having opposite spins to each other. In

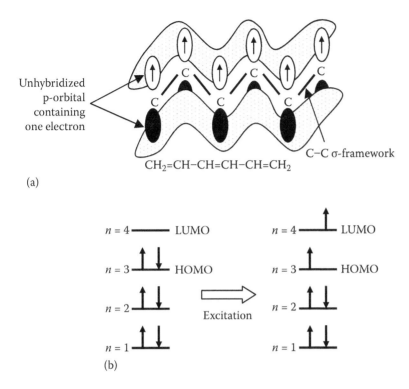

(a)

CH₂=CH–CH=CH–CH=CH₂

Unhybridized p-orbital containing one electron

C–C σ-framework

(b)

Figure 5.16 (a) Overlap of unhybridized p-orbitals on each carbon atom produces a π molecular orbital resulting in the delocalization of electron density along the hexatriene chain. (b) The particle-in-a-box model as applied to the hexatriene molecule results in an energy-level diagram showing electron pairs in three levels. Excitation of an electron to the $n = 4$ occurs by the absorption of energy.

Figure 5.16, the six free electrons in hexatriene enter the first three energy levels. Higher energy levels ($n > 3$) are empty when the molecule is in the ground state. The $n = 3$ level is known as the HOMO (highest occupied molecular orbital) level and the $n = 4$ level is known as the LUMO (lowest unoccupied molecular orbital) level. An electron in the $n = 3$ state can absorb a photon of energy and as a result enter the $n = 4$ level. This is known as the HOMO–LUMO electronic transition. Of course, the photon causing the excitation must have an energy that is equal to the HOMO–LUMO energy gap.

The higher-dimensional particle-in-a-box models discussed in Chapter 4 can also be used to represent electrons delocalized over a two-dimensional rectangular region (Equation 5.32) or a three-dimensional cubical region (Equation 5.33):

$$E_{2D} = \frac{h^2}{8m} \left(\frac{n_x^2}{a^2} + \frac{n_y^2}{b^2} \right)$$ (5.32)

$$E_{3D} = \frac{h^2}{8m} \left(\frac{n_x^2}{a^2} + \frac{n_y^2}{b^2} + \frac{n_z^2}{c^2} \right) \tag{5.33}$$

In these equations a, b, and c represent the length, width, and height of a cube (or just a and b for a rectangle), and the x, y, and z subscripts denote quantum numbers in the three different directions of a Cartesian coordinate system. These three quantum numbers independently assume values of 1, 2, 3, 4, and so on.

An interesting solution arises when dealing with an electron moving around a ring, another model that was first encountered in Chapter 4. The energy levels are given by Equation 5.34:

$$E = \frac{h^2}{8\pi^2 I} m^2 \tag{5.34}$$

$$I = m_e r^2 \tag{5.35}$$

I is the moment of inertia of the electron going around the ring (Equation 5.35), r is the radius of the ring, and m_e is the mass of the electron. The quantum number m can take on values of 0, ±1, ±2, ±3, and so on. This means that when m is 1, there are two energy levels with the same energy. We say that the electron in the $m = 1$ state is twofold degenerate. In general, $+m$ and $-m$ represent a doubly degenerate energy state.

This model can be used to describe the electronic structure of benzene. There are six π electrons in benzene, one in each unhybridized p-orbital on each carbon atom (Figure 5.17a). These six electrons are delocalized around the ring and are regarded as free electrons. Figure 5.17b shows an energy level diagram based on Equation 5.28. We can place two of them into the $m = 0$ level and four into the $m = ±1$ level. The first electronic transition would be the $1 \rightarrow 2$ transition, and the energy associated with this transition is given by Equation 5.36.

$$\Delta E = \frac{h^2}{8\pi^2 I} \left(2^2 - 1^2 \right) \tag{5.36}$$

Example 5.4   Estimating the Size of the Benzene Ring

Benzene absorbs light of wavelength ~250 nm. Estimate the radius of the benzene ring.

Solution   We first change the absorption wavelength into the corresponding energy using Equation 5.30.

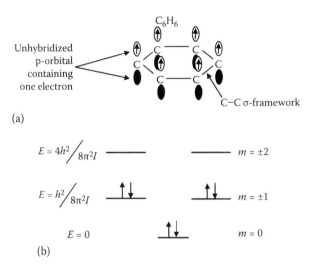

(a)

$$E = 4h^2 / 8\pi^2 I \quad \underline{\qquad} \qquad \underline{\qquad} \quad m = \pm 2$$

$$E = h^2 / 8\pi^2 I \quad \underline{\uparrow\downarrow} \qquad \underline{\uparrow\downarrow} \quad m = \pm 1$$

$$E = 0 \qquad \underline{\uparrow\downarrow} \qquad m = 0$$

(b)

Figure 5.17   (a) Overlap of unhybridized p-orbitals on each carbon atom produces a π molecular orbital resulting in the delocalization of electron density along the benzene ring. (b) The particle-in-a-box model as applied to the benzene molecule results in an energy level diagram showing electron pairs in three levels. The two energy levels corresponding to $m = \pm 1$ are doubly degenerate.

$$\Delta E = \frac{hc}{\lambda} = \frac{(6.626 \times 10^{-34}\ \text{Js})(2.998 \times 10^8\ \text{ms}^{-1})}{(250 \times 10^{-9}\ \text{m})} = 7.946 \times 10^{-19}\ \text{J}$$

Assuming that the $1 \rightarrow 2$ transition is associated with the absorption wavelength of 250 nm, we rearrange Equation 5.34 to determine the moment of inertia.

$$I = \frac{h^2(2^2 - 1^2)}{\Delta E(8\pi^2)} = \frac{(6.626 \times 10^{-34}\ \text{Js})^2(3)}{(7.946 \times 10^{-19}\ \text{J})(8\pi^2)} = 2.102 \times 10^{-50}\ \text{Js}^2$$

$$= 2.102 \times 10^{-50}\ \text{kgm}^2$$

(Note: 1 Js^2 = 1 kgm^2)

Finally, the radius can be estimated using Equation 5.35.

$$r = \sqrt{\frac{I}{m_e}} = \sqrt{\frac{2.102 \times 10^{-50}\ \text{kgm}^2}{9.109 \times 10^{-31}\ \text{kg}}} = 1.519 \times 10^{-10}\ \text{m} = 0.152\ \text{nm}$$

This is close to the experimentally measured value of around 0.25 nm.

The particle-in-a-box model can be used to describe the energy of an electron confined within a three-dimensional region of nanoscale dimension. This is known as quantum confinement, and examples involving quantum confinement in solid nanoparticles are given in Chapter 9.

5.4.2 Conjugation in organic molecules

The free electron model can be applied to conjugated organic molecules. In fact, the model can help us explain why the absorption wavelength increases as the length of the molecule increases. We will use the term *conjugation length* to describe the length of the alternating double- and single-bonded hydrocarbon chain in which the electrons are delocalized over. First let's review some background on the structure of conjugated organic molecules.

As briefly mentioned in the last section, a conjugated system has three sp^2 hybridized orbitals on every carbon atom, which form covalent bonds with nearby atoms. This accounts for the σ-bonding in the molecule. The leftover unhybridized p_z orbital combines with other p_z orbitals to form a delocalized π MO that spans the length of the molecule. For example, ethylene has two atomic p_z orbitals, ϕ_1 and ϕ_2. From these two atomic orbitals, two MOs, ψ_1 and ψ_2^*, are formed by taking linear combinations. The bonding MO, ψ_1, results from the in-phase combination of the wave functions of the two p orbitals, whereas the antibonding orbital, ψ_2^*, results from the out-of-phase combination. The overlap results in two new MOs—one bonding orbital with an energy lower than either of the original p orbitals and one antibonding orbital with an elevated energy. The relative energies of these MOs are illustrated in Figure 5.18. It must be pointed out that only the π MOs are shown in the figure. These MOs result from the overlap between unhybridized p-orbitals on the carbons. The MOs resulting from σ-bonds are not shown since this is not relevant to our discussion of conjugation.

In terms of shape, the bonding MO has electron density above and below the line connecting the two carbon atoms. There is no node between these atoms. In contrast, the antibonding MO has a node between the two carbons. The antibonding MO has electron density concentrated near each of the two carbon atoms, but zero electron density between these atoms. The presence of a node means no bond, thus the term antibonding MO.

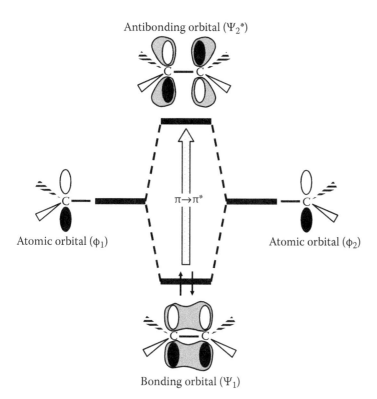

Antibonding orbital (Ψ_2*)

Atomic orbital (ϕ_1) $\pi \rightarrow \pi^*$ Atomic orbital (ϕ_2)

Bonding orbital (Ψ_1)

Figure 5.18 A partial MO energy-level diagram for ethylene emphasizing the π bonding (Ψ_1) and antibonding (Ψ_2) MOs. MOs are constructed by the linear combination of the AOs of ethylene, in this illustration the unhybridized p-orbitals (ϕ_1 and ϕ_2) on each carbon atom. The two electrons from each p-orbital are placed in the bonding MO (Ψ_1). The MOs due to σ-bonding are not shown.

The simple two p-orbital overlap scheme can be expanded to conjugated systems. We start by considering 1,3-butadiene, the simplest conjugated system. Because 1,3-butadiene has four atomic orbitals, four molecular orbitals must result. Figure 5.19 shows the atomic and molecular orbitals of 1,3- butadiene and the corresponding energy levels. Like ethylene, only the π MO are shown, which result from the overlap between unhybridized p-orbitals on the carbons. The MOs resulting from σ-bonds are not shown.

The important aspect of Figure 5.19 is the $\pi \rightarrow \pi^*$ transitions. Notice that the $\phi_2 \rightarrow \phi_3^*$ transition in 1,3-butadiene is much smaller than the $\phi_1 \rightarrow \phi_2^*$ transition in ethylene. As the number of p-orbitals increases in a conjugated system, the energy gap, ΔE, between the HOMO and the LUMO becomes progressively smaller. Figure 5.20 shows the MO energy levels of ethylene and the first three conjugated systems: 1,3-butadiene; 1,3,5-hexatriene; and 1,3,5,7-octatetraene. The vertical arrows emphasize the HOMO–LUMO transitions. Further, the energy difference between the

Figure 5.19 A partial MO energy-level diagram for buta-diene emphasizing the π bonding (Ψ_1 and Ψ_2) and antibonding ($\Psi_3{}^*$ and $\Psi_4{}^*$) MOs. MOs are constructed by the linear combination of the AOs of butadiene, in this illustration the unhybridized p-orbitals on each carbon atom grouped together and described as ϕ_1 and ϕ_2. The four electrons from each p-orbital are placed in the bonding MOs (Ψ_1 and Ψ_2). The MOs due to σ-bonding are not shown.

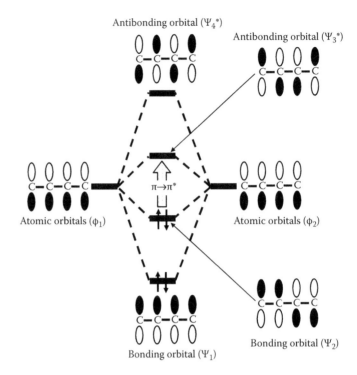

Figure 5.20 The MO energy-level diagram for a series of linear conjugated hydrocarbons. The figure shows how ΔE for the $\pi \rightarrow \pi^*$ transition changes as the conjugation length increases.

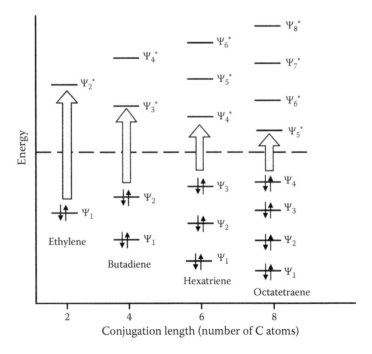

HOMO and LUMO determines the position of the absorption maximum (λ_{max}). Since wavelength is inversely proportional to ΔE, λ_{max} increases with the length of the chain of the conjugated carbon atoms.

In the extreme case of a conjugated polymer, the π bonding orbitals and the π^* antibonding orbitals form broad bands separated by a small ΔE. A long conjugated polymer is able to conduct electricity because electrons in the valence band (π) can be excited into the conduction band (π^*). There, they are free to move and carry charge. Because electrons in conjugated polymers must be excited in order to conduct electricity, conjugated polymers make excellent semiconductors.

Alan Heeger, Alan MacDiarmid, and Hideki Shirakawa shared the Noble Prize in Chemistry in 2000 for their discovery and development of conductive (conjugated) polymers in the late 1960s and early 1970s. They did much of their work on polyacetylene, the simplest conjugated polymer, with alternating single and double bonds in a linear carbon chain. Polyacetylene, however, is very susceptible to photo-oxidation, and more recent studies have focused on other, more stable conjugated polymers. Derivatives of polythiophene, polyaniline, polyfluorene, and poly(phenylene vinylene) or PPV have similar conductivities to polyacetylene but are much more stable against oxidative degradation. Polymers and other macromolecules will be discussed in Chapter 10.

5.4.3 Solvatochromism

The energy of electronic transitions in molecules and nanosystems are affected by the environment that surrounds those molecules. When a molecule undergoes an electronic excitation and electrons move into a previously unoccupied orbital, the charge distribution in the molecule changes. Such movement of charge may make a molecule or nanosystem more or less polar. Recall that more polar molecules have a more favorable potential energy of interaction with polar solvents due to energetically favorable dipole–dipole and dipole-induced dipole interactions. If an electronic transition causes a molecule to become more polar, its excited state will have a lower energy in a polar solvent than it would in a nonpolar solvent, and the wavelength at which it is excited will become longer (red-shifted) compared to the same molecule in a nonpolar solvent. This is called a **bathochromic shift**. Conversely, if an electronic transition causes a molecule to become less polar, the excited state will be less energetically favorable in a polar solvent than in a nonpolar solvent and the wavelength at which the molecule is excited will become shorter (blue-shifted). This is called a **hypsochromic shift**.

The general phenomenon of the wavelength at which a molecule or nanosystem is excited at shifting due to changes in solvent polarity is called solvatochromism. A simple example is shown in Figure 5.21; phenol blue undergoes a bathochromic shift since its first excited state resembles a zwitterion and is more polar than its ground state. Solvent polarity can be characterized via the dielectric constant or empirical polarity scales such as the ET_{30} parameter. While described here in the context of a solvent, the phenomenon of a molecule's environment affecting the wavelengths at which it absorbs light by stabilizing or destabilizing the ground or excited states can also be generalized to solids, molecules on surfaces, and molecules within other nanostructured systems.

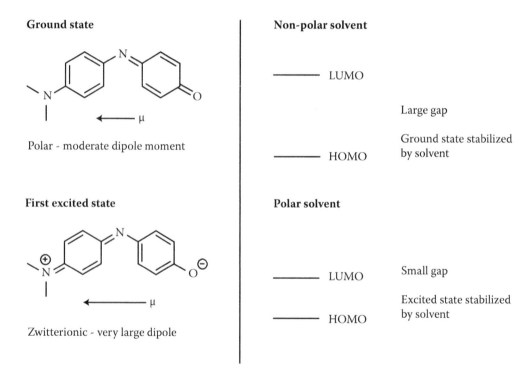

Figure 5.21   Left: first excited state of phenol blue is more polar than its ground state, as an electron is transferred from the aromatic amine on the left side of the molecule toward the ketone on the other end, forming a zwitterionic phenolate. Right: The energy of the LUMO (corresponding to the first excited state) for phenol blue is reduced in polar solvents, causing a bathochromic shift in the wavelength at which light is absorbed.

5.4.4 Aggregation and electronic structure

Molecular aggregates have electronic and spectroscopic properties that may be considerably different from the monomer. Consider a conjugated molecule that contains a dipole moment. We can represent this molecular building block as a rod shown in Figure 5.22, where the arrow indicates the direction of the dipole. Let's form a molecular aggregate in which individual monomers are arranged in a regular fashion. We can distinguish between two types of aggregation patterns called the *H*-type and *J*-type aggregation.

The *J*-type aggregate is a one-dimensional molecular assembly in which the dipole moments of the individual monomers are aligned parallel to the line joining their centers. This is sometimes referred to as the "end-to-end arrangement." In contrast, the *H*-type arrangement, while still a one-dimensional array, is one in which the dipole moments are aligned parallel to each other but perpendicular to the line joining their centers.

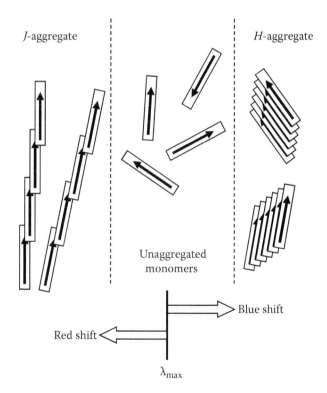

Figure 5.22 Noncovalent aggregation patterns in dipolar building blocks. The *J*-aggregate is a head-to-tail arrangement and the *H*-aggregate is a head-to-head arrangement between neighboring molecules.

This is sometimes referred to as the "face-to-face arrangement." Figure 5.22 illustrates the difference between a *J*-aggregate and an *H*-aggregate.

One of the most characteristic properties of *J*-type aggregation is that such materials absorb higher wavelength light with respect to the monomer absorption. We say that it is red-shifted in the absorption spectrum with respect to the monomer absorption (see Chapter 6). The absorption wavelength of the *H*-aggregate is a little lower (or blue-shifted) with respect to the monomer absorption wavelength. The energy shift of the absorption wavelengths of the aggregates has been explained by exciton theory. This theory will not be covered in detail in this text, but we can apply our understanding of conjugation length and the particle-in-a-box models to help understand this observation.

First assume that the electron motion is along the dipole only. A *J*-aggregate can be considered a line along which free electrons can move. Since the dipoles are arranged end-to-end, the electron motion can be considered as delocalized along the entire length of the aggregate. According to Equation 5.31, the length a is much greater than the length of an individual monomer. Therefore, ΔE is small and is inversely proportional to wavelength (Equation 5.30), and the aggregate absorbs light energy at a higher wavelength compared to a monomer. Conversely, an *H*-aggregate will have a short conjugation length and a larger ΔE value corresponding to the absorption of light of a smaller wavelength. The unaggregated monomer can be considered intermediate between these two extremes, and so the ΔE value corresponding to the excitation lies between the values for the *H*- and *J*-aggregate.

5.4.5 π–π stacking interactions

We end this chapter with a brief discussion of a very weak electron coupling interaction leading to aggregation known as the π–π stacking interaction. These interactions occur due to the presence of p-orbitals in conjugated ring systems such as benzene. The net effect of such an interaction is face stacking of planar rings as shown in Figure 5.23a for naphthalene. Although the effect is actually insignificant in small systems such as benzene, the interaction becomes stronger as the number of π-electrons increases. It should be pointed out, however, that even in these systems, electrostatic forces usually overcome π–π stacking interactions. Nonetheless, the interaction is particularly strong in planar polycyclic

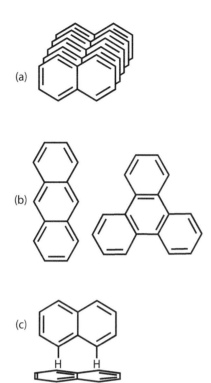

(a)

(b)

(c)

Figure 5.23 (a) π–π stacking between planar aromatic rings. (b) Structures of anthracene and triphenylene. (c) The edge-face interaction between two planar aromatic ring systems.

aromatic molecules containing many delocalized rings. Examples include anthracene and triphenylene Figure 5.23b). For large ring systems these interactions can be so significant that they may dominate the supramolecular chemistry and determine the overall structure of the aggregate. For example, the interaction determines the growth of organic crystals composed of such polycyclic molecules.

π-stacking interactions affect the properties of polymers, peptides, liquid crystals, and proteins. In biology, π-stacking occurs between adjacent nucleotides, and this adds to the stability of double-stranded DNA. A related phenomenon, called the edge–face interaction, is often observed in proteins where the hydrogen atom of one aromatic system points perpendicularly to the center of the aromatic plane of the other aromatic system Figure 5.23c). This type of interaction is thought of as related to the partial positive charge on the hydrogen atom connected to the aromatic ring.

End of chapter questions

1. $k_b T$ is often used to measure the strength of a given interaction, where k_b is Boltzmann's constant (1.38065×10^{-23} J K^{-1}). Coulombic interactions are generally among the strongest and the farthest reaching of the interactions that we have discussed. (a) Using the information given in Example 5.1, determine at what distance Na$^+$ and Cl$^-$ would have to be separated in order for $V(r)_{Na-Cl}$ to be equal in magnitude to $k_b T$ at room temperature (298K) in the gas phase ($\varepsilon = 1$). (b) Perform the same calculation in water and toluene (use the dielectric constants in Table 5.1) and discuss how this could have an impact on the solubility of NaCl in these two solvents.

2. (a) Imagine that a Ca^{2+} ion interacts with an H$_2$O molecule in the absence of any other molecules at a distance of 0.4 nm. Plot the interaction potential energy as a function of θ. Assume that the only interaction between the two species is an ion–dipole interaction and that the distance remains fixed. (b) At what angles does the magnitude of the potential energy reach a maximum? Why does this make sense? What do the positive and negative values of V represent?

3. Obtain an expression for the force between an ion and its induced dipole in an ion-induced dipole interaction. Is the force attractive or repulsive?

4. (a) Assume that two dipolar molecules (each with dipole moment μ) are in the same plane. Show that for all distances r, the interaction potential energy between the dipoles is smaller (more negative) if the two dipoles are oriented in a line rather than antiparallel to each other. (b) The answer from 4(a) is slightly misleading because it might lead us to believe that the dipoles always prefer to orient themselves in a line rather than anti-parallel. If the dipolar molecules are anisotropic in shape (oblong in the direction of the dipole), explain why the molecules might prefer to orient themselves in an antiparallel orientation rather than in a line.

5. Explain in your own terms why, in the absence of other interactions, a dipole-induced dipole interaction will always be attractive. Use diagrams if appropriate.

6. At what value of r (in terms of σ) does the Lennard-Jones potential reach its minimum value? Mathematically, what does σ represent?

7. The Onsager model treats the relationship between the dipole moment of a polar molecule within a liquid as the interaction of that dipole embedded in a spherical cavity (spherical molecule) within a dielectric composed of an infinite number of identical, freely rotating dipoles, such that the dipole induces a field on the dielectric and the average field of the dipoles in the dielectric partially counteracts the field from the dipole. This assumption of no interactions between individual molecules allows the dielectric response of a material to be estimated from single-molecule properties. However, consider the molecules listed in Table 5.1, where the Onsager model provides very good estimates in some cases and fails for others (particularly water). For each molecule listed in Table 5.1, (a) list what are likely to be the strongest intermolecular interactions it

will experience with other identical molecules based on its structure, and (b) briefly explain how these interactions and/or molecular shape could lead to the Onsager model producing a good or poor estimate of the dielectric constant.

8. Use the particle-in-a-box model to predict the values of ΔE and the corresponding values of λ_{max} for the conjugated linear hydrocarbons shown in Figure 5.21. Assume the C–C bond length is 154 pm and the C=C bond length is 134 pm.

9. β-Carotene is an organic molecule responsible for the red-orange color in plants and fruits. The molecular structure is shown below. Assuming that the length of the molecule is 2.94 nm, determine the value of λ_{max}. Does the result agree with the red-orange color observed for this molecule?

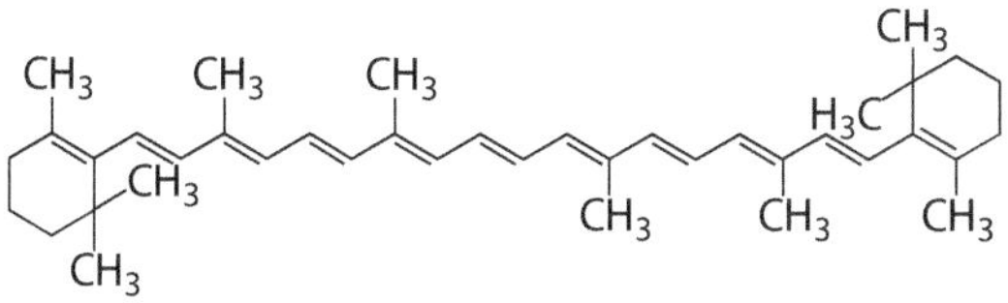

10. Two dyes are shown, each consisting of a dimethylamine group connected to an electron-withdrawing group by a conjugated bridge, forming a polar structure. Which sort of solvatochromic shift, if any, would you expect each dye to undergo? Explain your reasoning. Assume that the transition is a π–π^* transition and that the electrons move from an orbital that includes the amine lone pair to an orbital that includes the electron-withdrawing group at the other end of the molecule. It may be helpful to consider potential resonance structures for the molecule; see Figure 5.21 for an example.

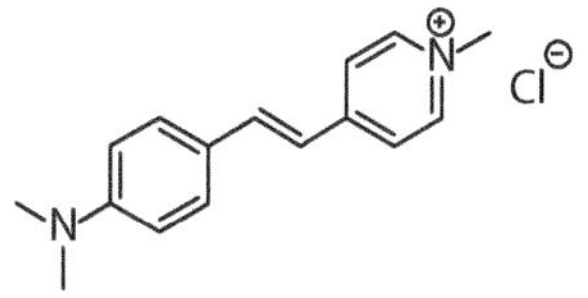

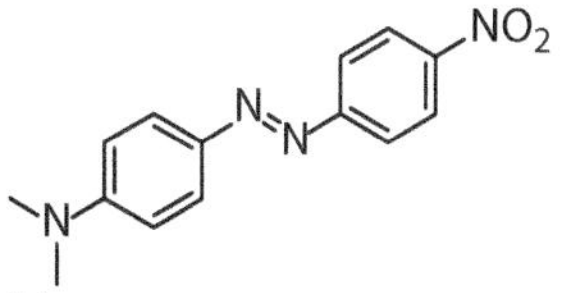

(a)

(b)

11. Discuss the possible types of aggregation patterns when the following water-soluble molecules self-assemble in the aqueous phase. Your discussion should include inter-molecular contributions from van der Waals interactions, electronic coupling effects, and predictions about the absorption properties of the resulting aggregates. How would molecules aggregate if they were forced at the interface of a hydrophobic surface and a water phase?

(a). $CH_3(CH_2)_{12}SO_4Na$

(b). $HO–CH=CH–CH=CH–CH_2–CO_2H$,

(c). $BrH_3N–CH=CH–CH=CH–CH_2–CO_2Na$

(d).

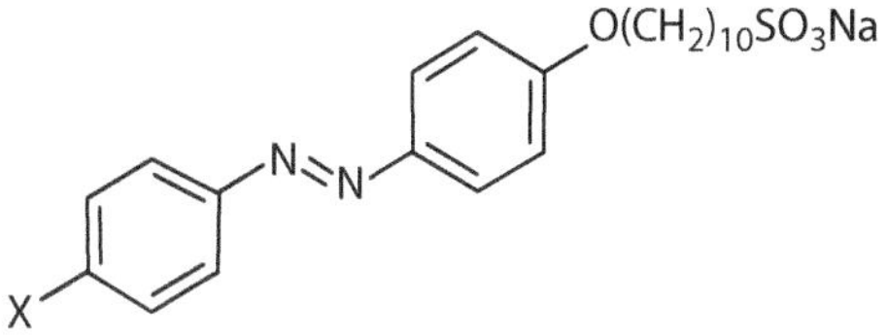

where X = OH or NO_2

(e).

12. The following molecules are identical except for the structure of dipolar headgroups. Place

the molecules in order of increasing head-group dipole moment and discuss the differences in the way these molecules aggregate into a monolayer at the air–water interface, if any difference exists.

References and recommended reading

Israelachvili, J. *Intermolecular and Surface Forces*, 3rd ed. 2011, Academic Press, San Diego, pp. 31–139. This is a classic textbook on the subject, but is probably for the more advanced student. Chapters 3 through 8 are particularly relevant to this text.

Kuhn, H.Kuhn, C. Chromophore coupling effects. In *J-Aggregation*, Kobayashi, T., Ed., 1996, World Scientific, Singapore, 1–140. This is an excellent book for those interested in H- and J-aggregation.

Lowe, J. P. *Quantum Chemistry*, 2nd ed. 1993, Academic Press, San Diego. This book is for the advanced student. Chapter 2 contains some excellent applications of the particle-in-a-box model.

Onsager, L. Electric moments of molecules in liquids. *Journal of the American Chemical Society*. 1936, 58: 1486–1493. A detailed derivation of the Onsager model and discussion of its assumptions is found in this 1936 paper.

Pavia, D., Lampman, G., and Kriz, G. *Introduction to Spectroscopy*, 2nd ed. 1996, Saunders College Publishing, Orlando, FL. The chapter on electronic spectroscopy provides some useful background on MO diagrams of organic molecules. This book is an excellent read for those interested in the kinds of electronic excitations discussed in Section 5.4.2.

Purcell, E. MMorin, D. J. 2013, *Electricity and Magnetism*, 3rd ed. Cambridge University Press, Cambridge, UK. This is an excellent general introduction to electrostatics and dynamics.

Reichardt, C. Solvatochromic dyes as solvent polarity indicators. *Chemical Reviews*. 1994, 94: 2319–2358. This paper provides a good review of solvatochromism, including common dyes and their solvatochromic shifts.

CHAPTER 6

Bulk Characterization Techniques for Nanomaterials

CHAPTER OVERVIEW

This chapter provides a broad survey of some of the common techniques used for characterization of the bulk properties of materials, including nanomaterials. These techniques can provide insight on properties such as the concentration or composition of molecules or nanoparticles, the thickness of nanofilms, or the arrangement of atoms in ordered structures such as crystals. However, all of these techniques typically average over a macroscopic region of a material to determine an ensemble property (analogous to the macroscopic thermodynamic properties discussed in Chapter 2) instead of directly examining a material's nanoscale dimensions. Many of these techniques are spectroscopic, involving the interaction of light at different frequencies with matter, and applying principles from Chapter 4. Several of techniques that will be discussed are typically encountered in undergraduate chemistry and materials labs, while others require more specialized equipment. Specific examples of the application of these methods to nanomaterials are found in Chapters 9 and 10.

6.1 SPECTROSCOPIC METHODS

Spectroscopy is the study of the ways in which electromagnetic radiation interacts with matter. Spectroscopy is an invaluable field of chemical analysis, and different spectroscopic methods are able to determine identity, concentration, and structural information of chemical compounds, among other useful information. Because the electromagnetic spectrum is so broad, different spectroscopic methods are employed for different regions of the spectrum. Figure 6.1 depicts a listing of the common regions of the electromagnetic spectrum as a function of wavelength and frequency. It also lists the most common spectroscopic methods employed in each region of the spectrum and the kinds of transitions those methods probe.

The following sections examine the application of a variety of spectroscopic techniques to the study of nanomaterials. In order to better

Figure 6.1 The electromagnetic spectrum and a listing of common spectroscopic methods used to interrogate those regions of the spectrum. Note that the energy of light increases at smaller wavelengths and higher frequencies.

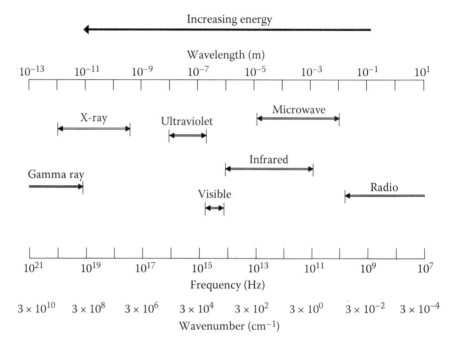

Spectroscopy	Typical Range	Transition Studied
X-ray absorption and diffraction	0.1–100 Å	Inner electron energy states
UV–vis absorption and fluorescence	200–800 nm	Outer (bonding) electron energy states
IR absorption and raman scattering	0.8–300 μm (14,000 to 30 cm^{-1})	Vibrational energy states
Nuclear magnetic resonance	0.6–10 m	Spin of nuclei in magnetic field

understand these methods, it is appropriate to begin with a brief discussion of the interactions between light and matter.

6.1.1 Interactions between light and matter

As described earlier (see Section 4.2.1), in many instances light is best thought of as an oscillating electromagnetic wave with a characteristic energy E. The energy of a photon of this electromagnetic radiation is a function of its frequency ν (or its wavelength λ), and can be calculated by Einstein's famous equation

$$E = h\nu = \frac{hc}{\lambda} \tag{6.1}$$

where h is Planck's constant and c is the speed of light. This equation was introduced in Chapter 4; because of its importance we repeat it here.

As previously discussed in Chapters 4 and 5, the energy states of molecules are quantized; molecules may only exist in a finite number of discrete states. These allowable energy states are the sum of several quantized aspects of the molecule, such as the energies of its electrons around their respective nuclei, the interatomic vibrations that exist in the molecule, and the rotations of the molecule around its center of mass. We can also say that a given molecule has quantized (or discrete) electronic, vibrational, and rotational energy states and that it can only exist at those energy states (or at a sum of those energy states). The lowest energy state of a molecule is termed the ground state and higher energy states are referred to as excited states. When a molecule gains energy by absorbing radiation or by transfer of energy between molecules from collisions (heat) or various electronic processes, it enters an excited state. Following this excitation, the molecule may relax from the excited state to a lower excited state or to the ground state. This relaxation is often accomplished by the emission of electromagnetic radiation (and consequently the frequency and the wavelength) of the emitted light is the exact difference between the upper and lower energy states. Using Einstein's equation (Equation 6.2), this process can be written as

$$\Delta E = E_1 - E_0 = h\upsilon = \frac{hc}{\lambda} \tag{6.2}$$

where ΔE is the difference in energy between the higher and lower energy states. Therefore, by measuring the wavelength of the emitted light, one can calculate the energy difference between the higher and lower states of the molecule.

On the other hand, when electromagnetic radiation is shone on a substance (rather than being emitted from it), several processes may take place (see Figure 6.2). Typically, the radiation is transmitted through the matter at a velocity that depends upon the refractive index (n) of the medium, which we previously discussed in Chapter 5. The consequences of light traveling at different velocities through different materials will be further discussed in the context of surface analysis techniques in Chapter 8. However, if the frequency (or energy) of the incoming light happens to match the exact difference between two of the allowed energy states in a

Figure 6.2 Energy diagrams depicting the transitions that occur during fluorescence (a) and phosphoresence (b). Note that the energies of the photons (either those being absorbed or those being emitted) are identical to the corresponding difference in energy states of the molecule.

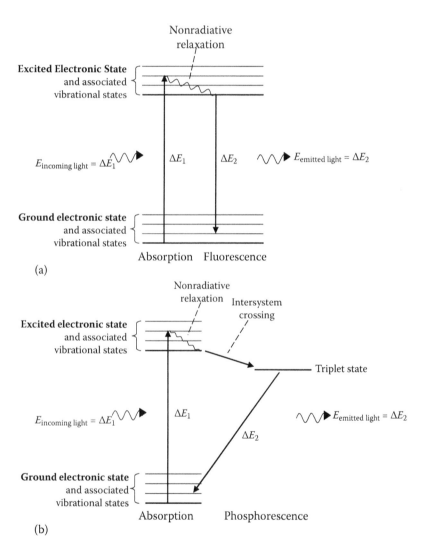

molecule, then the incoming light may be absorbed by the molecule. This outcome, in turn, promotes the molecule to an excited energy state and decreases the intensity of the light at that frequency according to the probability of photons of light interacting with the molecule or material. This entire process is called **absorption**, and the absorption of different frequencies of electromagnetic radiation corresponds to different types of energy transitions in a molecule. For example, absorption of light in the ultraviolet and visible region of the electromagnetic spectrum typically causes transitions between electronic energy states of a molecule. Alternatively, absorbed infrared radiation causes a transition between vibrational states of the molecule. It is important to emphasize that the incoming radiation is only absorbed by the molecule if the frequency of the radiation matches the exact energy difference between allowed energy states of the molecule. Otherwise, the radiation is only transmitted through the substance. Additionally, the transition must conserve spin and angular momentum and must involve a shift in the charge distribution in the molecule or nanosystem. The shift in electron density can be characterized by the overlap of two orbitals,

$$\rho_{overlap} = \phi_{ground}\phi_{excited} \tag{6.3}$$

where ϕ_{ground} represents the orbital from which an electron is being excited (e.g., the HOMO) and $\phi_{excited}$ represents the orbital to which that electron is being excited (e.g., the LUMO). For most transitions to have a significant probability of occurring, the shift in electron density must form a nonzero dipole, known as the **transition dipole**, which can be written in one dimension as

$$\mu_{ge} = -e \int \phi_{ground}(x)x\phi_{excited}(x)dx \tag{6.4}$$

where e is the elementary unit of charge.

If the transition dipole moment is zero, the transition is not allowed and has a zero or weak intensity. Note that while the transition dipole moment represents an induced polarization in the molecule as it interacts with the electric field of the light and absorbs a photon, it does not imply that either the ground or excited state must have a permanent dipole moment. Figure 6.3 shows the HOMO, LUMO, transition density, and transition dipole for butadiene. Note that while the charge distribution of both the ground and excited state are symmetric (no dipole moment), the transition density is not, and thus this transition would be allowed.

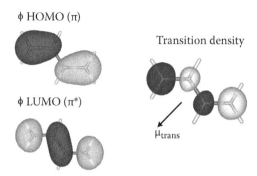

ϕ HOMO (π)

Transition density

ϕ LUMO (π*)

μ_trans

Figure 6.3   HOMO, LUMO, and HOMO–LUMO transition density and transition density for butadiene. The HOMO–LUMO transition leads to a shift in electron density that produces a substantial transition dipole, and is therefore dipole-allowed.

The relationship between the transition dipole and the strength of a transition (relative probability of light being absorbed) is quantified by the oscillator strength

$$f = \left(\frac{8\pi^2 m_e \Delta E}{3h^2 e^2} \right) \mu_{ge}^2 \tag{6.5}$$

where m_e is the mass of an electron. Oscillator strengths can span a very wide range of intensities; an oscillator strength on the order of 1 represents a very strong transition.

The concept of oscillator strength also allows us to connect the molecular or nanoscale property of the transition dipole to the bulk optical properties of a system. In the previous chapter, we discussed the refractive index in the context of molecular polarizability and dielectric screening, in which light is slowed down by the interaction of its electric field with the electric fields of permanent and induced dipoles in the matter that the light is passing through. Molecular polarizability represents a molecular response, while the refractive index represents a bulk response. However, as we have now observed, molecular polarization and optical transitions involve the formation of induced dipoles in a material.

As both absorption and refraction involve polarization of a material in response to an oscillating electric field, the refractive index can be generalized to encompass both phenomena by defining it as a complex number, the **complex refractive index**

$$\widetilde{n} = n + ik \tag{6.6}$$

Here, the real component, n, is the refractive index that we've previously introduced (and typically called the refractive index). The imaginary component, k, is known as the **extinction index** and represents absorption.

The complex refractive index is a function of the oscillator strengths of potential transitions in a system,

$$\tilde{n}(v)^2 = 1 + \frac{N_{\text{elec}}e^2}{4\pi^2 m_e \varepsilon_0} \sum_i \left(\frac{f_i}{v_i^2 - v^2 - i\gamma v_i} \right) \tag{6.7}$$

where N_{elec} is the density of electrons participating in transitions in the material, m_e is the mass of an electron, ε_0 is the permittivity of vacuum, v is the frequency of light, v_i is the frequency of each transition the material undergoes, and γ is a line-width parameter for each transition that can represent various sources of broadening, including quantum mechanical, differences in the environment of each molecule or nanosystem, interaction with other states (e.g., different vibrational levels of each electronic state), and so on. While beyond the scope of this book, a combination of some algebra and nonlinear fitting of peaks in optical spectra can be used to obtain the real and imaginary components of the refractive index of a material over a wide frequency range.

Essentially, what the complex refractive index describes is that when light at some frequency encounters a material, it always causes some sort of polarization in the molecules within that material. If the frequency does not correspond to a quantum mechanical transition, the interaction with the material only slows the light down (as represented by n) but does not transfer energy to the material. However, if the light is at a frequency that corresponds to a transition, the polarization of the electrons in the material becomes resonant with the light, electron density shifts location in the molecule, and the light is absorbed (as represented by k).

The nature of the frequencies at which a molecule or substance absorbs the incoming radiation and the amount of radiation absorbed at those frequencies provides valuable information about that molecule, such as the type and strength of different bonds that exist in the molecule. While all optical transitions follow similar rules, different types of transitions can be used to provide complementary information (e.g., geometry and concentration).

However, absorption is not the only important process involving light in molecules and nanomaterials. After the molecule has absorbed the incoming radiation, it ordinarily returns to the ground state after some short amount of time due to one of several relaxation processes. For some molecules, **fluorescence** is a common relaxation process in which the molecule relaxes by reemitting light, generally of a lower energy (longer wavelength) than the absorbed light. Fluorescence is typically observed in the ultraviolet–visible region of the electromagnetic spectrum and is depicted schematically in Figure 6.2. **Nonradiative relaxation** is the most common relaxation process, in which the excited molecule relaxes without reemitting any electromagnetic radiation. This process typically occurs in small steps by the conversion of the excited energy into kinetic energy through collisions with nearby molecules, producing heat. Nonradiative relaxation is also depicted in Figure 6.2.

Phosphorescence is another relaxation process that typically occurs in the UV–vis region. It is observed when an excited electron undergoes non-radiative intersystem crossing to a lower-energy excited state in which the electron in the excited state and the remaining electron in HOMO have the same spin (a triplet state). Relaxation is much slower since the electron needs to flip its spin again to return to the ground state while obeying the Pauli exclusion principle. When the excited electron then relaxes back to a state of lower energy, a photon is produced at a longer wavelength (lower frequency) than the absorbed light. Phosphorescence is a much rarer phenomenon than fluorescence and is consequently less important to our discussion of spectroscopic techniques.

In summary, molecules may exist only at discrete, quantized energy states. By examining the ways in which a given molecule interacts with light, one can gather information about its energy states, thus providing valuable insight into the identity of the molecule, the strength and type of its chemical bonds, and the concentration of the molecule in the substance being studied. For our elementary purposes, those observations serve as the basis of spectroscopy. Let's consider several important types of spectroscopy and their applications to the study of nanomaterials.

6.1.2 UV–visible spectroscopy

6.1.2.1 Principles of UV–visible spectroscopy

UV–vis spectroscopy uses the transmission of visible and/or ultraviolet light through a sample to determine the presence and/or the amount of material that absorbs light within the sample. As mentioned in our discussion of the interactions between light and matter, absorption of a photon in the ultraviolet and visible wavelength range (~190 nm to 800 nm) typically induces an electronic transition within the absorbing molecule, promoting it from a low-energy ground state electron configuration to an excited state. The region of a molecule that absorbs the light and contains the electrons that undergo the electronic transition is called the **chromophore**. For example, the amine group $-NH_2$ absorbs light at a wavelength of about 190 nm. The absorption causes an electron from the lone pair on nitrogen to be excited into an antibonding molecular orbital. This transition is denoted as n → σ*. For our purposes, we consider only the amount of light absorbed and the corresponding absorption wavelength (λ_{max}). The exact electronic nature of the transitions is less important. It should be noted that the absorption by chromophores in a given molecule depends on the exact electronic environment of the chromophore within the molecule and its surrounding environment (e.g., solvent or surface). As a matter of convention, we also note that electronic transitions are typically reported in terms of the wavelengths of light (generally nm) that excite the transition.

All light that encounters a material either passes through it (transmission), is reflected off the surface of the material, or is absorbed by the material. Therefore, if light is absorbed by a sample, the amount of light transmitted through the sample as measured by the intensity or radiant power of the beam decreases. The transmittance T of the sample is defined as the ratio of the intensity of the beam after passing through the sample (I) to the original intensity of the beam (I_o). The absorbance A of the sample is defined as the negative logarithm of transmittance, as shown in Equation 6.8:

$$A = \log\left(\frac{I_o}{I}\right) = -\log T \qquad (6.8)$$

For samples with relatively low concentrations of chromophores, the absorbance of the sample can be directly related to its concentration. This relationship is known as the Beer–Lambert law and is given in Equation 6.9:

$$A = c\,\varepsilon\,l \qquad (6.9)$$

where c is the concentration of the sample, l is the sample path length, and ε is a parameter known as the molar absorbitivity or extinction coefficient of the material. The molar absorptivity is an important parameter that is proportional to the probability of absorbing a photon for a particular molecule at a specific wavelength of light. Its value is specific to a particular chromophore in a particular environment and does not depend on concentration or path length.

The extinction coefficient, however, is closely related to another optical property that we have discussed—the refractive index. The extinction coefficient is proportional to k, as

$$\varepsilon(\lambda) = \frac{4\pi k}{\lambda c} \tag{6.10}$$

where c is the concentration of chromophores. It can also be related to the oscillator strength of a transition as

$$f = 4.3 \times 10^{-9} \int \varepsilon(\nu) d\nu \tag{6.11}$$

where the integral is over the absorption peak for which the transition is being calculated. The extinction coefficient can be calculated from the absorbance if the concentration of the material and the path length is known, allowing further connecting of molecular and bulk properties.

In addition to solutions, UV/vis spectroscopy can also be used to characterize solid materials. If the path length and ε of a sample are known, then the density of the sample's contents can be calculated from the absorbance value. Alternatively, if the content density is known, then the path length can be calculated. In the context of a thin film (Figure 6.4),

Figure 6.4 Transmission mode UV–vis absorption spectroscopy can be used to determine the thickness of a thin nanofilm. Beer's law is used just as with bulk phase measurements, but the thickness of the film replaces the path length of the sample cell in the equation.

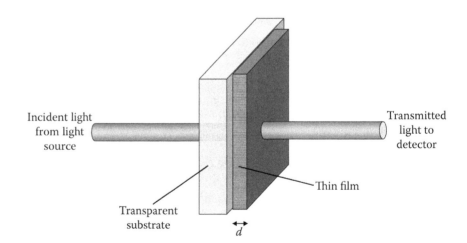

the path length is simply the film thickness (d), and in this manner the thickness of a nanofilm can be determined using UV–vis spectroscopy.

6.1.2.2 Setup of a UV–visible spectrophotometer

At a basic level, a UV–vis spectrophotometer consists of a light source, a dispersive device for selecting wavelengths of light, a sample holder, and a detector, along with several optical components. Many spectrophotometers use a double-beam configuration, in which a beam splitter allows the light beam to pass through a sample cell and a reference cell alternately throughout the measurement process. The beam is diverted into the reference cell by the splitter several times per second, and the absorbance of the reference cell is measured and automatically compared with the absorbance of the sample cell to correct for fluctuations or variations in the lamp intensity over time and improve the accuracy of the measurement. A schematic of a double-beam spectrophotometer is shown in Figure 6.5.

In practice, UV–vis spectrophotometer contain two different light sources, one to produce light in the UV range and one to produce visible light. Except in spectrophototometers capable of measuring multiple wavelengths simultaneously, the light sources are never run simultaneously. Instead, as the spectrophotometer measures the absorbance of a sample across a predetermined range of wavelengths, the light source switches from the visible source to the ultraviolet source. The light change is generally set to occur around 360 nm to allow for the highest light intensity at each wavelength.

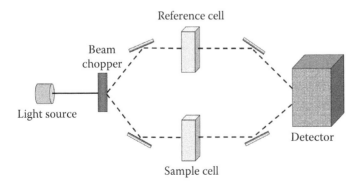

Figure 6.5 A schematic of a common double-beam spectrophotometer used to make bulk phase measurements.

6.1.3 The absorption of visible light by a nanofilm

As an example of the application of UV–vis spectroscopy to characterize the growth of a nanofilm, consider the anionic polymer PAZO whose structure is shown in Figure 6.6. The PAZO monomer unit contains two phenyl groups bridged by an azo group (–N=N–). This unit functions as a chromophore and absorbs light at about 360 nm due to a $\pi \rightarrow \pi^*$ transition. Thus, an aqueous solution of PAZO appears orange because of the broad absorption of light centered at 360 nm. The polymer PEI encountered previously contains amine groups that absorb light at about 190 nm. There is no absorption in the visible region and thus an aqueous solution of PEI is colorless. Both PAZO and PEI can be used to create multilayer films for a variety of applications. We will discuss these applications and

Figure 6.6 (a) The absorption spectrum of PAZO (structure shown in the inset). λ_{max} represents the wavelength at which the absorbance is a maximum. (b) A plot of absorbance at λ_{max} as a function of bilayer number. The bilayer represents a layer of polycation complexed with a layer of polyanion.

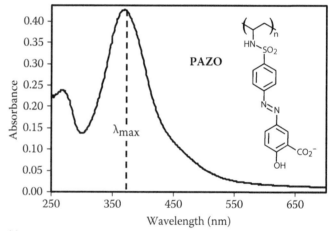

(a)

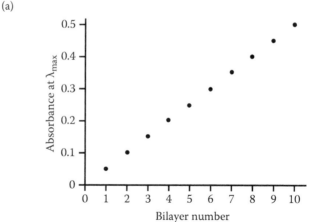

(b)

the methods used to fabricate the multilayers in Chapter 9. About 1-nm-thick layers of PEI and PAZO can be constructed on a glass substrate in a layer-by-layer fashion. These layers are held together by strong electrostatic interactions between the positively charged PEI and the negatively charged PAZO. Each PEI and PAZO pair in the film can be described as a "bilayer." Figure 6.6b shows a UV–vis spectrum of a PEI and PAZO multilayer film as a function of the bilayer number. Only the absorbance value at ~360 nm is shown. In this range, only the chromophores associated with PAZO are able to absorb light (centered at a wavelength of 360 nm). The PEI is essentially transparent at visible and longer UV wavelengths, and therefore the increase in the absorbance in Figure 6.6b reflects the increase in the amount of total PAZO in the film after each bilayer.

Example 6.1   Determining Absorbance Values

Estimate the slope of the line in the graph shown in Figure 6.6 and use it to predict an absorbance value of a film composed of 15 bilayers. How would you determine the concentration of PAZO in the film from the PAZO's molar absorptivity? What other technique would be useful in this determination?

Solution   A linear fit to the data yields a slope of ~0.02 (Figure 6.7). The intercept is close to zero. The slight negative number in the intercept is likely due to a baseline shift. Thus the equation of the line is $y = 0.02x$, where $y = A$. When $x = 15$, the absorbance (A) is 0.75. From Beer's law, $A = \varepsilon cl$, where l represents path length, or in this

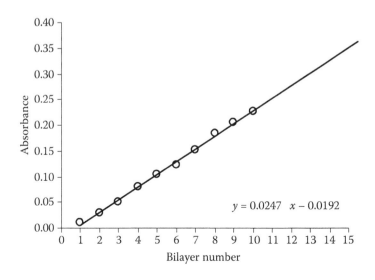

Figure 6.7   The absorbance (at λ_{max}) data of a polyelectrolyte multilayer film. The bilayer represents a layer of polycation complexed with a layer of polyanion.

$$y = 0.0247 \ x - 0.0192$$

case the thickness of the film. We can estimate l using techniques such as ellipsometry, DPI, or SPR (discussed in Chapter 8), and if we know ε (molar absorptivity), we can obtain the concentration, c.

6.1.4 Molecular fluorescence spectroscopy

6.1.4.1 Principles of fluorescence and fluorescence quantum yield

As described in Section 6.1.1 on the interactions between light and matter, fluorescence is the process whereby a molecule that has been excited by the absorption of radiation relaxes to a lower energy state by emitting a photon of lower energy than the photon that was absorbed. Not all molecules exhibit fluorescence; molecules that do fluoresce are called **fluorophores**. Each fluorophore has a characteristic absorption profile, which is identical to the UV–vis absorption spectrum for the molecule. For most fluorophores, fluorescence is typically observed in the ultraviolet to visible range of the electromagnetic spectrum, meaning that it corresponds to transitions in the electronic state of the molecule (much like UV–vis absorption spectroscopy). As with UV–vis spectroscopy, the wavelength of light (generally nm) is typically used to describe fluorescence spectra rather than wavenumber (cm^{-1}) or frequency (Hz) units.

Wavelengths that produce fluorescence when absorbed are referred to as excitation wavelengths. However, not all excitation wavelengths cause the fluorophore to fluoresce to the same degree or even to produce fluorescent light of the same wavelength. Therefore, for each excitation wavelength, the fluorophore has a characteristic fluorescence emission profile, which is the range and intensity of wavelengths that are produced when that excitation wavelength is used to excite the fluorophore. Alternatively, for each emission wavelength there also exists an excitation profile, which is the range of wavelengths that produce fluorescence of that emission wavelength and the intensities of emission associated with each of those excitation wavelengths. Both emission and excitation profiles provide useful information about the fluorophore and its local environment, so most fluorometers can be set to determine either the emission spectrum at a set excitation wavelength or the excitation spectrum at a set emission wavelength. It should be noted that because fluorescence-emitted photons are typically of lower energy (i.e., longer wavelength) than their excitation photon counterparts, the emission spectrum for a given fluorophore usually occurs at longer wavelengths than its excitation spectrum. Typical emission and excitation spectra for an optically active polymer are shown in Figure 6.8. The polymer is a

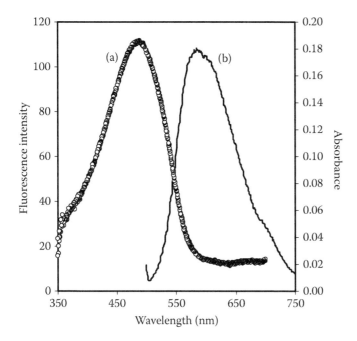

Figure 6.8 Typical absorption (a) and fluorescence emission (b) spectra of PPV, an optically active polymer. The fluorescence intensity is in arbitrary units. The molecules are excited at ~475 nm (the wavelength corresponding to the peak absorbance) and emits light at ~575 nm (the wavelength corresponding to the peak fluorescence).

derivative of poly(phenylene vinylene) (PPV), and such materials are being developed for applications in nanophotovoltaics.

As discussed previously (Section 6.1.1), there are several alternative relaxation processes that may occur in lieu of fluorescence, such as non-radiative relaxation and intersystem crossing. Furthermore, reabsorption of the fluorescent photon by a neighboring fluorophore may occur, especially at higher concentrations of fluorophore. Therefore, it is convenient to define a parameter that represents how often a given fluorophore successfully emits a fluorescent photon. Fluorescence quantum yield (Φ_f) is the usual measure of such a parameter and is defined as the ratio of photons absorbed to photons emitted for a given fluorescing species. The correct determination of Φ_f is especially important in the characterization of certain emissive nanomaterials such as those designed for photoluminescent devices.

Determination of Φ_f for a given material A can normally be accomplished in a rather straightforward manner by comparison with a standard material of known quantum yield Φ_{std} according to

$$\Phi_{fa} = \Phi_{std}\frac{g_a n_a^2}{g_{std} n_{std}^2} \qquad (6.12)$$

In this equation n_a and n_{std} are the refractive indexes of the solvents in which the material and the standard were dissolved, respectively. g_x is effectively defined as the derivative of the total fluorescence intensity of a fluorophore with respect to its absorbance of the excitation light (i.e., a measure of how the total fluorescence intensity changes with respect to absorbance). For low concentrations of the fluorophore, the relationship is usually linear (i.e., g_x is constant), so g_x can be calculated by integrating the total fluorescence spectrum for a certain excitation wavelength and then plotting integrated fluorescence intensity versus absorbance of the excitation wavelength for a number of different concentrations.

6.1.4.2 Setup of a fluorometer for bulk phase and thin-film fluorescence measurements

Figure 6.9 shows the typical setup of a fluorometer. The detector is usually placed at ~90° from the incident light beam to avoid collecting incident light; light emitted through fluorescence is typically uniform in all directions. However, some incident light will still be collected due to nondirectional **Rayleigh scattering**, which is discussed later in the chapter. Optical filters can also be placed between the sample and the detector in order to ensure that only the fluorescence is detected. The fluorescence emission spectrum is gathered by shining a beam of a certain excitation wavelength on the sample and then scanning the output (emission) wavelength range with the detector while recording the intensity of fluorescence at each emission wavelength. Alternatively, the excitation spectrum can be obtained by setting the detector to monitor a specific emission wavelength and recording the intensity of fluorescence while scanning the input (excitation) wavelength range.

Figure 6.9 Schematic of a typical fluorometer.

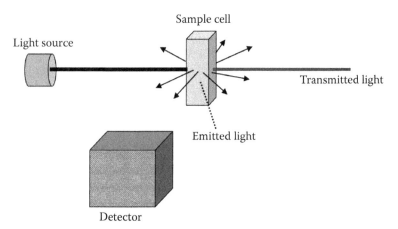

Light source

Sample cell

Transmitted light

Emitted light

Detector

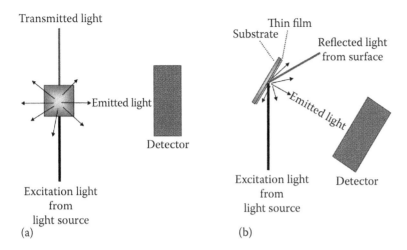

Transmitted light

Emitted light

Detector

Excitation light
from
light source

(a)

Thin film
Substrate

Reflected light
from surface

Emitted light

Excitation light
from
light source

Detector

(b)

Figure 6.10 Methods for monitoring the fluorescence of samples in a cuvette in solution phase (a) or in a thin film (b). The detector is strategically placed so that it will intercept the maximum amount of fluorescent light without picking up any of the transmitted or reflected light.

In some circumstances the intensity of the emission is too low to be recorded without amplification. In these situations it is best to use an instrument that uses a **photomultiplier tube** for detecting light. It is worth noting that photomultiplier tubes should be used only in cases of low-power radiation; otherwise, they will be damaged. A photomultiplier tube contains a photocathode surface that emits electrons when exposed to radiation. It also has multiple other electrodes, called dynodes. Each photoelectron that strikes a dynode causes many more electrons to be emitted. This creates an avalanche effect that results in millions of electrons for each initial photoelectron to be collected at an anode. The resulting current is measured in terms of voltage and is related to the wavelength being scanned.

For bulk phase measurements, the solution phase sample is generally placed into a fluorescence cuvette and fluorescence measurements are made at 90° to the incident beam. For thin films, a substrate on which the thin film has been built is placed so that the incident light shines on the surface of the substrate at a large angle of incidence (i.e., more parallel to the surface) and fluorescence is detected at a different angle than the angle of reflection. Each of these two setups is depicted in Figure 6.10.

6.1.5 Vibrational spectroscopy methods

6.1.5.1 Introduction to vibrational modes

When a molecule absorbs infrared (IR) radiation, it is excited to a higher energy state, as described in Section 4.5. As with the absorption of other types of electromagnetic radiation, the IR absorption process is quantized and the molecule can absorb only particular frequencies of IR radiation. The absorption of IR radiation corresponds to a change in energy of approximately 20 kJ/mol. This is the amount of energy required to cause covalent bonds to stretch, bend, and twist, particular combinations of which are called the vibrational modes of a molecule. Only those frequencies of IR radiation that match the natural vibrational frequencies of the covalent bonds of the molecule lead to IR absorption by the molecule. These frequencies are called **vibrational modes**. The energy absorbed leads to an increase in the amplitude of the vibrational motions of the bonds in the molecule. Furthermore, in order for a molecule to absorb IR radiation, it must undergo a change in dipole moment during the course of the vibration. The change in dipole moment is due to the motion of atoms in response to the oscillating electric field of the infrared light shining on the sample. Just as how electronic transitions only occur with frequencies of light with the same energy as the difference between electronic states, energy transfer, and consequently, IR absorption, occurs when the frequency of the light is the same as a frequency of bond vibration. For vibrational transitions, the transition dipole moment is the change in the dipole moment as the atoms in the molecule vibrate. Therefore, symmetric diatomic molecules that have no dipole moment, such as N_2, H_2, and O_2, do not absorb IR radiation, since their vibration cannot produce a nonzero transition dipole. The frequency of a vibrational mode for a bond is given by

$$\bar{\nu} = \frac{1}{2\pi c}\sqrt{\frac{k}{\mu}} \tag{6.13}$$

which is essentially the same as Hooke's law used to describe a spring undergoing harmonic oscillation. In this equation, the constant k is called the force constant of the bond and its units are typically N/m. The magnitude of k gives a direct measure of the stiffness of a covalent bond and μ is known as the reduced mass. For a simple diatomic molecule containing two atoms of mass m_1 and m_2, the reduced mass is given by

$$\mu = \frac{m_1 m_2}{m_1 + m_2} \tag{6.14}$$

Molecules that have more than two atoms have multiple vibrational modes. The number of vibrational modes is given by 3N-6 for a free, non-linear molecule in the bulk, where N is the number of atoms in the molecule. The subtracted degrees of freedom represent rotation or translation of the molecule. Linear molecules have 3N-5 vibrational modes since they have one fewer rotational degree of freedom due to their symmetry. As a simple example, consider the structure of H_2O. The central oxygen atom is sp^3 hybridized, meaning that the molecule is bent with single bonds between the oxygen and hydrogen atoms. The molecule has a net dipole moment and is thus able to absorb IR radiation. The molecule is non-linear and composed of three atoms, so it has $9 - 6 = 3$ vibrational frequencies. Figure 6.11 shows three possible ways that H_2O can vibrate. When the two O–H bonds are simultaneously increasing and decreasing in length we say that the vibration is symmetric. If one O–H bond length increases while the other decreases, then we have an asymmetric vibration. The molecule can also undergo a bending motion. All three modes of vibration lead to changes in the dipole moment and therefore result in the absorption of IR radiation at characteristic frequencies. Since it is easier to bend the molecule than to stretch it, the bending mode has a relatively small k value. The implication is that, of the three observed IR absorption frequencies, the bending mode corresponds to the lowest frequency. If we replace the H atoms in H_2O with deuterium atoms (D_2O), then the reduced mass of D_2O is greater than that of H_2O. According to Equation 6.13, all three IR absorption frequencies for D_2O are smaller than those for H_2O.

Let's consider the simple molecule CO_2. The central carbon atom is sp hybridized, meaning that CO_2 is a linear molecule with double bonds between the carbon and oxygen atoms. It is composed of three atoms, so it has $9 - 5 = 4$ vibrational modes. Since the molecule is linear and symmetric around the central C atom, CO_2 has a zero net dipole moment (the dipole moments from the two C = O bonds cancel each other out).

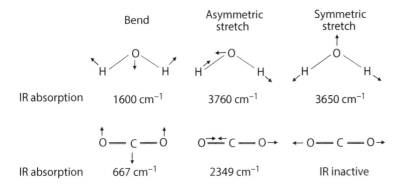

Figure 6.11 The vibrational modes of H_2O and CO_2 and the IR absorption associated with each vibrational mode. The symmetric stretch in CO_2 does not produce a net change in the dipole moment of the molecule and is therefore IR inactive. The bending mode is doubly degenerate.

When this molecule vibrates symmetrically, as shown in Figure 6.11, the net dipole moment remains zero. Thus, the symmetric vibration of CO_2 does not lead to the absorption of IR radiation. However, the asymmetric and bending modes of CO_2 produce a net dipole moment, and therefore they lead to the absorption of IR radiation. Therefore, it will have two absorption frequencies, one corresponding to the asymmetric mode, and the other corresponding to the doubly degenerate bending mode.

Organic molecules usually contain a variety of functional groups. Each functional group absorbs a specific IR frequency that is characteristic of that group. For example, the carbonyl group (C = O) absorbs around 1700 cm^{-1}. The exact value depends on the local environment of the functional group, such as the identity of other atoms that are attached to the carbonyl functional group. Table 6.1 shows a number of functional group vibrations and their characteristic IR absorption frequencies. As a matter of convention, we note that IR spectroscopists typically report IR values in wavenumbers (cm^{-1}), a unit that is directly proportional to frequency ν and defined as

$$\text{wavenumber} = \frac{\nu}{c} = \frac{1}{\lambda} \qquad (6.15)$$

where c is the speed of light (in cm s^{-1}) and λ is the wavelength of light.

Table 6.1
A Listing of Common Functional Groups and Their IR Frequencies

Function Group	Group Frequency (cm^{-1})
–C–H (stretch)	2850–2960
=C–H (stretch)	3000–3100
≡C–H (stretch)	~3300
C=C (stretch)	1620–1680
C≡C (stretch)	2100–2260
–O–H (alcohols, H-bonded, stretch)	3200–3600
–O–H (carboxylic acids, H-bonded, stretch)	2500–3000
–N–H (stretch)	3300–3500
–N–H (bend)	~1600
C=O (stretch)	1670–1820
C≡N (stretch)	2220–2260
–S–H (stretch)	2550–2600
–S–S– (stretch)	470–620
Si–O–Si (stretch)	1020–1095
Si–O–C (stretch)	1080–1110
–N=N– (stretch)	1575–1630

By plotting absorption or transmission versus incident IR frequency, one can map out the exact absorption frequencies for a sample containing the molecule of interest. This infrared spectrum provides a fingerprint of the molecular structure of the molecule. Several specialized applications of infrared spectroscopy for characterizing molecules confined to surfaces will be discussed in Chapter 8.

6.2 LIGHT SCATTERING METHODS

6.2.1 Scattering and absorption

In the opening discussion of the interactions between light and matter (see Section 6.1.1), we focused on processes such as absorption, emission, and fluorescence. However, another interaction called light scattering occurs commonly when light encounters nanoscale matter. In order to understand the scope and utility of Raman spectroscopy, a brief introduction to light scattering is provided.

When light passes through a solution or suspension, much of the light is transmitted directly through the solution, but some of the light is scattered in different directions. To better understand this scattering process, it is useful to model the scattering of light. Suppose that when a photon interacts with a molecule, the molecule is promoted to a virtual excited state as shown in Figure 6.12. This virtual state is short-lived, so we expect

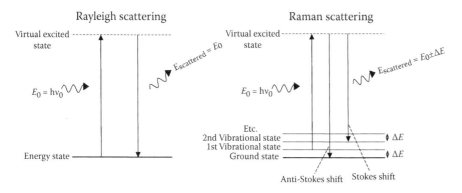

Figure 6.12 Energy diagrams modeling the Rayleigh and Raman scattering processes. A photon interacts with a molecule, promoting it to a nonquantized virtual state. The molecule quickly returns to a lower energy state, emitting a scattered photon. If the molecule returns to the same state in which it began, Rayleigh scattering has occurred. If the molecule returns to a higher or lower vibrational energy state, then Raman scattering has occurred.

that the molecule almost immediately returns to its ground state, reemitting the photon. The photon is reemitted in a random direction. It is important to note that this model of the scattering process should not be confused with absorbance and reemission of the photon. The incoming photon is not absorbed by the molecule according to our model. The promotion to the virtual excited state is a result of a momentary interaction with the photon and the entire process is not quantized (meaning it can happen for any energy of light). Absorbance only occurs when the frequency of the incoming light exactly matches the difference between allowed energy states of the molecule.

6.2.2 Rayleigh and Raman scattering

Most commonly, light is scattered elastically (or without losing or gaining any energy), in a process described as **Rayleigh scattering**. From Einstein's equation (Equation 6.1), we know that the energy of light is directly related to its frequency (or wavelength). Therefore, since Rayleigh scattering is elastic, the scattered light has the same frequency (or wavelength) as before it was scattered. However, due to the wave nature of light, Rayleigh scattering is a wavelength-dependent process, meaning that some wavelengths are scattered to a greater extent than others. Indeed, it is this aspect of Rayleigh scattering that accounts for the blue color of the sky. As light from the sun interacts with particles in the atmosphere, it undergoes Rayleigh scattering. Blue light is scattered more than the other wavelengths of light, because Rayleigh scattering is strongest for the shorter wavelengths of light that are closer in size to the air molecules responsible for scattering, and so it appears as though the sky is a blue color. Rayleigh scattering is depicted schematically using our model of light scattering in Figure 6.12.

Raman scattering occurs when light is scattered inelastically by a molecule. In other words, Raman scattering happens when the scattered light is of a higher or lower energy after it has been scattered than before. This increase or decrease in energy is generally due to a change in the vibrational energy of the molecule. According to our model of light scattering, this type of Raman scattering might occur when the molecule relaxes from the virtual excited state to a vibrational state that is higher or lower in energy than the state at which the molecule was previously. In this case, the reemitted (or scattered) photon is a slightly different frequency (or wavelength) than before it interacted with the molecule. The shift in the frequency of the scattered light from its original value directly matches the energy difference between the vibrational states of the molecule as in infrared absorption. However, Raman scattering is a distinct process from infrared absorption—the vibrational modes are the same, but the rules for when a scattering event can occur are different. Instead of requiring a change in dipole moment, Raman scattering requires a change in the polarizability (see Chapter 5) of the molecule. This allows vibrational frequencies that cannot be excited using infrared spectroscopy, such as the stretching of homonuclear diatomics such N_2 or the symmetric stretching mode of CO_2, to be excited.

In our diagram (see Figure 6.12), Raman scattering is represented as follows. Suppose the molecule happened to be in the first vibrational state above the ground state at the time of interaction with the photon. The molecule would be excited to the virtual excited state and then immediately relax, producing a scattered photon as described previously. Suppose, however, that the molecule would relax to either the second vibrational state or to the ground state rather than the first vibrational state. In this case, the reemitted photon would have either gained or lost energy according to the exact difference in energies between the vibrational states, represented by ΔE in the figure. And, because the energy of light is a function of its frequency, the higher or lower energy light has a frequency that is slightly shifted from its original value and Raman scattering has occurred. Not only has the Raman scattered light changed direction, but it has also shifted its frequency.

6.2.3 Raman spectroscopy

Because the frequency shifts of Raman scattered light are a direct result of changes in vibrational states of the molecule, it seems reasonable to assume that one could gather similar information to IR spectroscopy by simply measuring the shifts in frequency of Raman scattered light. Indeed, this is the basic idea behind Raman spectroscopy, and frequency shifts in a Raman spectrum are directly analogous to IR absorption frequencies. In practice, shifts in frequency of Raman scattered light are monitored and those shifts are matched to specific bond vibrations within the molecule being studied. At a basic level, this allows one to identify the functional groups of the molecule being studied as well as their local chemical environment. Combining data from a Raman experiment and an infrared experiment on a particular molecular system can provide even more detailed information about the structure of the molecule.

As a matter of naming convention, one should note that a frequency shift of Raman scattered light to a lower energy (or lower frequency) is called a Stokes shift. A frequency shift to a higher energy (or higher frequency) is called an anti-Stokes shift. Typically, Stokes shifts are more probable than anti-Stokes shifts, so their signals are stronger. As might be expected, Raman scattering is much less likely to occur than Rayleigh scattering for a given molecule and a given wavelength of light. Therefore, a Raman spectrum that maps intensity versus frequency of the scattered light has a centralized and strong peak that represents the Rayleigh scattering of the incident light. The Rayleigh peak is symmetrically surrounded by a distribution of much smaller peaks that represent the Stokes lines at the lower frequencies and the anti-Stokes lines at the higher frequencies. This peak is often blocked from the detector by a filter.

6.2.4 Light scattering by nanoparticles

As discussed in Chapter 5, intermolecular interactions can drive molecules to aggregate into colloidal particles such as micelles. Like molecules, these nanoparticles can scatter light, and the manner in which they scatter light can provide information about the concentration and size of particles in the system. In addition to Rayleigh (inelastic) and Raman (elastic) scattering as discussed above, these particles can also undergo two other types of inelastic scattering, known as Mie scattering and geometric scattering. The type of scattering depends on the size of the particle interacting with electromagnetic radiation. Consider the unitless parameter, x, defined as,

$$x = \frac{2\pi r}{\lambda} \tag{6.16}$$

where r is the radius of the particle and λ is the wavelength of the incident light. Rayleigh scattering occurs when $x \ll 1$ or when the particles are small compared to the wavelength of light. Mie scattering occurs when the particles are the same size as the wavelength of light (i.e., $x \approx 1$) and geometric scattering occurs when the particles are relatively large (i.e., $x \gg 1$).

6.2.5 Determining particle size using scattered light

The intensity, angular distribution, and polarization of the scattered light depend on factors such as the shape and size of the particles, as well as on the interactions between them. Light scattering experiments can thus provide useful structural information (particle shape and size) and inter-particle interactions on colloidal systems. The measurements are generally instantaneous, noninvasive, and allow representative sampling of poly-disperse samples. However, the presence of small particle impurities, particularly those that have a tendency to scatter the light, can impose serious errors in the measurement.

In practice, a collimated beam of light of a given wavelength (λ) and intensity (I_o) passes through a solution containing the dispersed nanoparticles (Figure 6.13). The intensity of the scattered light is then measured as a function of the angle (ϕ) between the incident beam and the scattered beam.

We will begin by discussing how light scattering can be used to determine the aggregation number of a micelle. Micelles—spherical aggregates of amphiphilic molecules—were briefly mentioned in Chapter 1 and will be discussed further in the Chapter 7. Micelles have diameters typically on the order of a few nanometers. The wavelength of visible light is about two orders of magnitude greater. Let's consider a beam of visible light passing through an aqueous solution containing spherical micelles. The solution can be described in terms of two refractive indices. As previously dis-cussed, the refractive index quantifies interaction between light and matter, with the real component (n) representing the effective speed of light through the material and the imaginary component (k) representing absorption. Here, we are examining two materials that (ideally) do not absorb any of the incident light and have different polarizabilities, and thus, different refractive indices. Therefore, the system can be characterized by the real components of the refractive index of the ran-domly dispersed nanospherical micelles (n_{micelle}) and of the continuous solvent (n_{solvent}). These two refractive indices have different values, and

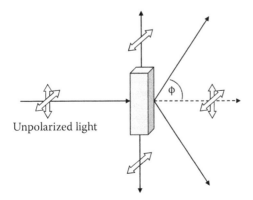

Unpolarized light

Figure 6.13 The scattering of unpolarized light through a sample. The intensity of the scattered light is measured as a function of the angle (ϕ) between the incident beam and the scattered beam.

the average refractive index of the solution will therefore vary with the local concentration of micelles. This variation will cause light to be scattered. The intensity of the scattered light depends on the intensity and wavelength of the incoming light, the solution refractive index increment (i.e., how n varies with concentration, dn/dc), and the average number density of micelles in the solution (N). These parameters can be determined experimentally and are used to obtain the optical constant K_o (Equation 6.17):

$$K_o = 2\pi^2 \frac{n^2}{\lambda^4 N} \left(\frac{dn}{dc}\right)^2 \tag{6.17}$$

In order to understand how the intensity of light varies with the scattering angle ϕ, we need to describe the incoming unpolarized light as being composed of two mutually perpendicular polarized components (electric fields oscillating in a specific direction; polarization of light will be discussed in detail in Chapter 8) (Figure 6.13). For small scattering angles ($\phi \sim 0$), these components will contribute equally to the scattered intensity. At very large scattering angles, one of the two polarized components contributes to a greater degree to the scattering. In fact, when $\phi = 90°$ the component polarized along the direction of the scattered beam has no contribution to the scattered intensity. By measuring the intensity of scattered light as a function of ϕ, we can determine a quantity known as the Rayleigh ratio (Equation 6.18).

$$R_\phi = \frac{d^2}{1 + \cos^2\phi} \frac{I}{I_o} \tag{6.18}$$

In this equation, d is the distance between the sample and the detector. The Rayleigh ratio will be different for a colloidal solution compared to the pure solvent. In fact, it can be shown that the difference between these two Rayleigh ratios is given by Equation 6.19.

$$\Delta R_\phi = \frac{2\pi^2 n^2}{\lambda^4} \left(\frac{dn}{dc}\right)^2 RTc \sqrt{\frac{d\Psi}{dc}} \tag{6.19}$$

The term $RTc\sqrt{d\Psi/dc}$ is known as the concentration fluctuation factor and describes the free-energy cost in creating an inhomogeneity in micelle concentration. For micellar systems, the concentration fluctuation factor causes a dramatic change in the scattering intensity as we go from monomers to predominantly micelles. Therefore, one would expect that ΔR_ϕ would be related to both the critical micelle concentration (CMC, see page 244) of the surfactant and the concentration of the surfactant monomers (c_m). The exact dependence is shown in Equation 6.20:

$$\Delta R_\phi = \frac{K_o(c_m - CMC)}{10^3/M + 2B(c_m - CMC)} \tag{6.20}$$

M represents the molecular weight of the micelle and B is a constant known as the second virial coefficient. The sign of B provides information on the intermolecular interactions between micelles. A negative value indicates a net attraction between the micelles and a positive value indicates a net repulsive interaction. A value of zero indicates an "ideal" micellar solution in which there are intermicellar interactions.

Example 6.2   Determining the Aggregation Number of a Micelle

Consider values of $K_o/\Delta R_\phi$ recorded as a function of surfactant monomer concentration. How would you determine the aggregation number of the micelle?

Equation 6.20 can be rearranged to give

$$\frac{K_o(c_m - CMC)}{\Delta R_\phi}10^{-3} = \frac{1}{M} + 2 \times 10^{-3}B(c_m - CMC)$$

Thus, a plot of the left-hand side of this equation versus ($c_m - CMC$) will give a straight line with an intercept equal to $1/M$ and a slope equal to 2×10^{-3} B.

6.2.6 Dynamic light scattering

Dynamic light scattering (DLS), also known as photon correlation spectroscopy, is a method used to determine the size distribution of particles in a solution. This is a powerful technique that can accurately report the sizes of particles present across several orders of magnitude. Furthermore, measurements are typically easy to make; this makes DLS ideal for many applications from measuring the size of micelles and other nanoparticles as a function of concentration to assaying a protein solution for the presence of aggregates.

DLS works by shining a laser at a sample solution and observing the scattered light. If the particles in the solution are small compared to the wavelength of the light ($x < 10$, from Equation 6.16), then this phenomenon will proceed primarily through Rayleigh scattering. In this process, particles absorb a photon and then reemit a photon shortly thereafter. However, the emitted photon is sent in a random direction. Thus, even if the entire incident light comes from one direction, the emitted light is radiated in all directions (scattered). This is an elastic process, meaning that if light of a particular wavelength is absorbed, the scattered light will have exactly the same wavelength. With this in mind, if one conducted an experiment shining a 632-nm HeNe laser at a solution of interest, one might naively expect the scattered light to look like that shown in Figure 6.14a. This figure depicts all scattered light being detected at exactly 632 nm.

In actuality, however, scattered light is observed in a distribution centered on the expected wavelength (Figure 6.14b). The reason for this is that the particles in solution are undergoing Brownian motion. This means that at any given moment, some particles are traveling toward the detector and have their emitted light blue-shifted relative to the incident

Figure 6.14 (a) In a simplistic model all scattered light is detected at one wavelength (the wavelength of the incident light). (b) The actual observed scattered light; the wavelengths are distributed around the expected wavelength due to Doppler shifting of the light from Brownian motion of the particles.

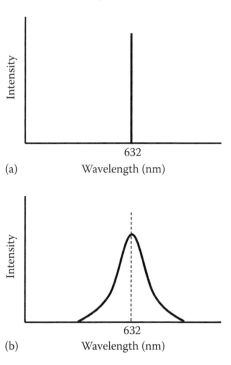

light. Similarly, some particles are moving away from the detector and have their emitted light red-shifted. On average, however, the particles are not moving relative to the detector and thus the distribution's center lies on the wavelength of the incident light.

The observed wavelength of scattered light is not the only thing affected by Brownian motion. The total observed intensity also fluctuates in time. This is because Brownian motion causes the distance between any two adjacent atoms to vary in time. The light emitted from these atoms experiences interference; the distance between the atoms determines whether this interference is constructive or destructive. Thus, as this distance varies for a certain pair of atoms, the net light emitted from them will fluctuate in intensity. Since there are so many particles in solution, the net intensity of scattered light approximately evens out and is never very far from the average intensity. However, it does not perfectly average out, and the net intensity of signal in solution can be seen to fluctuate.

Critically, the intensity at any given point in time depends on the location of all of the particles in solution. Thus, if one takes a measurement at time zero and another measurement very quickly after that, the intensity of the second measurement will be very similar to that of the first measurement so long as the particles have not had sufficient time to diffuse far from their original positions. If, on the other hand, the particles have had enough time to move significantly, the intensity will be essentially random. This concept can be expressed as a correlation. That is, if the second measurement is likely to be more similar than average to the first measurement, that is a positive correlation. If, however, the second measurement is likely to be less similar than average to the first measurement, that is called negative correlation. A correlation of zero means there is no relation between the measurements and the first intensity gives the experimenter no information about the second intensity. In a DLS experiment, as the time delay between the two intensity measurements increases from zero to infinity, the correlation decreases from unity (perfect correlation) to zero (no correlation).

We can take this data and plot correlation versus time delay for solutions of different particles (Figure 6.15). This graph gives a time scale for how fast a given particle diffuses. The faster it diffuses, the more quickly it will reach zero correlation as the particles need less time to move to different positions from where they started. These correlation graphs can be used to calculate the diffusion coefficient, D, which we introduced in Chapter 3.

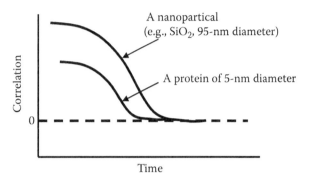

Figure 6.15 A plot of correlation versus time delay between subsequent measurements for two different particles. As expected, the larger particles stay correlated for a longer period of time, indicating that they diffuse more slowly.

The hydrodynamic radius of a particle can be obtained from its diffusion coefficient, and thus, from dynamic light scattering data via rearrangement of the Stokes–Einstein equation:

$$r_H = \frac{k_B T}{6\pi\mu D} \qquad (6.21)$$

In this equation r_H is the hydrodynamic radius, k_B is the Boltzmann constant, T is temperature, η is the solution viscosity, and D is the diffusion coefficient. As previously discussed, the larger the particle, the slower it diffuses, as shown in Figure 6.15; this is why the diffusion rate and diameter have an inverse relationship. The number of particles observed at each diameter can be plotted versus diameter to obtain a graph such as that in Figure 6.16. This figure demonstrates how a complex correlation curve with two distinct exponential regions can be

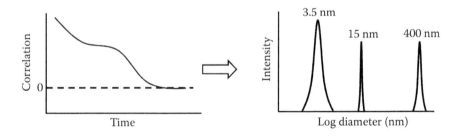

Figure 6.16 A plot of signal intensity versus particle size along with the correlation plot from which such a graph is derived.

transformed into an easy-to-interpret graph with two peaks at the corresponding particle sizes.

There are several things that can affect diffusion speed besides particle size that need to be controlled for (or at least considered). The first is the size of the hydration shell around each particle. Since it is the hydrodynamic radius of each particle that is measured, if two particles have equal size, but one is more strongly coupled to the surrounding media, that particle will diffuse more slowly and register as larger. Second, the ionic strength of a solution affects the coupling of the media to the particles, so ionic strength must typically be set at a standard level. Third, the surface morphology of a particle can affect diffusion speed. If a particle has long, comblike appendages on its surface, they will slow its diffusion and make it appear larger than a smooth sphere of equal size. Finally, changing particle geometry can affect diffusion speed. If a particle shifts from a more compact spherical formation to a more extended conformation, it will be slowed down and again appear larger despite not changing mass at all. This last effect is of particular interest in protein analysis because it can sometimes detect protein morphology shifts between different forms.

One important limitation to DLS is that the signal intensity received from a given particle is proportional to the sixth power of the particle's diameter. Thus, if one has a solution containing a 1:1 mixture by molar concentration of 10-nm particles and 100-nm particles, the peak area corresponding to the larger particles will be a million-fold larger than that for the smaller particles despite the solution containing an equal number of each. This problem only worsens as the size difference increases and serves to limit the range of particles that can be accurately analyzed in a single solution.

Dynamic light scattering is an excellent method for determining the sizes of particles present in a solution. It is able to accurately size particles from the nanometer scale to above the micron scale so long as any one particular solution does not contain widely varying particle sizes. This technique has found utility in applications ranging from colloidal science to proteomics, and has become an integral part of many nanoscience laboratories. Few other methods of particle sizing present such an attractive combination of ease of use and wide region of sensitivity.

6.3 X-RAY SPECTROSCOPY

Photons at energies above those that interact with valence electrons or covalent bonds can also be used to obtain structural and compositional information about nanoscale systems. X-ray spectroscopy is a collection of techniques that use x-ray excitation in order to probe the electronic structure of molecules in a material. These techniques are particularly useful methods for determining a material's composition, probing the ordering of individual atoms in a crystal, or observing nanoscale molecular processes. Each technique is based on the measurement of absorption, emission, or scattering of electromagnetic radiation caused by x-rays.

6.3.1 Absorption

When x-rays pass through a thin layer of matter, the intensity of the x-rays is diminished as a result of absorption and scattering. The effect of scattering can be ignored in wavelength regions at which significant absorption occurs. Each element has its own absorption spectrum with well-defined x-ray absorption peaks that can be used to identify it. In the x-ray region of light, enough energy is available to ionize the molecules that the photons interact with. When x-ray energy corresponding to the binding energy of a core (nonvalence) electron is absorbed, the core electron is ejected from the atom. This results in an excited ion. There is a higher probability of this happening when an x-ray beam with energy equal to the binding energy of the core electron is used. As the x-ray beam increases in energy away from the core electron binding energy, the probability of the corresponding wavelength being absorbed diminishes, and so the amount of the x-ray beam absorbed will decrease. If the x-ray beam is too low in energy, the corresponding wavelength will not be appreciably present and will not be able to eject the core electron. This will cause an abrupt decrease in the amount of the x-ray beam absorbed. The wavelength of x-rays ranges from about 10 to 10^{-6} nm, but conventional x-ray spectroscopy generally uses only wavelengths between 2.5 and 0.001 nm because this range contains the x-rays with energies corresponding to core electron-binding energies, which differ between elements.

6.3.2 Fluorescence

The excited ion, which is a result of the x-rays' ejecting of a core electron, will fluoresce through transitions of electrons in higher energy levels to the vacancy left by the ejected core electron. These transitions allow the excited ion to return to its more stable ground state. This fluorescence is measurable and is often used in concert with nanoscience techniques such as fluorescence recovery after photobleaching (FRAP), fluorescence resonant energy transfer (FRET), and fluorescence interference contrast microscopy (FLIC), which are discussed in Chapters 9 and 10. X-ray fluorescence is also commonly used in dyes. These fluorescing dyes can be attached to target molecules of interest that can be excited and observed in conditions where dyes in the visual wavelengths of light are not usable. This is often the case for in vivo studies when there is a layer between the observer and the dyed molecule that allows for x-rays to pass through but does not allow visible light to pass through.

6.3.3 Diffraction

When x-rays pass through matter, the radiation interacts with electrons in the matter in such a way that the path of the x-rays can be altered. This scattering effect of matter on x-rays is known as diffraction. In a crystal or any ordered sample, the x-rays scatter in ways that produce higher-intensity areas and lower-intensity areas, also termed constructive and destructive interference, respectively. By comparing these high- and low-intensity areas, it is possible to determine the architecture and the ordering in matter at the nanoscale. It is, of course, slightly more complicated in practice, because when an x-ray beam strikes an ordered crystal, which can be thought of as multiple layers of atoms, each subsequent layer of atoms scatters some of the beam and lets the remainder of the beam through. So that this diffraction can take place in a way that allows us to relate it to the materials structure, the space between layers of atoms in the material must be about the same distance as the wavelength of the radiation used to probe it and the atoms comprising the system must be highly ordered with few defects.

In order to use the high- and low-intensity areas that result from shining x-rays through an ordered crystal, a way to relate them to the structure must be known. Fortunately, in 1912, W. L. Bragg determined this relationship. The Bragg equation for constructive interference is as follows:

$$2d \sin \theta = n\lambda \qquad (6.22)$$

In this equation d is the lattice spacing, or the distance between atoms in the subsequent layers. θ is the incident angle between the beam of x-rays and the plane of the atom layer. λ is the wavelength of the x-rays (see Figure 6.17). Lastly, n is an integer related to the order of the reflection.

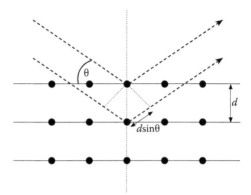

Figure 6.17   X-ray diffraction from crystal in which atoms are arranged in a primitive cubic lattice with separation *d*. The two beams shown will be in phase only in directions where their path-length difference 2*d* sin θ equals an integer multiple of the wavelength λ. As a result the incoming beam is deflected by an angle 2θ, producing a *reflection* spot in the diffraction pattern.

End of chapter questions

1. Ethidium bromide is a common dye used for studying DNA since the fluorescence intensity of ethidium bromide increases greatly when it is bound to DNA compared to when it is in solution.

a. Ethidium bromide has an extinction coefficient of 5680 M^{-1} cm^{-1} at 478 nm in water. Calculate the concentration of a solution measured to have an absorbance of $A = 0.8$ in aqueous solution.

b. Would you expect ethidium bromide to undergo any sort of solvatochromic shift (discussed in Chapter 5)? If so, what type?

c. Ethidium binds by inserting between DNA bases such that its aromatic system is stacked with the aromatic systems of the DNA bases. Develop a hypothesis using the concepts you have learned in Chapters 5 and 6 as to why its fluorescence intensity is enhanced when it is bound. Searching the internet for structures of the DNA–ethidium complex and fluorescence spectra may be helpful in formulating an answer.

2. We previously discussed how the HOMO-LUMO transition for butadiene is allowed since it has a significant transition dipole moment.

The HOMO–LUMO+1 transition density is shown below. Light regions represent increased density, dark regions represent decreased density.

a. Sketch the LUMO+1 of butadiene; the discussion of applications of the particle-in-a-1D-box model in Chapter 5 may be beneficial.

b. Would you expect the HOMO–LUMO+1 transition to be dipole-allowed? Why or why not?

3. Acetylene, also known as ethylyne (C_2H_2), is the simplest stable linear hydrocarbon. Two of its vibrational modes (symmetric C–C stretch and asymmetric C–H stretch) are shown below.

$$\overset{\longleftarrow}{H}-C\equiv C-\overset{\longrightarrow}{H}$$

$$\overset{\longleftarrow}{H}-C\equiv C-\overset{\longleftarrow}{H}$$

a. How many total vibrational modes does ethylene have?

b. Of the two vibrational modes shown, which do you expect to be IR active? Which do you expect to be Raman active?

4. Describe how you would use information given about molecular fluorescence spectroscopy to calculate Φ_f for a set of data.

Hint: Use an appropriate equation and integrate it to find g_x and thus calculate Φ_f.

5. What is the concentration fluctuation factor of a surfactant that forms micelles with very large molecular weight? Express your answer in terms of the second virial coefficient.

6. (a) Explain in words why the distribution of Rayleigh scattered light is a broad peak instead of a sharp signal.
(b) If a laser emitting light at 700 nm is sent through a dispersion and the Rayleigh scattered light is centered around 640 nm, what can be deduced about the direction of motion of the particles in solution?
(c) Does the shift mentioned in part (b) correspond to a red shift or blue shift?

7. (a) If the hydrodynamic radius of a particle doubles, what happens to its diffusion speed?
(b) If the radius is halved, the temperature is quadrupled, and the viscosity of solution increases by half, by what factor does the diffusion speed change?
(c) In a solution of hexane, would one expect to observe multiple signals? Explain why or why not.

References and recommended reading

Berne, B. J. *Dynamic Light Scattering: With Applications to Chemistry, Biology, and Physics*. 2000, Dover Publications, Mineola, NY. This book explains DLS and how Maxwell's equations lead to the intensity of the scattered radiation. This is an advanced graduate-level textbook.

Fowles, G. R. *Introduction to Modern Optics*, 2nd ed. 1975, Dover Publications, New York. This undergraduate textbook provides a good introduction to optics, including the concept of the complex refractive index/dielectric function.

Skoog, D. A., Holler, F. J. and Crouch, S. R. *Principles of Instrumental Analysis*, 7th ed. 2017, Brooks/Cole Boston.

This undergraduate textbook includes descriptions of both the physical principles and instrumentation used in many types of chemical and materials analysis.

Thulstrup, E. W. and Michel, J. *Elementary Polarization Spectroscopy*. 1989, VCH Publishers, New York. While principally focused on techniques involving polarized light, which will be introduced in Chapter 8, this undergraduate textbook provides an excellent introduction to optical transitions and their relationship to structural and electronic properties of materials, including ordered materials such as liquid crystals.

CHAPTER 7

Fundamentals of Surface Nanoscience

CHAPTER OVERVIEW

Surfaces and interfaces occur everywhere in nature, from the membranes surrounding biological cells to the multitude of particles within the vast expanses of the oceans. Surfaces play a key role in nanoscience since they are often used as platforms for the growth of nanomaterials. Therefore, understanding nanomaterials requires understanding important elements of surface science. **Interfaces** represent the two-dimensional plane between two distinct regions of matter, whether in different phases (e.g., the interfaces between glass and water) or in the same phase but of different compositions (e.g., oil and water). **Surfaces** are a subtype of interface where one of the bulk phases is a gas, usually air. Physical and chemical processes occurring at interfaces tend to be very different from corresponding processes in the bulk phase as such processes are confined to a region with one nanoscale dimension (i.e., a 2D nanomaterial). For instance, the hydrophobic effect influences chemistry at surfaces and is often responsible for the formation of films of nanoscale thickness. In this chapter, a consideration of hydrophobicity and the surface energy of solids and liquids will lead to a discussion of contact angles and wetting phenomena. This naturally leads to a discussion of self-assembled monolayers and adsorption phenomena. An understanding of how intermolecular interactions influence the adsorption and aggregation of molecules into nanostructures (such as micelles) is provided by considering the amphiphilic nature of surfactant molecules.

7.1 FUNDAMENTALS OF SURFACE SCIENCE

7.1.1 Surface energy of solids and liquids

As discussed in Chapter 5, molecules of the same type tend to experience a net attractive interaction. In liquids, this cohesive force keeps molecules close to each other, and each molecule is symmetrically surrounded by others such that *averaged over the bulk*, the constituent molecules have no preferred orientation or direction of motion and the system as a whole has no net force on it. The picture is considerably different on the surface since molecules here are not surrounded symmetrically by others (Figure 7.1). At the surface, a molecule experiences cohesive forces from others in the bulk beneath it, but (comparatively) negligible interactions above it. This asymmetry in force results in a net inward pull on the surface-bound molecule normal to the surface. This is the molecular basis of **surface tension**, which is defined as the force acting parallel to the surface and at right angles to a line of unit length on the surface. The experimental measurements of surface tension are described in Chapter 8.

The existence of surface tension is the reason why many liquids, such as water, tend to spontaneously contract and minimize the surface-area-to-volume ratio, which in turn maximizes the number of interactions between molecules *within* the liquid. To this end, such liquids adopt a spherical geometry in the absence of all external forces such as gravity. One can determine the work done in expanding the surface of a liquid

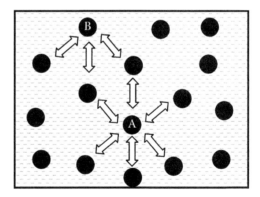

Figure 7.1 Bulk and surface interactions between molecules in a pure phase (e.g., H_2O). Bulk molecule A is surrounded symmetrically by its neighbors. Surface-bound molecule B is surrounded asymmetrically by its neighbors.

against the surface tension forces. This work actually represents the surface free energy of the expansion of the interface.

7.1.2 Surface free energy of adsorbed monolayers

The measurement of surface tension is perhaps the oldest method used to characterize a nanofilm floating on the surface of water. One class of molecules that can form nanofilms on water, and which will be discussed further later in the chapter, are known as **amphiphiles**, and are capable of self-assembling into nanostructures of remarkable complexity. Amphiphilic molecules, such as lipids and surfactants, have a tendency to accumulate on the surface of water. The driving forces behind this self-assembly process involve a combination of factors, such as the lack of solubility of the hydrocarbon portion of the molecule in water and the tendency of the charged or polar headgroup to point toward the aqueous phase. The net result is the formation of a film composed of a monolayer of highly oriented molecules. The closely packed monolayer is in dynamic equilibrium with individual molecules in the bulk aqueous phase, and so the packing density of the film increases as the concentration of molecules in the bulk phase increases. However, structural and thermodynamic constraints often limit the packing density, and a saturated monolayer is usually formed when the bulk concentration is sufficiently high. The thickness of such a film is approximately the length of the amphiphilic molecule (nanoscale). Due to this short dimension, the monomolecular film is often described as a two-dimensional nanoassembly.

Figure 7.2 shows a monolayer of the surfactant sodium dodecyl sulfate (SDS) confined to the air–water interface. Later in the book, we show that nanofilms at the air–water interface are of vital importance to biological processes and that their properties can be exploited in a number of important applications. Thus, it is often crucial to know how well the film is packed within both the monolayer and the cross-sectional area occupied by each molecule on the liquid surface. These factors often determine the physicochemical properties of the film.

Surface tension measurements provide a simple but powerful method for determining a variety of characteristics of adsorbed monolayers, including the density of the monolayer. Surface tension measures the stability of a surface. If the surface is relatively unstable, it has a large surface tension value and is considered a "high energy surface." For example, liquid water has a high surface energy because the molecules prefer not to be on the

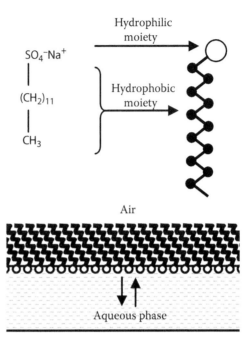

Figure 7.2   A monolayer of the anionic surfactant sodium dodecylsulfate (SDS) at the aqueous–air interface. The monolayer is in equilibrium with SDS molecules in the bulk aqueous phase. The hydrophobic moieties of SDS are pointing away from the aqueous phase and the polar headgroups are buried in the aqueous phase. This orientation of the SDS molecules on the surface stabilizes the air–water interface.

surface, but rather to be surrounded completely by other water molecules and interact with them through energetically favorable intermolecular hydrogen bonding.

A more rigorous way to define surface tension is to say that it is the free energy required to transport a molecule from the bulk phase to the surface and hence expand the area of the surface (dA). This surface free energy (dG_{surf}) is given by

$$dG_{surf} = \gamma dA \tag{7.1}$$

where the proportionality constant γ is the surface tension. The units of surface tension are energy per unit area, $J\ m^{-2}$ (or since energy can be thought of as applied force multiplied by the distance moved, the units are also newtons per meter, $N\ m^{-1}$). Astute readers may have observed that these dimensions are the same as the Hooke's law force constant.

Thus, we can interpret surface tension as a two-dimensional analog to a spring—it is the measure of the resistance of a surface to increase its area.

Example 7.1 Determining the Work Done in Expanding a Liquid Film

Consider a simple experiment in which a film is withdrawn from a soap solution as shown in Figure 7.3. By noting that work done is force times displacement and that this work is the surface free energy given by Equation 7.1, show that surface tension is $F/2L$. Prove that the units of surface tension are N m^{-1}.

Solution   Work must be done to pull up the frame and create a film (Figure 7.3b).

Work done ($dw = F\,dh$) is equal to the surface free energy ($\gamma\,dA$), where A is the area of the surface. The soap film has two sides or surfaces, so in this example area A is actually $2A$. Also, A is the length L times the distance moved dh.

Therefore, $F\,dh = \gamma\,2dA = \gamma\,2L\,dh$.

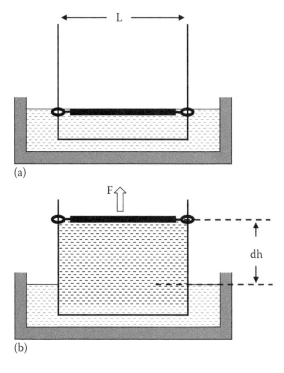

(a)

(b)

Figure 7.3 A film being withdrawn from a soap solution to a height dh by a force F, using a wire frame of length L. (a) Apparatus prepared for producing a film. (b) Film created by moving the bar above the solution surface.

Rearranging gives

$$\gamma = \frac{F}{2L} .$$

Since force is measured in N, the units of γ are N m^{-1}.

The larger the surface tension, the greater the resistance to increase the surface area, which in turn fundamentally depends on the interplay between various intermolecular interactions occurring on the surface. Table 7.1 lists the surface tension values of a few common liquids at three different temperatures.

Water has one of the highest known surface tension values (about 72 mN m^{-1} at room temperature) for the reasons mentioned above. However, when amphiphilic molecules are present in water, they tend to aggregate on the surface and lower the surface tension. For example, adding a milligram of SDS to approximately 200 mL of pure water lowers the surface tension by about 30 mN m^{-1}. This decrease in surface tension occurs because the amphiphilic molecules orient themselves at the interface such that they expose their water-insoluble hydrophobic tails to the air while keeping the polar head group buried in the aqueous phase, thereby freeing up water molecules that would otherwise be forced to remain unfavorably at the air–water interface (Figure 7.2). These conditions stabilize the surface and consequently lower the surface tension.

Table 7.1

Surface Tension Values of Some Pure Liquids

Liquid	γ/mN m^{-1}		
	25°C	50°C	75°C
Water	71.99	67.94	63.57
1-Decanol	28.51	26.68	24.85
Mercury	485.48	480.36	475.23
Ethanol	21.97	19.89	–
Bromine	40.95	36.40	–
Pyridine	36.56	33.29	30.03
Toluene	27.93	24.96	21.98
Benzene	28.22	25.00	21.77

Source: *CRC Handbook of Chemistry and Physics, Internet Version 2007*, 87th ed., Taylor & Francis, Boca Raton, 2007. With permission.

The degree to which the surface tension decreases depends on the structure of the amphiphile and the packing density of the resulting monolayer. If the concentration of the amphiphiles is low, then we can expect that the number of molecules on the surface is also relatively small. In a way, this dilute surface approximates a gaseous phase in which individual molecules are far apart and are free to move over the surface in a random fashion. As the concentration increases, however, the packing of the amphiphilic molecules at the surface becomes denser and the surface tension consequently decreases. This behavior continues with rising concentration until the point at which a saturated monolayer is formed. Beyond this concentration the surface tension value does not change. The phase behavior of amphiphiles at the air–water interface is discussed in Chapter 8.

7.1.3 Contact angles and wetting phenomena

When a drop of water is placed on a planar solid surface, it may at one extreme completely spread to cover the entire surface, or at the other extreme form a spherical droplet on the surface. These situations represent either complete wetting or complete dewetting. Usually the degree of wetting is intermediate between these extremes and depends largely on the interfacial energy (or surface tension) between the liquid and the solid surface. The contact angle is the angle at which the liquid–vapor interface meets the solid surface (Figure 7.4). If the planar surface is horizontal and the droplet is not moving, this angle is called the **static contact angle** (Figure 7.4a). If the liquid is in motion because the surface is tilted, then we can identify two **dynamic contact angles**, the advancing contact angle and the receding contact angle (Figure 7.4b). This picture is similar to a raindrop running down the surface of a window. Usually the advancing contact angle is much larger than the receding angle, and the difference between the two values is called contact angle hysteresis.

Figure 7.4(a) shows a nonwetting drop (conventionally called a "sessile drop") on a planar solid surface making a contact angle, θ, which must be greater than zero. The various interfaces are described by their surface tensions: the liquid–vapor tension (γ_{LV}), the solid–vapor tension (γ_{SV}), and the solid–liquid tension (γ_{SL}). The Young equation provides a relationship between these various surface tensions and the static contact angle:

$$\gamma_{SV} = \gamma_{SL} + \gamma_{LV} \cos \theta \qquad (7.2)$$

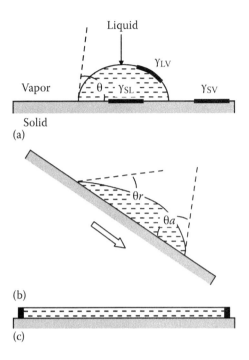

Figure 7.4 A liquid drop on a solid surface. (a) The drop is nonwetting with a contact angle θ. (b) The drop is running down the solid with advancing (θ_a) and receding (θ_r) contact angles. (c) Complete spreading. The solid dark lines in (a) highlight the various interfaces with different surface tensions: the liquid–vapor tension (γ_{LV}), the solid–vapor tension (γ_{SV}), and the solid–liquid tension (γ_{SL}).

Since γ_{LV} and θ can easily be measured (see Section 8.1), the value of $\gamma_{SV} - \gamma_{SL}$ can be determined. If the liquid completely wetted the solid (θ = 0°), then the value of $\gamma_{SL} = 0$. This is a hypothetical situation in which there is no "tension" between these two phases. In fact a plot of cosθ versus γ_{LV} is linear with a negative slope. The line can be extrapolated to the value cosθ = 1 and the corresponding surface tension measured (on the *x*-axis). Cosθ = 1 corresponds to a contact angle of 0°, or complete wetting. The corresponding surface tension is called the critical surface tension (γ_c). Since under these conditions of complete wetting $\gamma_{SL} = 0$, the Young equation tells us that $\gamma_c = \gamma_{SV}$.

It is worth noting that surfaces can be hydrophobic (water hating) or hydrophilic (water loving). The surface of glass, for example, is considered hydrophilic due to being terminated by polar hydroxyl groups and Teflon hydrophobic due to being terminated with nonpolar fluorocarbons. Water will not spread on hydrophobic surfaces but will instead form a droplet with a relatively large contact angle. Conversely, water will spread on a hydrophilic surface since it has a strong affinity for that surface, forming a thin film with a very small contact angle.

7.1.4 Nanomaterials and superhydrophobic surfaces

If the contact angle of a sessile drop approaches 180°, the drop essentially adopts its spherical geometry on the surface and moves around much like a frictionless bearing on the surface. Figure 7.5 shows a spherical droplet of water on a hydrophobic surface. In order for a water droplet to behave this way, the surface needs to be one that minimizes the area of the solid-water interface, or a **superhydrophobic** surface. Creating such surfaces has been a fascinating challenge in nanoscience.

Superhydrophobic surfaces are not new technology by any means—scientists have been experimenting with superhydrophobicity for nearly a century and they have been entranced by the superhydrophobicity of natural materials like the lotus leaf for far longer. However, new developments as well as new potential applications for their use have spurred the popularity of superhydrophobic surfaces. It is important, therefore, to understand the basic types of superhydrophobic surfaces, their specific characteristics, and the potentials for the use and development of each

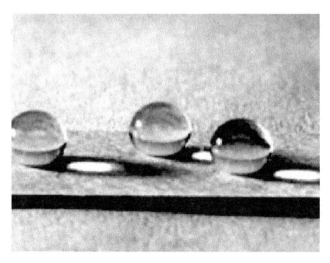

Figure 7.5 A sessile drop on a superhydrophobic surface. (Reproduced with permission from Ma, M., Hill, R. M., Lowery, J. L., Fridrikh, S. V., Rutledge, G. C. "Electrospun Poly(Styrene-block-dimethylsiloxane) Block Copolymer Fibers Exhibiting Superhydrophobicity." *Langmuir* 2005, 21:5549–5554.)

method of fabricating these surfaces. All of these concepts are outlined in the popular review article, "Progress in Superhydrophobic Surface Development," by Paul Roach, Neil Shirtcliffe, and Michael Newton, 2008.

First, a basic understanding of the mechanism behind superhydrophobicity is important in understanding its applications. Superhydrophobic surfaces each have two distinct states of wetting, each governed by separate equations and having separate characteristics. These states are illustrated in Figure 7.6. The first type was outlined by Wenzel in 1936 and describes a wetting state in which water rests on a surface whose morphology has been altered so that in a given area, water is in contact with more surface than if the surface were completely flat. **Wenzel states** are described by the equation

$$\cos \theta^W = r \cos \theta \qquad (7.3)$$

where θ is the contact angle on an unmodified surface, θ^W is the Wenzel contact angle (on the rough surface), and r is the ratio of the actual surface area of the substrate to the projection of that surface onto a horizontal surface. In other words, r is the ratio of the actual surface area to what the surface area would be if the substrate were completely smooth. This equation essentially states that Wenzel wetting increases the contact angle of a drop of water by creating more hydrophobic surface with which the water can interact. However, if the surface is hydrophilic ($\theta < 90°$), then Wenzel wetting actually increases the hydrophilic properties of the surface. Also, because water is present between perturbations, water in a Wenzel state is less likely to roll off the substrate than on a flat surface of the same material. Thus, Wenzel wetting relies on two factors—an already hydrophobic substrate and an increase in the surface area of that substrate.

Figure 7.6   The various wetting states on rough surfaces: (a) Wenzel state, (b) Cassie–Baxter state, and (c) intermediate wetting.

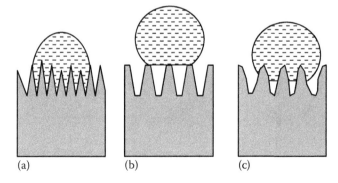

(a)   (b)   (c)

The other major type of wetting was discovered by Cassie and Baxter in 1944 and describes a wetting state in which water rests on nano- or micro-sized "pins" with air spaces in between. Unlike Wenzel wetting, **Cassie–Baxter states** have a smaller area of substrate in contact with the water than a planar surface and rely on the air gaps in between regions on the surface of the substrate to form a superhydrophobic surface. In fact, Cassie–Baxter surfaces can be fabricated with hydrophilic substrates and still display superhydrophobicity. Unlike Wenzel surfaces, water in Cassie–Baxter states rolls off surfaces much more easily than on a flat surface of the same material. These properties are reflected in the equation governing Cassie–Baxter wetting:

$$\cos \theta^C = \varphi_s(\cos\theta) + (1 - \varphi_s)\cos\theta_X \qquad (7.4)$$

where φ represents the fraction of surface present at the top of the protrusions (where the water is in contact with the substrate), $(1-\varphi)$ represents the fraction of air gaps, and θ_X is the contact angle over the air gaps, which is approximated at 180°, and θ^C is the Cassie–Baxter contact angle (on the rough surface). Thus, there are two ways to increase superhydrophobicity of a Cassie–Baxter surface—either increase the value of θ by increasing the inherent superhydrophobicity of the substrate, or decrease φ_s by making bigger air gaps.

Interconversion between these two states is possible, but there is an energy barrier to overcome much like any transition state between two energy minima. This phenomenon comes into play when Cassie–Baxter surfaces display Wenzel wetting, which is possible if water falling from a great distance (i.e., rain) is forced into the crevices between roughness features. The transition from one state to another is important when discussing some methods of creating superhydrophobic surfaces.

The final important feature of superhydrophobic surfaces that must be mentioned before a more thorough description of the specifics is the advantage of multiple-scale roughness, in which perturbations range from the nanoscale to the microscale. Recent study has shown this kind of surface to increase the ease with which drops roll off the surface (in other words, decreasing the contact-angle hysteresis), the prevention of conversion from Cassie–Baxter to Wenzel states, and the tendency to convert from Wenzel to Cassie–Baxter states. This property is beneficial for a variety of reasons and will be revisited later.

Because the review article "Progress in Superhydrophobic Surface Development" specifically outlines the major methods of creating superhydrophobicity, we will refrain from repeating a list of specific methods and focus on unifying themes and methods of fabrication as well as the most salient features of each major method. One such feature belongs to the first method described by this article—fibrous hydrophobic surfaces. These surfaces can range from carbon nanotubes bound to cotton fibers to nanoscale polymer structures created by electrospinning. Fibrous hydrophobic surfaces are versatile, as fibrous materials are inherently rough, and coating with hydrophobic materials or nanorods can increase this roughness. Very high contact angles have been reported for fibrous hydrophobic surfaces. Some researchers have even reported surfaces having self-cleaning properties that are dirt- and oil-resistant while producing high contact angles and low hysteresis. This technology appears to be promising for the development of water- and contamination-resistant clothing that would essentially wash itself and is already in use in a variety of spill- and stain-resistant clothing brands.

Another interesting feature that is unique to a specific fabrication method is the ability of certain hydrophobic semiconductors to be crystallized into a superhydrophobic surface that exhibits superhydrophobicity in the dark but superhydrophilicity in the light. One likely explanation for this phenomenon is that superhydrophilicity is generated by the excitation of electrons on the surface, which suggests that there are surface properties that can be changed by applying a voltage. Instead of excitation due to ultraviolet (UV) radiation, a voltage could be used, thus allowing for materials that can be dried instantly at the flick of a switch or wetted completely with another flick.

Several overarching themes appear in this article as well. One of the most notable is the generation of fractal solids as superhydrophobic surfaces, which may be achieved in many ways. One common approach is crystal growth, in which a material is cooled or condensed into a fractal solid. Fractal solids can be formed from both organic and inorganic materials. Another common method is the diffusion-limited growth process, in which a material deposits so quickly onto a substrate that the growth of the film is limited by the flow of material over the substrate. This method, by randomness, creates small disturbances on the surface of the flat substrate, which then cause more particles to attach to these perturbations. The process continues until large bumps form, which then have small disturbances on their surfaces that create subbumps, and so on. The greatest benefit of these kinds of assemblies is that being fractal, these surfaces have multiple-scale roughness. The main bumps can act as Wenzel-type surfaces, while the smaller bumps can act as Cassie–Baxter surfaces. As is expected, very low hysteresis and very large contact angles have been reported on fractal solids. Another important benefit of fractal solids is their ease of fabrication—because these surfaces are usually created by chaotic movement of material, they require less specialized machinery and can be made with a variety of materials. In fact, fractal superhydrophobic surfaces have been reported using semiconducting polymers, giving these particular surfaces the ability to switch properties under the influence of UV light or an applied voltage.

One very common downside of fractal solids, however, is their tendency to be opaque. Their multiscale roughness is very beneficial to their superhydrophobic properties, but because these features range from nanoscale to microscale, they interact with visible light. However, if transparent fractal solids could be synthesized, they could very well be used for windshield coatings on cars or as protective layers for photoactive devices such as LEDs or photovoltaic cells.

This problem is not present in another type of superhydrophobic surface fabrication—uniform nanostructures. These structures can be generated in many ways, from carbon nanotube growth to lithographic etching. These structures are the opposite of fractal patterns in that although they are sometimes random in arrangement, they are generally uniform in size. These materials can be easily generated using laser lithography as well as chemical etching to produce surface morphologies of a specific predetermined shape and pattern. As well, these surfaces can be easily "decorated" with additional coatings and features to produce increased hydrophobicity. Structures of this type have been reported to reach contact angles up to 178° using dodecanoic acid–coated, cobalt hydroxide nanopins. As well, significant (168°), although not quite as substantial, Wenzel contact angles were reported using decorated carbon nanotubes. Perhaps most importantly, these kinds of structures are of one size and thus do not often absorb in the visible range. This feature makes these kinds of structures more viable for the kinds of surfaces previously discussed. The drawback of these kinds of materials is that they are much less effective at repelling water than fractal structures. They lack multiscale roughness, and even methods that generate more random arrays are less hydrophobic than fractal surfaces.

What has emerged in recent years has not been the development of one dominant method of constructing superhydrophobic surfaces. Rather, many different methods have developed, each with its own benefits and drawbacks. As more applications for superhydrophobic surfaces appear—clothing, photoactive devices, windows, and more—the drive is for a variety of superhydrophobic materials that can be customized to fit these multiple needs.

7.2 ADSORPTION PHENOMENA: SELF-ASSEMBLED MONOLAYERS

A molecule approaching a surface may experience a net attractive force and consequently become trapped or confined at the surface. Such a species is called the adsorbate, and adsorption is the physical process by which adsorbate molecules accumulate onto a solid surface. Desorption is the opposite process in which molecules leave the surface and enter the bulk phase. The solid surface in question is referred to either as the **substrate** or **adsorbent**. The former term is typically used for planar surfaces and the latter term often refers to high-surface-area porous solids. The sticking of a reactant molecule to a surface was first proposed by Michael Faraday in 1834 as the initial step of a surface-catalyzed reaction. In this section we are interested in the kind of adsorption that results in a monolayer and has a thickness on the order of nanometers. The surface itself does not need to be flat but may be rough or even porous. The extent of adsorption depends not only on the types of intermolecular forces involved (van der Waals, electrostatic, hydrogen bonding), but also on the surface area; the greater the surface area of the substrate, the greater the extent of adsorption.

The best adsorbents are those with large total surface areas, such as silica gel (SiO_2, surface area >1000 m^2/g) and activated carbon. Silica gel is commonly used in chromatographic columns to enhance the separation of solute mixtures by taking advantage of the different degrees of adsorption of the various components. Another interesting example of an adsorption phenomenon occurs in polar stratospheric clouds in the upper atmosphere, in which highly porous particles of ice act as substrates for the adsorption of gases such as HCl.

In this and the next section we focus on the adsorption of molecules from solution or gaseous phase onto a solid substrate. However, it is important to appreciate that adsorption is a surface phenomena and can occur at any interface. For example, molecules may adsorb from an aqueous solution to the aqueous–air interface. Liquid–liquid interfaces, such as the boundary between an oil and water phase, also represent regions at which adsorption may occur. Surface adsorption plays a central role in the formation of nanomaterials. Adsorbents present themselves as platforms for the self-assembly of molecules into nanostructures. Specific examples in which nanomaterials are synthesized this way are presented in Chapter 10. Solid substrates can be chemically modified so that adsorption can be selective. This modification is known as surface functionalization and may be as simple as oxidizing a metal surface to render it hydrophilic and change the surface energy so that polar molecules are spontaneously adsorbed to the surface.

Many solids have the property of adsorbing large quantities of gases and solutes from liquid solutions. This process is generally very specific both with respect to the adsorbent and the material adsorbed and driven largely by thermodynamic considerations. Adsorption is usually an exothermic process and can be divided into two kinds: **chemisorption** (chemical adsorption) and **physisorption** (physical adsorption). In general, if adsorption is specific and if large amounts of heat are liberated (greater than about 50 $kJmol^{-1}$), the adsorption process is referred to as chemisorption (first proposed by Irving Langmuir in 1916). In this process, bonds are broken in the adsorbate molecules and new covalent bonds are formed between the adsorbent and the adsorbate until a complete monolayer has been established. The resulting substrate-adsorbate bond strengths range from 200 to 500 $kJmol^{-1}$. The chemisorbed monolayer is irreversibly bound to the solid surface and changes the surface properties of the solid substrate to resemble those of the exposed portion of the adsorbate. Thus, chemisorption is an excellent method of chemically functionalizing a solid surface. As an example of chemisorption, consider the molecule octadecanethiol (ODT), shown in Figure 7.7. The molecule contains 18 carbon atoms (17 methylene groups and one methyl group) terminated by a thiol group (SH). The thiol group is extremely reactive toward gold, resulting in a strong Au–S covalent bond. Thus, by placing a gold-coated substrate into a solution of ODT in a solvent such as chloroform, molecules of ODT spontaneously chemisorb to the Au surface and form a tightly packed monolayer within hours. A few features regarding this process are worth noting. First, adsorption is rapid, irreversible, and stops after a complete monolayer is formed. Second, in addition to the strong Au–S bonds that are formed, the strong van der Waals interactions between the neighboring alkyl chains in the ODT monolayer allow the molecules to pack very tightly and force the

Figure 7.7 The formation of a self-assembled monolayer by the chemisorption of octadecane thiol on a gold surface.

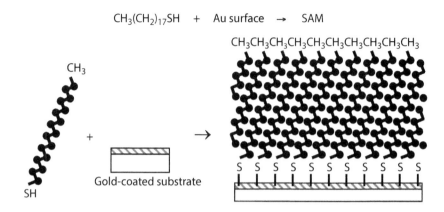

hydrocarbon backbone into an all-trans conformation (Figure 7.7). The resulting film is solid in nature, has a thickness almost the length of a fully extended ODT molecule, and is referred to as a self-assembled monolayer (SAM). Finally, the SAM renders the solid surface hydrophobic because the surface now contains a very high density of closely packed methyl groups. Consequently, a drop of water on this surface has a large contact angle (>110°).

If adsorption is nonspecific and if only small amounts of heat are liberated, comparable to the heat of vaporization of the adsorbed material, the process is physisorption. The interaction between the adsorbent and adsorbate is much weaker (~20 kJmol^{-1}) in a physisorbed film compared to a chemisorbed film. It should be noted that although physisorption is typically nonspecific, thermodynamic considerations remain very important for physisorption processes. As an example, we consider the physisorption of a long-chain alcohol, such as dodecanol, from an aqueous solution onto a solid hydrophobic surface. Let the hydrophobic surface be composed of an ODT monolayer chemisorbed to a gold surface. The adsorption process is driven by a strong tendency of the alcohol to avoid water (dodecanol is only sparingly soluble in water) and to associate with any accessible hydrophobic surface. Thus, the dodecanol adsorbs to the ODT SAM (Figure 7.8). The resulting dodecanol monolayer is closely packed with the OH headgroup orientated toward the water phase. Like the ODT SAM, the alkyl chains of dodecanol are closely associated with each other due to favorable van der Waals interactions, and form a solid-like SAM on the surface of the hydrophobic film. Unlike the ODT, the adsorption of dodecanol is reversible. The molecules in the film are in dynamic equilibrium with the dodecanol molecules in the aqueous bulk phase.

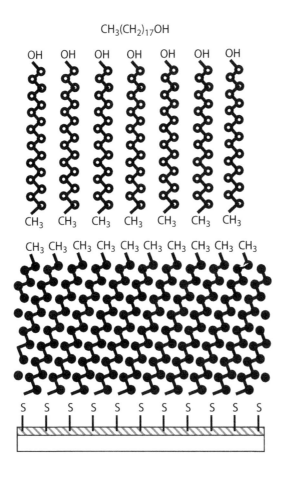

$CH_3(CH_2)_{17}OH$

Depleting the number of dodecanol molecules in the bulk phase reduces the packing density in the monolayer at the solid–aqueous interface.

Adsorption characteristics such as temperature, bulk concentration dependence, and reversibility are often used to distinguish chemisorption and physisorption. The energetic difference between the two adsorption processes is nicely illustrated in the one-dimensional Lennard-Jones potential energy curve for the adsorption of a diatomic molecule on a planar surface. $V(x) = 0$ at large distances corresponds to zero interaction between the substrate and adsorbate molecule at infinite separation. As the molecule approaches the surface (going from point A to B in Figure 7.9), there is a negative attractive potential between the surface and the adsorbate molecule. The potential reaches a minimum at distance B in the case of physisorption and at distance C in the case of chemisorption.

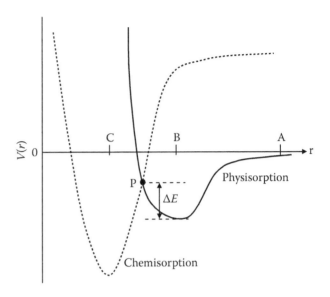

Figure 7.9 A one-dimensional Lennard-Jones potential energy curve for the chemisorption and physisorption of a molecule on a planar surface. A, B, and C represent various distances between the surface and the molecule.

These distances represent the corresponding equilibrium substrate-adsorbate bond lengths for the two processes. At distances less than B for physisorption (or less than C in the case of chemisorption), the attractive interaction lessens until $V(x) = 0$, below which repulsive interactions start to become more important. The minimum potential (at distances B and C) for chemisorption is at a smaller substrate-adsorbate distance compared to physisorption because the former processes lead to a shorter bond distance between the substrate and adsorbate. Furthermore, the potential well for chemisorption is deeper than for physisorption due to a stronger substrate-adsorbate bond strength than that formed in physisorption, with the former process involving the breaking and making of covalent bonds. Sometimes a molecule can be trapped in a physisorbed state before being chemisorbed. In this case the physisorbed molecule is a precursor to chemisorption. The two potential energy curves shown in Figure 7.9 cross at a distance represented by point P. This is the point where the physisorbed precursor can "cross over" to the chemisorbed state. ΔE represents the activation energy (kinetic barrier) in going from the physisorbed state to the chemisorbed state.

The adsorption capacity of solid surfaces is determined from measurements of the mass (or moles) of material adsorbed and the area available for adsorption per unit mass of adsorbent. Usually the amount of solute adsorbed from solution per unit mass of adsorbent depends on the solute bulk concentration up to its saturation point. As well, the amount adsorbed per unit mass of adsorbent for a given solute bulk concentration decreases with increasing temperature. If the adsorption is physical, it is reversible and the solute leaves the surface of the adsorbent to reestablish equilibrium when the adsorbent is removed from a solution in equilibrium and placed in a solution of lower concentration. Such reversibility may not be shown if the process is chemisorption. The following section describes how adsorption capacity can be measured and how information, such as monolayer coverage and substrate surface area, can be extracted by measuring the adsorption capacity.

7.2.1 Simple adsorption isotherms

An adsorption isotherm is a mathematical relationship between surface coverage (i.e., the fraction of the adsorbent surface covered by the adsorbate) versus the bulk concentration of the adsorbate. The "isotherm" term means that these measurements are carried out at constant temperature. Figure 7.10 shows a few simple adsorption isotherms. The line that trails off to a constant surface coverage (Figure 7.10a) represents a situation in which adsorption stops because the surface is saturated by the adsorbate. This typically corresponds to monolayer coverage. In

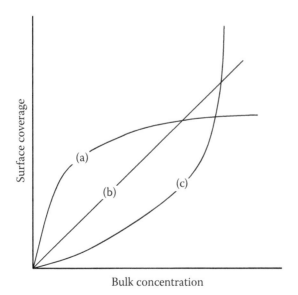

Figure 7.10 Adsorption isotherms describing (a) Langmuir adsorption, (b) multilayer adsorption, and (c) cooperative adsorption.

Figure 7.10(b), adsorption seems to continue indefinitely as a function of bulk concentration. Adsorption in this case does not stop after a saturated monolayer has formed; instead, adsorption continues and multilaycrs are formed on the surface. In Figure 7.10(c), adsorption seems to increase exponentially with bulk concentration. This situation occurs if the presence of the adsorbate at the surface promotes further adsorption due to favorable intermolecular interactions. This kind of adsorption is known as cooperative adsorption. It may or may not eventually level off at very high bulk concentrations.

If the bulk phase is a gas, then the adsorption isotherm can be represented by a plot of surface coverage as a function of gas pressure at a constant temperature. Adsorption isotherms allow us to determine some important characteristics of the adsorption process such as equilibrium constants, the number of adsorption sites available for adsorption on the substrate, and the enthalpy of adsorption.

The Langmuir adsorption isotherm, presented by Langmuir in 1918, is the simplest adsorption isotherm. The Langmuir adsorption isotherm assumes that there are a finite number of identical adsorption sites on the solid substrate and that each site is taken up by a single adsorbate molecule. During the adsorption processes, these sites are taken up until a point of saturation is reached in which all sites have been taken up by adsorbate molecules. This point represents complete monolayer coverage, at which further adsorption ceases. Furthermore, the Langmuir adsorption isotherm assumes that there are no interactions between the adsorbate molecules during the adsorption process and that the enthalpy of adsorption is independent of surface coverage. Figure 7.10(a) represents a Langmuir adsorption profile.

By representing the solid surface by $S(s)$ and the adsorbate molecules in a gas phase by $A(g)$, we can write the following equilibrium between the adsorbate and the substrate:

$$A(g) + S(s) \underset{k_d}{\overset{k_a}{\rightleftharpoons}} AS(s) \tag{7.5}$$

The constants k_a and k_d are the rate constants for the adsorption and desorption process, respectively. At equilibrium, the forward and reverse rates are the same, and so from basic kinetics,

$$k_a[A][S] = k_d[AS] \tag{7.6}$$

or

$$\frac{k_a}{k_d} = \frac{[AS]}{[A][S]} = K \tag{7.7}$$

where K is the equilibrium constant for the adsorption process. The adsorbate surface contains a certain concentration of sites available for adsorption. Let β be the concentration of sites per square meter and let θ be the fraction of surface sites occupied by the adsorbate molecules. The concentration of occupied sites is therefore given by $\theta\beta$, and the concentration of free available adsorption sites is given by $\beta - \theta\beta = (1 - \theta)\beta$. The rate of desorption (v_d) is proportional to $[AS]$, which in turn is proportional to the number of occupied surface sites. Furthermore, the rate of adsorption (v_a) is proportional to $[A][S]$, which in turn is proportional to the number of available sites and the number density of molecules in the bulk phase. Thus,

$$v_d = k_d \theta \beta \tag{7.8}$$

$$v_a = k_a(1 - \theta)[A] \tag{7.9}$$

At equilibrium, these two rates are equal, so

$$k_d \theta \beta = k_a(1 - \theta)[A] \tag{7.10}$$

Rearranging gives

$$\frac{1}{\theta} = 1 + \frac{k_d}{k_a[A]} = 1 + \frac{1}{K[A]} \tag{7.11}$$

Note the similarity to the Scatchard model. For a gas phase adsorption process, $[A]$ represents the concentration (number of molecules per unit volume) of the adsorbate gas. Assuming that the adsorbate can be modeled as an ideal gas, the concentration can be expressed as a pressure by use of the ideal gas equation, $PV = nRT$.

$$[A] = \frac{\text{number of molecules}}{V} = \frac{nN_A}{V} = \frac{nN_AP}{nRT} = \frac{N_AP}{RT} = \frac{P}{kT} \tag{7.12}$$

where P is the pressure of the adsorbate gas, and $k\ (= R/N_A)$ is the Boltzmann constant. Equation 3.11 can be rearranged as Equation 7.13.

$$\frac{1}{\theta} = 1 + \frac{kT}{KP} = 1 + \frac{1}{aP} \tag{7.13}$$

where $a = K/kT$. Equation 7.13 is the Langmuir adsorption equation. A plot of $1/\theta$ versus $1/[A]$ or $1/P$ yields a straight line with a slope equal to the reciprocal of the equilibrium constant (or kT/K).

The fraction θ can be related to the following ratio:

$$\theta = \frac{V}{V_m} \qquad (7.14)$$

In this equation, V is the volume of the gas adsorbed to the surface and V_m represents the volume of gas corresponding to monolayer coverage. Complete monolayer coverage corresponds to $\theta = 1$, and $V = V_m$. Equation 7.14 can be substituted into Equation 7.13, giving Equation 7.15:

$$\frac{1}{V} = \frac{1}{aPV_m} + \frac{1}{V_m} \qquad (7.15)$$

Example 7.2 The Adsorption of Nitrogen onto a Mica Surface follows the Langmuir Adsorption Model

The following data were collected at 273.15 K. Determine the value of a, V_m, and the total number of surface sites. What would the concentration of surface sites be if the mica substrate was a square with 2-cm sides?

$P/10^{-12}$ torr	2.50	1.32	0.48	0.30	0.20
$V/10^{-8}$ m³	3.40	2.92	2.00	1.54	1.25

Solution According to Equation 7.15, a plot of $1/V$ versus $1/P$ will have a slope of $1/aV_m$ and an intercept of $1/V_m$. Figure 7.11 shows this plot. A linear fit to the data yields the equation $y = 1 \times 10^{-5} x + 3 \times 10^7$.

The intercept on the y-axis is equal to $3 \times 10^7 = 1/V_m$, and so $V_m = 3.3 \times 10^{-8}$ m³.

From the slope $1 \times 10^{-5} = 1/aV_m$, and so $a = 3.0 \times 10^{12}$ torr^{-1}.

To determine the total number of sites on the mica surface, we first need to determine the number of molecules corresponding to V_m.

First, we know that under these conditions, 1 mol of gas occupies 2.24×10^{-2} m³.

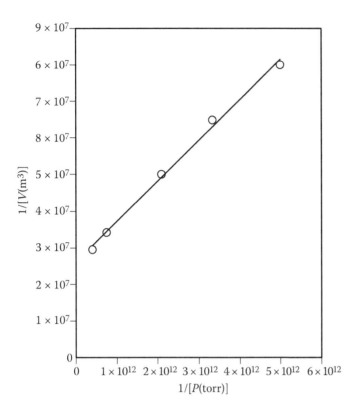

Figure 7.11 A plot of $1/V$ versus $1/P$ for the adsorption of N_2 on mica at 273.15 K.

The number of moles corresponding to V_m is

$$\frac{3.3 \times 10^{-8}\text{m}^3}{2.24 \times 10^{-2}\text{m}^3\text{mol}-1} = 1.47 \times 10^{-6}\text{mol}$$

The number of molecules is simply N_A times the number of moles.

$$\left(6.022 \times 10^{23}\text{mol}^{-1}\right)\left(1.47 \times 10^{-6}\text{mol}\right) = 8.85 \times 10^{17}\text{molecules}$$

Now each N_2 molecule occupies a single site on the mica surface. Thus, there must be 8.85×10^{17} sites on the surface.

For a square of length 2 cm, the concentration of surface sites is simply the ratio of the total number of sites to the total surface area of the square.

$$\frac{8.85 \times 10^{17}\text{molecules}}{(0.02\text{m})^2} = 2.21 \times 10^{21}\text{m}^{-2}$$

7.2.2 Other useful adsorption isotherms

The Langmuir isotherm makes a number of assumptions that may not apply to all adsorbates and adsorbents. For instance, strong intermolecular interactions may cause deviations from Langmuir adsorption behavior. Many systems display multilayer adsorption. The Brunauer Emmett Teller, or BET, isotherm is useful for modeling such behavior. The key assumptions made in the BET isotherms are that physisorption on a solid is infinite, that there are no interlayer interactions in the multilayer film, and that each layer can be described by the Langmuir model. The BET isotherm takes the form shown in Equation 7.16

$$\theta = \frac{V}{V_m} = \frac{cz}{(1-z)[1-(1-c)z]} \tag{7.16}$$

The term $z = P/P^*$, where P^* is the vapor pressure above the adsorbate before any adsorption occurs. The constant c is related to the enthalpy of adsorption of the first layer (ΔH_1) and the enthalpy of vaporization of the subsequent layers (ΔH_{vap})(Equation 7.17)

$$c = \exp\left(\frac{\Delta H_1 - \Delta H_{vap}}{RT}\right) \tag{7.17}$$

Figure 7.12 shows how the form of the isotherm changes for different values of c. When $c \gg 1$, the isotherm simplifies to Equation 7.16 and is

Figure 7.12 The effect of changing the parameter c on the BET isotherm.

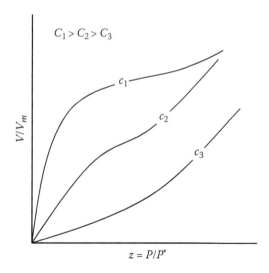

useful in describing surfactant adsorption (Section 7.3).

$$\frac{V}{V_m} = \frac{1}{1 - z} \qquad (7.18)$$

There are many useful isotherms, each one modeling a specific kind of adsorption. For example, the Temkin isotherm, $\theta = c_1 \ln(c_2 P)$, where c_1 and c_2 are constants, describes adsorption for systems in which ΔH_1 changes with pressure or concentration of the adsorbate. The Freundlich isotherm, $\theta = c_1 P^{1/C2}$, describes adsorption involving strong intermolecular interactions. In Section 7.3.2, the Gibbs' adsorption equation is introduced, which allows adsorption at the solid–aqueous interface to be modeled reasonably well.

7.3 SURFACTANT CHEMISTRY

The word surfactant is short for surface-active agent. Surfactants describe a class of molecules that have a tendency to adsorb to surfaces and interfaces and lower the interfacial tension. For instance, the air–water interface has a surface tension of 72.8 mN/m. The addition of 10% NaOH to the water increases this value to 78 mN/m. However, adding a typical surfactant to the water at relatively low concentrations (~mM) can lower the surface tension to 20 mN/m. Interestingly, the surface tension between an aqueous surfactant solution and an oil such as heptane can be as low as 1 mN/m.

To rationalize the effect surfactants have on surface tension, we need to consider the structure of these molecules. Surfactants are amphiphilic, which means that one part of the molecule is soluble in a specified fluid (the lyophilic part) and the other part of the molecule is insoluble (the lyophobic part). If the fluid is water, then the soluble part is called the hydrophobic part, and the insoluble part is the hydrophilic part. Thus, the structure of a surfactant has a region that is largely nonpolar, typically hydrocarbon or fluorocarbon in nature, and another region that is polar, charged, and interacts strongly with water. These two regions are commonly referred to as the hydrophobic tail (or chain) and the hydrophilic head group, respectively. Figure 7.13 shows the structure of some common surfactants, including some naturally occurring ones such as phospholipids (DMPC).

Figure 7.13 The molecular structure of some common surfactants, with the hydrophobic and hydrophilic moieties indicated. Note that the hydrocarbon chains are shown in all-trans conformation. This conformation is rarely adopted in micellar structures or in adsorbed surfactant films at interfaces. The surfactants shown are sodium dodecylsulfate (SDS), didodecyldimethylammonium bromide (DDAB), zwitterionic dodecyl-N,N-dimethyl-3-ammonio-1-proponate-sulfonate (DDAPS), nonionic $C_{12}E_3$, cationic. *Gemini* surfactant, and the zwitterionic lipid molecule 1,2-dimyristoyl-sn-glycero-3-phosphocholine (DMPC).

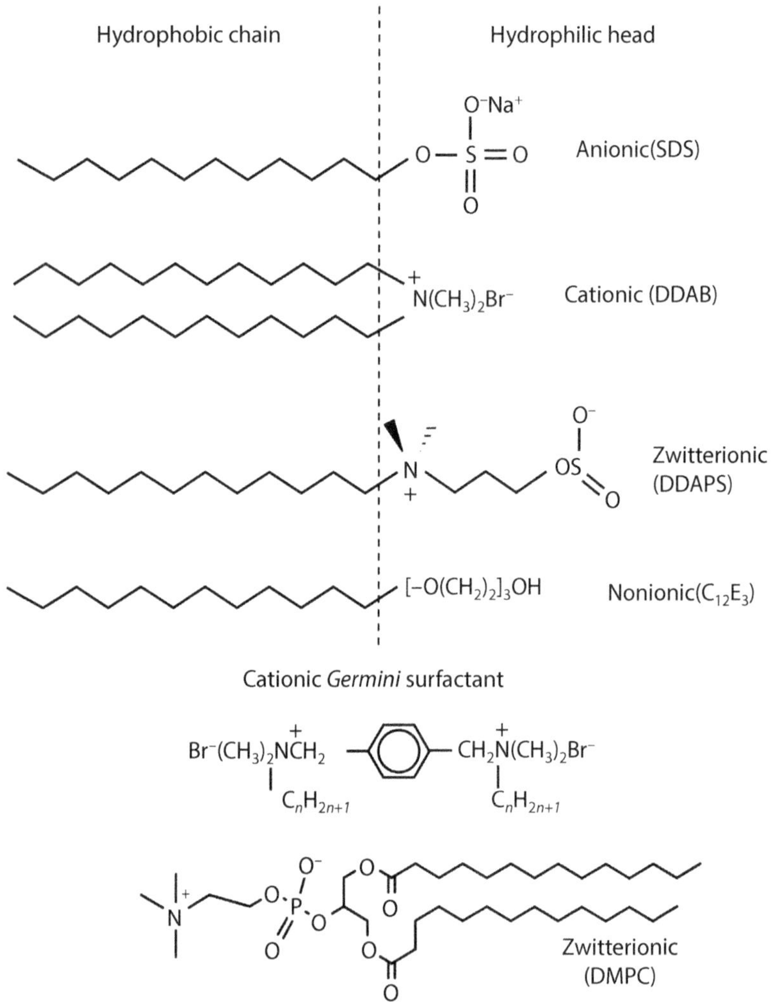

Hydrophobic chain | Hydrophilic head

Anionic(SDS)

Cationic (DDAB)

Zwitterionic (DDAPS)

Nonionic($C_{12}E_3$)

Cationic *Germini* surfactant

Zwitterionic (DMPC)

The amphiphilic nature of surfactant molecules gives them some interesting properties in water. The hydrophobic region is insoluble and the surfactant molecules are forced to accumulate on the surface and expose these nonpolar chains away from the water and toward the air. As Figure 7.14 illustrates, the number density of the molecules at the surface increases with bulk concentration. As with other adsorbates on surfaces, each surfactant has a concentration for which no more surfactant molecules can pack at the interface and a saturated monolayer is formed. The packing density of surfactant molecules within the monolayer will depend on the intermolecular interactions between neighboring head groups and the hydrophobic interactions between the tail groups. The formation of

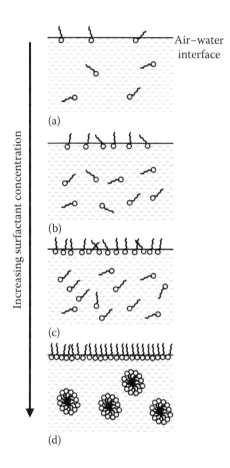

(a)

(b)

(c)

(d)

Increasing surfactant concentration

Air–water interface

Figure 7.14 Surfactant molecules in the bulk aqueous phase are in equilibrium with a monolayer at the air–water interface. The population of the molecules increases with increasing bulk concentration (a)–(c). At the CMC (d) a saturated monolayer is formed and micelles are present in the bulk phase.

the monolayer represents a thermodynamically favorable situation: polar head groups are buried in the aqueous phase and the hydrophobic tails are avoiding contact with water. This scenario explains why surfactants lower the surface tension of the air–water interface. It is easy to see that surface tension decreases as the surfactant concentration increases, but then levels off to a constant value when a saturated monolayer is formed. In Chapter 8, we see how the surface tensions of surfactant solutions are measured.

7.3.1 Micelle and microemulsion formation

What happens if we continue to increase the surfactant concentration in an aqueous solution beyond that required to form a monolayer? The molecules are unable to adsorb to the air–water interface. Instead, they self-assemble in the bulk phase into nanostructures or aggregates known as micelles, which were briefly mentioned in Chapter 6. The shape and size of these micelles depends on the structure of the surfactant. For example, SDS forms spherical micelles; each micelle is 4 nm in diameter and is composed of about 60 molecules. The number of molecules making a micelle is known as the aggregation number.

The concentration at which a full monolayer is present at the interface and micelles begin to form in the bulk solution is known as the critical micelle concentration, or CMC. The process of micellization is in some ways similar to precipitation, but the precipitate itself has a very narrow size distribution and is stable and soluble in water. This property is because the micelle structure is such that the hydrophobic chains are aggregated in the core while the polar head groups form the exterior part of the structure (Figure 7.14d). The delicate balance between inter-head group interactions and the geometry of the surfactant molecule result in a micelle with a particular shape, size, and aggregation number.

The CMC of a particular surfactant solution is very sensitive to impurities and other physical conditions. For example, increasing the temperature of an aqueous surfactant solution increases the CMC. Thermal agitation makes it more difficult for the molecules to self-assemble into micelles, so a higher concentration is required to reach the CMC. The addition of salt decreases the CMC of ionic surfactant solutions because the added ions screen the charged head groups of the surfactant, thus making it easier to form micelles. As the chain length of the surfactant increases, the CMC decreases due to the reduction in solubility of the surfactant. Table 7.2 lists the CMCs of some common surfactants.

In a nonpolar solvent, surfactant molecules can self-assemble to form "reverse micelles." In this situation, the molecules are aggregated in such a way that the hydrophobic moieties form the exterior of the micelle and the polar regions of molecules form the core of the micelle. These micelles are thermodynamically stable in a nonpolar solvent such as hexane.

Micelles in aqueous solution are capable of solubilizing small amounts of oil added to the aqueous surfactant solution above its CMC. As a result, the micelles swell with oil and increase in size (Figure 7.15). This swollen micellar phase is thermodynamically stable and is known as a microemulsion. In essence, tiny oil droplets are solubilized in water. Microemulsion phases are generally made by putting an aqueous surfactant phase in contact with an oil phase. The two phases don't mix completely, but a small equilibrium amount of the oil will enter the

Table 7.2

The CMC Values of Various Surfactants

Surfactant	CMC (mol dm^{-3})
Dodecylammonium chloride	1.47×10^{-2}
Dodecyltrimethylammonium bromide	1.56×10^{-2}
Decyltrimethylammonium bromide	6.5×10^{-2}
Sodium dodecyl sulfate	8.3×10^{-3}
Sodium tetradecyl sulfate	2.1×10^{-3}
Sodium decyl sulfate	3.3×10^{-2}
Sodium octyl sulfate	1.33×10^{-1}
$CH_3(CH_2)_9(OCH_2CH)_6OH$	3×10^{-4}
$CH_3(CH_2)_9(OCH_2CH)_9OH$	1.3×10^{-3}
$CH_3(CH_2)_{11}(OCH_2CH)_6OH$	8.7×10^{-5}
$CH_3(CH_2)_7C_6H_4(OCH_2CH)_6OH$	2.05×10^{-4}

Source: Holmberg, K., Jönsson, B., Kronberg, B., Lindman, B. 2003. *Surfactants and Polymers in Aqueous Solution*, 2nd ed., John Wiley & Sons, Chichester, West Sussex, England. With permission.

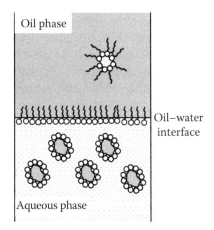

Figure 7.15 When an aqueous surfactant phase above its CMC is brought in contact with a pure oil phase, the micelles are able to solubilize small oil droplets. The aqueous phase is a microemulsion. A monolayer of the surfactant likely populates the oil–water interface. The diagram also shows a "reverse" micelle in the oil phase containing a water droplet.

aqueous phase and turn it into a microemulsion. In fact, a complex equilibrium is established between the oil phase, a monolayer at the oil-water interface, and the aqueous micellar solution. It is also possible for reverse microemulsion aggregates to be present in the oil phase.

The amphiphilic nature of surfactants is the reason why these molecules self-assemble into nanostructures, both as micelles and microemulsions, and as monolayers. The ability to adsorb at interfaces and lower surface tension is the main reason why surfactants are used in almost all detergent and cleaning products. Microemulsions, with their ability to solubilize oil, have shown promise in tertiary oil recovery. Furthermore, manipulating the molecular structure while still preserving the amphiphilic nature allows the possibility of introducing a desirable functionality into the surfactant molecules and then exploiting its ability to self-assemble into novel materials.

7.3.2 The determination of surface excess: The CMC and the cross-sectional area per molecule

Surface tension measurements of aqueous amphiphilic solutions allow one to determine the bulk concentration at which a saturated monolayer is reached. This concentration also corresponds to the CMC if a surfactant is used. Furthermore, one can use surface tension measurements to determine the cross-sectional area occupied by each molecule on the surface. Let's first consider a series of aqueous solutions of the cationic surfactant cetyltrimethylammonium bromide [$(C_{16}H_{33})N(CH_3)_3Br$], or CTAB. CTAB is an effective antiseptic agent against bacteria and fungi and is used in hair conditioning products. After measuring the concentration of each solution, we can generate a surface tension versus concentration plot, as shown in Figure 7.16. The surface tension values drop from pure water ($\sim$72 mN m^{-1}) to $\sim$40 mN m^{-1} when the bulk aqueous concentration of CTAB is 1 mM. The decrease in surface tension in this range is due to the increase in packing density as the bulk concentration is increased. Beyond 1 mM, the surface tension remains constant at 40 mN m^{-1}, which indicates that the packing density of CTAB molecules at the air–water interface is not increasing as the bulk concentration increases. At 1 mM and beyond, the surface has been saturated with CTAB molecules and a close-packed, highly oriented monolayer is present on the surface.

What happens to the excess surfactant molecules in the bulk phase when the concentration exceeds that required to form a complete monolayer on the surface (1 mM in the case of CTAB)? They are unable to remain as individual molecules in solution and instead self-assemble into nanoscale micelles. This is shown in Figure 7.16, where a break point in a plot of the surface tension versus surfactant concentration marks the CMC. The CMC of an aqueous solution of CTAB is 1 mM.

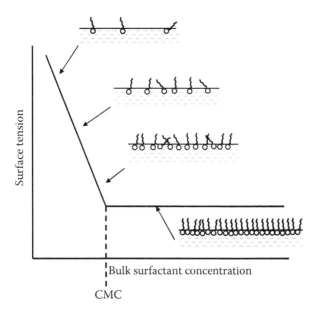

Figure 7.16 The drop in surface tension of air–water interface with increasing surfactant concentration. The intersection of the horizontal (constant surface tension) and the sloped lines corresponds to the CMC of the surfactant.

Surface tension measurements also allow the determination of a quantity known as surface excess. Consider an aqueous solution containing surfactant molecules. Let's assume that we can determine the concentration of molecules in a two-dimensional plane somewhere in the bulk phase below the air–water interface. We can then compare this bulk phase concentration to the concentration of molecules on the actual surface of the solution. Since we are considering a surfactant solution, we expect the surface concentration of surfactant molecules to be much higher than the bulk phase concentration. The difference between these two concentrations is known as the surface excess. The value tells us how much extra surfactant we have on the surface compared to a two-dimensional plane in the bulk phase. Surface excess is, to a good approximation, a measure of the concentration of molecules on the surface.

For nonionic amphiphiles, the Gibbs' adsorption equation relates surface tension (γ) to surface excess (Γ) at a given temperature (T), and is given by

$$\Gamma = -\frac{1}{RT}\frac{d\gamma}{d\ln C} \qquad (7.19)$$

where C is the surfactant concentration in solution below the CMC and R is the molar gas constant in J K^{-1} mol^{-1}. The surface excess is obtained from the slope of a plot of the surface tension versus the logarithm of the concentration. For ionic amphiphiles, a slightly different equation is used (Equation 7.20) because the presence of the counterion has to be taken into account and the surface-surfactant-counterion system as a whole must be electrically neutral.

$$\Gamma = -\frac{1}{2RT}\frac{d\gamma}{d\ln C} \tag{7.20}$$

Figure 7.17(b) shows a plot of the surface tension values versus the logarithm of the concentration for aqueous CTAB solutions. Below the CMC, the change in the slope reflects the different surface excess values at the various concentrations. At (i), the slope (or the value of $d\gamma/d\ln C$) is relatively small, indicating a small surface excess. At the higher concentration indicated by (ii), the derivative $d\gamma/d\ln C$ becomes larger, and according to Equation 7.19, the surface excess becomes larger. At higher

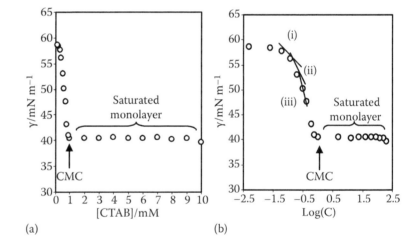

(a)

(b)

Figure 7.17 The surface tension of aqueous CTAB solutions versus (a) concentration and (b) logarithm of the concentration. The surface tension values were obtained at 20°C using the Wilhelmy plate method. Beyond the concentration corresponding to the CMC, the surface tension values become constant, indicating a saturated monolayer at the aqueous–air interface.

concentrations like point (iii), the surface excess increases even more. The largest slope occurs at the CMC and the maximum possible surface excess is achieved, thus indicating the presence of a saturated monolayer at the aqueous–air interface. Equations 7.19 and 7.20 cannot be applied beyond the CMC.

Example 7.3 Calculating Surface Excess

Consider a concentrated aqueous solution of *n*-decanol. A plot of the surface tension versus the logarithm of the concentration gives a slope of –3.23 mN/m. Use dimensional analyses to determine the units of surface excess and then calculate the surface excess. Using this value, determine the cross-sectional area per molecule of *n*-decanol at the air–water interface.

Solution   Since the logarithm term in Equation 7.19 is unitless, the derivative $d\gamma/\ln C$ has the same units as surface tension. The units of R are $J\,K^{-1}\,mol^{-1}$ and we know that $1\,N = 1\,Jm^{-1}$. Thus, according to Equation 3.19, the units of Γ are

$$\frac{1}{\left(JK^{-1}mol^{-1}\right)\cdot K}\,Nm^{-1} = \frac{1}{\left(JK^{-1}mol^{-1}\right)\cdot K}\,Jm^{-1}m^{-1} = mol\,m^{-2}$$

The surface excess of the *n*-decanol solution is

$$\Gamma = -\frac{1}{RT}\frac{d\gamma}{d\ln C} = -\frac{1}{\left(8.314\ JK^{-1}mol^{-1}\right)(298\ K)}\left(-3.23\times10^{-3}Nm^{-1}\right)$$

$$= 1.30\times10^{-6}mol\,m^{-2}$$

Assuming the solution of *n*-decanol is close to saturation, we have close to a complete monolayer of the alcohol at the aqueous–air interface. The cross-sectional area per adsorbed molecule is inversely proportional to the adsorbed amount. If the surface excess is expressed in mol/m², then the area per molecule, σ, is

$$\sigma\left(m^2/molecule\right) = \frac{1}{N_A\Gamma}$$

σ is usually expressed in nm^2/molecule. Since N_A = 6.023 × 10^{23} mol^{-1}, we have

$$\sigma\left(nm^2/molecule\right) = \frac{\left(10^9\,\dfrac{nm}{m}\right)^2}{6.023 \times 10^{23}\,\dfrac{molecules}{mol} \cdot \Gamma}$$

$$= \frac{1.6603 \times 10^{-6}\,\dfrac{nm^2 \cdot mol}{m^2 \cdot molecules}}{\Gamma}$$

$$= \frac{1.6603 \times 10^{-6}\,\dfrac{nm^2 \cdot mol}{m^2 \cdot molecules}}{1.30 \times 10^{-6}\,\dfrac{mol}{m^2}}$$

$$= 1.28\ nm^2/molecule$$

Thus, each n-decanol molecule occupies an area of 1.28 nm^2 on the surface of water.

End of chapter questions

1. Calculate the work done when the surface of water increases by 50 nm^2.

2. Surfactant molecules on the surface of water, if sufficiently dilute, can be described as a two-dimensional gas phase. If there are no intermolecular interactions between the surfactant molecules, the gas can be considered ideal. Instead of the familiar $PV = nRT$, this ideal will obey a two-dimensional ideal gas equation $\Pi\sigma = RT$, where Π is the "surface pressure" due to the surfactant molecules, and σ is the surface area per molecule. Derive an expression for the reversible isothermal work due to expansion of this two-dimensional gas. What is the work done when the gas expands and the area per molecule increases from 20 nm^2 to 40 nm^2?

Clue: The work can be determined by solving the following integral:

$$W = -\int \Pi\, d\sigma$$

3. The modified van der Waals equation is a more realistic equation describing a monolayer of lipid molecules at the air–water interface (see Question 2). This equation can be described as

$$\Pi = \frac{KT}{\sigma - \beta} - \frac{\alpha}{\sigma^2}$$

where the surface pressure (Π) is a function of the independent variables temperature (T) and the surface area per lipid molecule (σ), i.e., $\Pi\,(T,\sigma)$. K, α, and β are constants.

Derive an expression describing the reversible isothermal work due to the expansion of a lipid monolayer obeying this equation of state.

4. Charcoal is an excellent adsorbate for organic molecules. The amount of dodecanol adsorbed on this material from a toluene solution was measured at room temperature. The following data gives the equilibrium amount adsorbed on charcoal and the corresponding equilibrium concentration of dodecanol in the bulk phase.

Bulk concentration (mol dm^{-3})	0.010	0.035	0.061	0.104	0.149
Amount adsorbed (µmol g^{-1})	24.0	50.3	70.0	81.2	90.8

a. By means of a graph, show that the data fit the Langmuir adsorption model. Calculate the area occupied by each adsorbed dodecanol molecule at saturation coverage. Take the adsorption area of the charcoal to be 100 m^2 g^{-1}.

b. How would the adsorption isotherm change if the dodecanol was adsorbed from a slightly more polar solution compared to toluene?

5. This question concerns the application of Young's equation, which considers the equilibrium state of a drop of liquid on a solid surface in terms of the various surface tensions. Consider a drop of oil (an n-alkane) on the surface of solid polytetrafluotoethylene (PTFE). As expected, the oil spreads to some degree and forms a drop on the surface. The following oils were placed on the solid surface and the corresponding contact angles measured. The surface tension of each oil is also given.

Oil (n-alkane), n	cosθ	γ (mN/m)
6	0.95	18
8	0.87	22
12	0.78	25
16	0.72	27

Estimate the critical surface tension (γ_C) corresponding to complete wetting. Show that its value is simply γ_{SV}. Do you think it is true that all liquids with a surface tension less than γ_C will spread on PTFE? If so, explain.

6. The Gibbs adsorption equation can be used to obtain values of surface excess at various concentrations. The following surface tensions were measured for aqueous solutions of n-pentanol at 20°C (see graph below). A polynomial function fit to the data is also shown along with the equation representing the best fit.

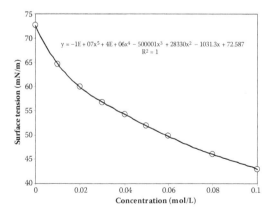

a. Calculate the surface excess concentrations and the average area occupied by each adsorbed molecule for bulk concentrations of 0.01, 0.02, and 0.04 mol dm^{-3}.

b. Plot a surface pressure versus area per molecule (Π–σ) curve for the adsorbed *n*-pentanol monolayer and compare it with the corresponding curve for an ideal gaseous film. You should say something about how the molecules are interacting.

   Clue: The compressibility of the film is

   $$Z = \frac{\Pi \sigma}{kT}$$

   This should be equal to 1 at all surface pressures for an ideal gas.

7. Consider the following plot of surface tension versus SDS concentration. The squares represent the data from a purified sample of SDS (>99.99%, recrystallized three times from ethyl acetate), and the circles show the data from a batch used as received from a popular vendor (>99.0%). The latter sample contains trace amounts of the corresponding alcohol (dodecanol).

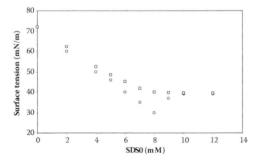

a. Estimate the approximate CMC of pure SDS. Explain how this CMC is affected by (i) a decrease in temperature, (ii) an

increase in ionic strength, (iii) an increase in pH, and (iv) the addition of a *polycation*.

b. With the aid of diagrams, explain the observation of the minimum in surface tension for the contaminated SDS sample. Explain clearly why the data differs from the purified sample.

8. In a paper in the *Journal of Chemical Education*, Bresler and Hagen reported an interesting laboratory project to investigate surfactant adsorption (Bresler, M. R., Hagen, J. P. *J. Chem. Edu.* 2008, *85*, 269–271). This problem will go through some of the steps these authors took to obtain an expression for the surface tension of a surfactant solution.

a. Rearrange the Gibb's adsorption equation (Equation 7.19) to the form

   $$d\gamma = -\frac{RT\Gamma}{C}dC$$

b. The surface excess Γ can be assumed to follow the Langmuir adsorption isotherm, especially when considering a nonionic surfactant. Express Equation 7.13 in the form

   $$\theta = \frac{K[A]}{1 + K[A]}$$

   and argue that

   $$\sigma\Gamma = \frac{K_{ad}C}{1 + K_{ad}C}$$

   where σ is the cross-sectional area of the surfactant molecules at the surface, and K_{ad} is the equilibrium constant for adsorption. What assumption has been made in saying that surfactant adsorption follows the Langmuir adsorption isotherm?

c. Solve

$$\sigma\Gamma = \frac{K_{ad}C}{1 + K_{ad}C}$$

for the surface excess, and then substitute the result into

$$d\gamma = -\frac{RT\Gamma}{C}dC$$

Integrate the resulting expression to obtain the Szyszkowski equation,

$$\gamma = \gamma_0 - \frac{RT}{\sigma}\ln(1 + K_{ad}C)$$

where γ_0 is the surface tension of pure water.

9. The Szyszkowski equation in Question 8 can be used to obtain the standard Gibb's energy of adsorption (ΔG_{ad}^o). By plotting surface tension versus bulk concentration below the CMC and then curve fitting the data, one can obtain K_{ad}. This value is related to the standard Gibb's energy of adsorption by the equation $\Delta G_{ad}^o = -RT\ln K_{ad}$. The following data was collected for a nonionic surfactant.

$C/\mu mol L^{-1}$	0.0	2.5	5.2	8.0	13.0	17.5	21.0	31.5
γ/mNm^{-1}	72.8	53.6	49.7	45.2	42.6	40.8	40.7	40.8

a. Plot γ versus $\ln C$ and determine the CMC of the surfactant.

b. Plot γ versus C and determine the parameters γ and K_{ad} using nonlinear curve fitting.

c. Determine the standard Gibb's energy of adsorption.

d. The standard Gibb's energy of micellization is given by $\Delta G_{mic}^o = RT\ln CMC$. Determine ΔG_{mic}^o.

10. Berberan-Santos commented on Bresler and Hagen's paper (Questions 8 and 9) in a letter published in the same journal (*J. Chem. Ed.* 2009, *86*, 433). Berberan-Santos pointed out that the Szyszkowski equation can be reduced to the following equation if the surfactant concentration is sufficiently low:

$$\gamma = \left(\gamma_0 + \frac{\Delta G_{ad}^o}{\sigma}\right) - \frac{RT}{\sigma}\ln C$$

Derive this equation by considering a dilute surfactant solution. Use the data given in Question 9 to plot γ versus $\ln C$. From the linear plot obtain the values of σ, ΔG_{ad}^o, and K_{ad}. Berberan-Santos stated that nonlinear fitting is preferable in general (Question 9), but may not be mandatory. Discuss this in the context of the data presented in Question 9.

11. This question concerns the stability of spherical "nanobubbles." Consider a bubble, like that formed when a soapy film on a ring is blown. The bubble has an internal pressure P_1, and a radius r. P_0 is the external pressure (e.g., the pressure of the surrounding air). At equilibrium, the bubble is stable and $dG/dr = 0$, where dr is the infinitesimal decrease in bubble radius. If $P_1 > P_0$, work must be done to ensure $dr = 0$. The change in Gibb's energy due to the change in surface area is approximately equal to

$$dG = -8\pi r \, dr\gamma + \Delta P \, 4\pi r^2 dr$$

where γ is the surface tension of the bubble and $\Delta P = P_1 - P_0$. The first term in the above equation is a measure of change in Gibb's energy due to surface tension, and the second term describes the mechanical work done against the pressure difference across the bubble surface.

Use the equilibrium condition $dG/dr = 0$ to derive the Laplace equation:

$$\Delta P = \frac{2\gamma}{r}$$

The Laplace equation essentially gives the pressure inside a stable bubble.

a. The Young–Laplace equation can be applied to nonspherical bubbles. In this expression, R_1 and R_2 represent the two principal radii of curvature. Show that the Young–Laplace equation reduces to the Laplace equation for a spherical bubble.

b. Calculate the Laplace pressure (ΔP) in units of bars for bubbles of radius 1 nm, 2 nm, 10 nm, and 1000 nm. Comment on how ΔP changes with the size of these "nano-bubbles."

c. Using your understanding of Laplace pressure, explain why small boiling chips are often added to hot reaction mixtures.

References and recommended reading

Clint, J. H. *Surfactant Aggregation*, 1992, Springer Science + Business Media, New York. This book contains some excellent chapters on micelles and microemulsion. It is easy to follow and is highly recommended for anyone interested in surfactant science.

Evans, D. F. and Wennerström, H. *The Colloidal Domain*, 2nd ed., 1999, Wiley-VCH, New York. pp. 99–153, 217–295. This is one of the finest books written on colloidal systems and includes a thorough discussion of surfactant chemistry, monolayers, and microemulsions. Chapters 3 and 5 are particularly relevant to the intermolecular interactions in nanomaterials.

Hartland, S., Ed. *Surface and Interfacial Tension: Measurement, Theory, and Applications*, 2004, Surfactant Science Series, Volume 119. Marcel Dekker, Santa Barbara, CA. This book provides a rigorous treatment of interfacial tension, film stability, and wetting phenomena. This book is recommended only for the serious student.

Holmberg, K., Jönsson, B., Kronberg, B., and Lindman, B. *Surfactants and Polymers in Aqueous Solution*, 2nd ed., 2003, John Wiley & Sons, Chichester, West Sussex, England. This is one of the most thorough and comprehensible books on surfactant chemistry. It is light on mathematical rigor but provides very thorough descriptions and examples.

Roach, P., Shirtcliffe, N. J., and Newton, M. I. "Progress in Superhydrophobic Surface Development," *Soft Matter*, 2008, 4, 224–240. This is an excellent review article on superhydrophobic surfaces and their relation to hard nanomaterials. The review contains many examples and references to methods used to construct superhydrophobic surfaces.

Rosen, M. J. *Surfactants and Interfacial Phenomena*, 3rd ed., 2004, John Wiley & Sons, New York. This book is an excellent source of data on many surfactants, including industrial synthetic surfactants, "green" surfactants, and amphiphilic molecules relevant to living systems.

Tóth, J., Ed. *Adsorption: Theory, Modeling, and Analysis*, 2002, Surfactant Science Series, Volume 119. Marcel Dekker, Santa Barbara, CA. This book contains many papers describing adsorption isotherms, including the adsorption behavior of biological molecules. The book is mathematically rigorous and is recommended only to those seriously interested in adsorption phenomena.

CHAPTER 8

Surface Characterization and Imaging Methods

CHAPTER OVERVIEW

This chapter provides a broad survey of some of the common techniques used to characterize interfacial properties and properties of nano-materials on surfaces. As discussed in Chapter 7, surfaces and interfaces have many unique properties due to their symmetry (functioning as a 2D nanosystem) and the ability for particles at interfaces to simultaneously encounter two different environments The methods introduced in this chapter range from traditional surface science tools, spectroscopic methods, and gravimetric techniques to more specialized characterization approaches such as nonlinear optical methods and interferometic techniques. The chapter focuses on the principles behind the techniques and the interpretation of data. Specific examples of the application of these methods to nanomaterials are found in Chapters 9 and 10.

8.1 SURFACE TENSIOMETRY: THE SURFACE TENSIOMETER

There are a number of methods used to determine the surface tension of liquids. One important method is to study the movement of the fluid up through a capillary tube, and this is probably the most accurate method of determining surface tension values. The basic setup is shown in Figure 8.1. From this setup, the surface tension of a particular fluid of density ρ can be calculated directly using the height h to which the liquid rises through the narrow capillary of radius r as described by Equation 8.1.

$$\gamma = \frac{rh\Delta\rho g}{2\cos\phi} \tag{8.1}$$

Figure 8.1 A fluid moving through a narrow capillary tube. The distance h the fluid travels depends on the surface tension (γ) and the contact angle (ϕ). The distance h also depends on the diameter $2r$ of the capillary.

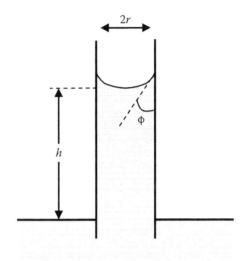

The term $\Delta\rho$ represents the difference between the density of the liquid (e.g., bulk water) and the density of the vapor (usually air). The angle ϕ is the contact angle the liquid makes against the capillary surface, and g is the acceleration due to gravity. For a narrow capillary the contact angle approaches zero, so the $\cos\phi$ term in Equation 8.1 is generally set equal to 1. The surface tension then depends on the fluid density, the radius of the capillary, and the height the fluid travels up the tube. Other methods for determining surface tension include measuring the volume of a drop detached from a narrow tube, analyzing an image of a pendant drop, and observing a jet of liquid emerging from a nozzle of elliptical cross section. A thorough treatment of these various methods can be found in Volume 119 of the surfactant science series, *Surface and Interfacial Tension: Measurement, Theory, and Applications.*

The Wilhelmy plate method is another common technique used to determine surface tension values and is the one to which we confine our attention for the remainder of this section. The Wilhelmy plate method involves measuring the forces acting on a "plate," usually a very thin piece of platinum or paper, at the liquid–air interface (Figure 8.2).

If a plate with dimensions l = length, w = width, t = thickness, and of density ρ_P is immersed to a depth h into a fluid of density ρ_L, then the forces acting on the plate are its weight, the upthrust on the submerged part of the plate due to buoyancy, and the surface tension of the liquid on the plate. The total force on the plate can be written as

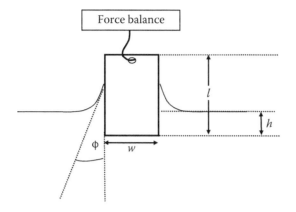

Force balance

$$F = (\rho_P lwt)g - (\rho_L hwt)g + 2(w + t)\gamma \cos \phi \qquad (8.2)$$

where g is the acceleration due to gravity. In this expression, $(\rho_P lwt)g$ is the weight of the plate, $(\rho_L hwt)g$ is the buoyant upthrust, and the surface tension contribution to the force is $2(w+t)\gamma\cos\phi$, where ϕ is the contact angle that the liquid makes on the plate as indicated in Figure 8.2.

Example 8.1 Measuring the Density from the Force Acting on a Plate

A clean plate of width w and thickness t is submerged into an aqueous surfactant solution. The solution has a surface tension value of γ mN/m. The force acting on the plate is measured as a function of immersion depth h. At $h = 2$ cm, the force is F_1, and at a depth of $h = 4$ cm, the force is F_2. Show how this information can be used to estimate the density of the surfactant solution.

Solution We can substitute the values of h and F into Equation 4.2 for the two cases and write

(a) $(\rho_P lwt)g - 0.02(\rho_L wt)g + 2(w+t)\, \gamma\cos\phi = F_1$

(b) $(\rho_P lwt)g - 0.04(\rho_L wt)g + 2(w+t)\, \gamma\cos\phi = F_2$

Subtracting these two equations gives $0.02(\rho_L wt)g = F_1 - F_2$.

Thus, by measuring the difference in force for a plate of a given width and thickness, ρ_L can be obtained.

The total force on the plate can be measured accurately by connecting the plate to a sensitive force balance. However, in order to accurately calculate the surface tension from the total force using Equation 8.2, we would need to know (also very accurately) the dimensions of the plate as well as the densities of both the plate and the liquid. Fortunately, the expression of the measured total force in Equation 8.2 can be simplified through careful manipulation of the procedure used to measure the total force on the plate, as follows.

First, before making any measurements and prior to submerging the plate into the liquid, the force balance is zeroed. Zeroing the balance eliminates the weight term from Equation 8.2. The plate is then submerged into the fluid and slowly raised until its lower edge is level with the liquid surface, thus eliminating the upthrust term because $h = 0$. In this position, Equation 8.2 is reduced to

$$F = 2(w + t)\gamma \cos \phi \qquad (8.3)$$

Using Equation 8.3, therefore, we can determine the surface tension provided that we know the contact angle the liquid makes to the plate. This is often not necessary, however, because the expression can be further simplified by realizing that as we slowly pull the plate away from the liquid surface, ϕ decreases and becomes zero (or $\cos\phi = 1$) just before detachment. Thus, if we measure the force just before detachment of the plate from the liquid, we can determine the surface tension from the dimensions of the plate:

$$\gamma = \frac{F}{2(w + t)} \qquad (8.4)$$

In order to obtain an accurate surface tension value, it is important to eliminate sources of contamination and work at a constant temperature ($\pm 0.5°C$), since surface tension is temperature-dependent. Furthermore, in order for Equation 8.4 to be used, a zero contact angle is required to exist between the plate and the liquid. This is usually achieved by using precisely cut paper plates, as the liquid can completely wet the paper, thus ensuring a zero contact angle. Platinum plates, even when cleaned thoroughly, can have their surfaces contaminated by the liquid during the first immersion, and their contact angle can vary during subsequent immersions. Flame-cleaning the plate is usually recommended when working with platinum plates.

Surface tension experiments have been pivotal to our understanding of the properties of monolayers. The technique provides direct information on packing density and the area occupied by a molecule in a monolayer. Although the area per molecule gives a qualitative indication of the thickness of the film, surface tension cannot be used to give the absolute film thickness or the absolute mass of the monolayer. Fortunately, film thicknesses as low as a fraction of a nanometer can be measured using optical methods such as ellipsometry (Section 8.3) or dual-beam polarization interferometry (Section 8.5). Gravimetric techniques also exist that can measure the mass of a monolayer on solid surfaces. The quartz crystal microbalance is a popular method used to determine the mass of a monolayer.

8.2 QUARTZ CRYSTAL MICROBALANCE

All gravimetric analyses rely on the determination of the mass of a material, or in our context, the mass of a nanofilm that has been deposited on a solid support. Gravimetric analysis is generally a precise analytical method when performed with well-calibrated balances. However, when working with nanofilms, the mass in question may well be as low as a few nanograms per square centimeter, and traditional methods for weighing such samples are not possible. The next few sections discuss methods that allow the direct measurement of the mass and the thickness of a nanofilm (or, more often, the measurement of mass-related and thickness-related parameters). These methods are the quartz crystal microbalance with dissipation monitoring (QCM-D), ellipsometry, dual beam polarization interferometry (DPI), and surface plasmon resonance (SPR). These techniques are routinely used in many research laboratories and can measure the changes in mass and thickness during nanofilm growth.

First, let's focus on the **quartz crystal microbalance (QCM)**. QCM is a powerful technique that can measure the mass of material as small as a few nanograms adsorbed to a surface. What makes this technique particularly appealing is that modern QCM instruments allow one to follow the mass deposition process as a function of time. In other words, the formation of a thin nanofilm, such as a model membrane, can be observed in real time. The quartz crystal microbalance is based on the piezoelectric characteristics of quartz, so in order to properly understand how QCM operates, let us first discuss the piezoelectric effect.

8.2.1 The piezoelectric effect

Since early times it has been known that an electric field could be induced in certain types of crystals if they underwent a change in temperature (i.e., were heated or cooled). This phenomenon was named pyroelectricity.

A Greek scientist, Theophrastus, first noticed the pyroelectric effect in 340 B.C. in experiments with tourmaline, a crystal silicate containing various metallic elements. In 1707, Johann Georg Schmidt showed that by heating or cooling tourmaline (as well as several other classes of crystals of proper geometries), a positive or negative electric field could be induced in the crystal.

In 1880, scientist-brothers Jacques and Pierre Curie discovered that they could also induce an electric field in certain classes of crystals, not by heating or cooling them, but by applying a mechanical stress. In addition, they realized that by changing the direction of the mechanical stress (i.e., a compression or expansion) they could control the sign of the electric field created. Their prior knowledge of pyroelectricity led them to this discovery, and they dubbed the phenomenon the **piezoelectric effect**, from the Greek *piezein*, meaning "to push." Today this is known as the *direct piezoelectric effect*—the ability of a crystal to produce an electric field in response to mechanical stress.

In subsequent years, the Curie brothers and other scientists began mapping out the crystalline structure requirements for piezoelectricity. They found that crystals possessing piezoelectric qualities have neither an absolute center nor a plane or axis of symmetry perpendicular to the axis of electric activity (the plane on which the electric field arises). There are currently 21 known classes of crystals that exhibit the piezoelectric effect. Three representative materials with crystal structures that exhibit piezoelectricity are quartz, cane sugar, and bone.

Another important advance in the understanding of piezoelectricity came with the discovery that a piezoelectric crystal would be physically deformed (e.g., expand or contract) in response to an applied voltage. This effect was effectively the opposite of the direct piezoelectric effect and was subsequently named the *converse piezoelectric effect*. It was also discovered that the converse piezoelectric effect could be used to induce vibrations along a crystal. By applying a precise alternating current, the deformations of the crystal could be rapidly changed between expansion and contraction, causing the crystal to oscillate.

The piezoelectric effect is the foundation of a host of technologies. One of the original applications was a piezoelectric device used during the development of sonar in which piezoelectric quartz crystals were used as transducers to detect echoes returning from underwater objects. Piezoelectric devices are also commonly used in microphones—the air pressure fluctuations caused by a sound distort the piezoelectric device in the microphone, producing an electrical signal. The converse piezoelectric effect is also used for the extraordinarily fine and accurate adjustment of devices such as the needle in a scanning tunneling microscope.

8.2.2 QCM principles

A quartz crystal microbalance operates on the principles of piezoelectricity. As we've discussed, a piezoelectric quartz crystal will oscillate at a specific resonant frequency in response to an AC voltage. (In fact, the resulting resonant frequency is so reliable that quartz crystals are used in many precision time-keeping devices.) This resonant frequency is dependent on the size and mass of the quartz crystal. If the mass of the quartz crystal is altered (such as when a molecule adsorbs to its surface), the resonant frequency at which the crystal is oscillating shifts slightly (Figure 8.3a). If the material forms a thin and rigid layer on the surface of the crystal such that motion of the crystal and adsorbate are strongly coupled, the change in mass on the crystal's surface is proportional to the shift in resonant frequency, as given by the Sauerbrey equation:

$$\Delta m = -\frac{C \cdot \Delta f}{n} \tag{8.5}$$

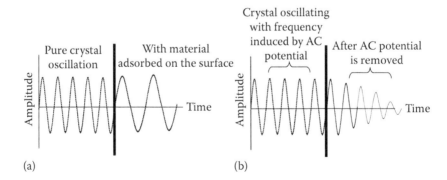

(a)   (b)

Figure 8.3   (a) The resonant frequency of a piezoelectric quartz crystal in a QCM system will oscillate more slowly when a material is adsorbed to its surface. Thus, a decrease in the resonant frequency is observed during an increase in mass adsorbed to the surface. (b) Dissipation is a measure of how quickly the QCM crystal stops oscillating after the AC potential is removed. As such, it describes the relative thickness and rigidity of thin films adsorbed to the surface of the quartz crystal. The decaying oscillations of the crystal after the circuit is broken obey an exponentially decaying sinusoidal curve. The numerical data of the decay can be fitted to Equation 8.7, which allows for the calculation of a time constant τ, which in turn can be used to calculate the dissipation of the crystal using Equation 8.8.

where m is mass, f is frequency, n is the overtone number of the crystal ($n = 1, 3, 5, 7, \ldots$), and C is a constant that depends on the specific quartz crystal.

A quartz crystal microbalance detects minute changes in mass by applying an AC potential across a quartz crystal to induce its resonance frequency and then monitoring the changes in that resonant frequency that result from the adsorption of molecules to the crystal's surface. Since those resonant frequency shifts are proportional to mass under the conditions described above, and because even very small shifts in the resonant frequency can be detected with modern electrical equipment, QCM operates as a sensitive balance or gravimetric device (hence the name microbalance). By monitoring the shifts in frequency as a function of time, QCM is able to monitor the formation of a thin nanofilm in real time. It should be noted, however, that the Sauerbrey relation shown in Equation 4.5 is often not exact when applied to surfaces at the solid–liquid interface because it was developed for oscillations in vacuum or air and only applies to rigid masses attached to the crystal. It generally under-estimates the mass adsorbed to the crystal surface under a liquid phase. Therefore, resonant frequency shifts detected using QCM should be considered a mass-related parameter, not an absolute measurement of mass adsorbed to a surface. More complex models incorporating additional parameters can improve estimation of mass for thicker and floppier/less strongly coupled films.

Example 8.2   What Are the Detection Limits of QCM?

A typical QCM-D instrument uses 5-MHz quartz crystals (meaning the crystal's fundamental resonant frequency is ~5 MHz). These crystals have a Sauerbrey constant of $C = 17.7$ ng Hz^{-1} cm^{-2}. If the QCM-D instrument is capable of detecting changes in resonant frequency of ~0.1 Hz, what is the detection limit of a typical QCM-D instrument at its fundamental resonant frequency ($n = 1$)?

Solution   We utilize the Sauerbrey relation to find

$$\Delta m = -\frac{C \cdot \Delta f}{n} = -\frac{(17.7 ng \cdot Hz^{-1} \cdot cm^{-2}(\sim \pm 0.1 Hz)}{1} = \sim \pm 2 ng \cdot cm^{-2}$$

Therefore, the detection limit of a typical QCM-D instrument is ~2ng cm^{-2}. Indeed, the QCM is an incredibly sensitive balance!

8.2.3 QCM and dissipation (D)

The quartz crystal microbalance with dissipation monitoring (QCM-D), as compared with traditional QCM, offers additional information called dissipation about the materials adsorbed to the QCM crystal surface. Dissipation is a measure of the ability of the adsorbed material to release or dissipate the energy of the oscillating QCM-D crystal. As such, it provides insight into characteristics of the adsorbed film such as its density, thickness, and viscosity. Collectively, these properties are sometimes referred to as viscoelastic properties. Dissipation is classically defined as

$$D = \frac{E_{\text{dissipated}}}{2\pi \, E_{\text{stored}}} \tag{8.6}$$

where $E_{\text{dissipated}}$ is the energy lost during one oscillation cycle and E_{stored} is the energy stored in the crystal. In essence, measuring both a frequency shift and dissipation provides complementary information about a system, similar to how optical methods can obtain complementary information from scattering/refraction and absorbance.

In practice, dissipation is determined by monitoring the time decay of the quartz crystal's oscillation when the AC potential is removed. The decay in the crystal's oscillations is an exponentially decaying sinusoidal of the form

$$A(t) = A_o e^{-t/\tau} \sin\left(2\pi \, ft + \varphi\right) \tag{8.7}$$

where t is time, τ is the decay constant, f is frequency, and φ is the phase angle. By numerically fitting the observed decay of the crystal's oscillations to Equation 8.7, the time constant τ can be obtained, from which dissipation D can be calculated as

$$D = \frac{1}{\pi \cdot f \cdot \tau} \tag{8.8}$$

In other words, dissipation can be thought of as a measure of how quickly the crystal stops oscillating when the electrical circuit is broken (Figure 8.3b). If the adsorbed thin film on the crystal's surface is thick and "floppy" (or not very rigid), then it is decoupled from the crystal's oscillations and efficiently dissipates the energy of the crystal. Consequently, the crystal stops oscillating quickly and a high value of dissipation is reported. Conversely, if the film is thin and rigid, it oscillates together with the crystal and does not dissipate the crystal's energy effectively. Hence, it takes a longer amount of time for the crystal to stop oscillating and a low dissipation value is reported. In this way, dissipation is a measure of the "floppiness" of the adsorbed film (or better, its lack of rigidity).

By using a suitable mathematical model, the dissipation and frequency shifts resulting from the adsorption of a nanofilm to a QCM-D crystal surface can be used to calculate the viscoelastic properties of the film. Viscoelasticity is the property of materials that exhibit both viscous and elastic characteristics when undergoing some kind of deformation. Viscous materials resist flow when a stress is applied. Elastic materials when stretched very rapidly return to their original state once the stress is removed. Viscoelastic materials have elements of both of these properties. While the amount of valuable information obtained using such a mathematical model is impressive, caution should be observed in its use as the model inevitably requires the input of one known parameter (such as thickness, density, etc.) in order to calculate the other viscoelastic properties. If this required parameter is unknown or not known precisely, then the resulting calculated viscoelastic properties are unreliable. For these reasons, it is often helpful to perform a QCM-D experiment in conjunction with measurements taken from ellipsometry, SPR, or DPI.

8.2.4 Modern QCM-D setup

Traditional QCM was developed in the 1960s as a method to detect the adsorption of gas molecules to surfaces, and has been used for decades for monitoring the formation of thin films from the gas phase. With more recent advances, the capabilities of QCM have been extended to detect surface adsorption at the solid–liquid interface, and QCM-D has become a valuable tool in the characterization of thin films under solution.

A typical QCM-D setup involves the quartz crystal being mounted in a flow cell with electrodes mounted on either side of the crystal, as shown in Figure 8.4. The QCM-D crystal is generally prepared with an active sensor surface such as gold, hydroxyapatite, or SiO_2. The resonant frequency of the QCM crystal is monitored as a function of time during the exposure of the crystal to a given solution under continuous flow conditions, typically in the range of 0.100–0.300 mL/min. If the solution contains materials that have some sort of affinity for the QCM crystal surface, they adsorb preferentially, increasing the mass of the quartz crystal and

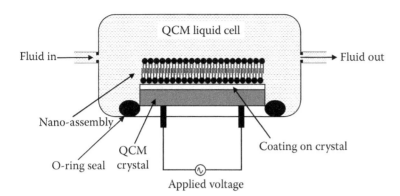

consequently producing a negative shift in the measured frequency. If the adsorbed material is sufficiently well behaved (i.e., is relatively thin and rigid), then the Sauerbrey relation can be applied, allowing us to calculate the mass of the adsorbed thin film.

QCM-D has a wide variety of applications. It is widely used in a biophysical context to detect interactions between biomacromolecules, such as between an enzyme and its substrate. It is also commonly used to examine the characteristics of polymer thin film formation as well as film behaviors under different conditions. Another application is determination of the effectiveness of different detergents to clean a given surface.

8.3 ELLIPSOMETRY

There are many approaches to determining the thickness of a nanofilm. As discussed in Chapter 6, measuring the absorbance of light through a nanofilm is one way to determine its thickness. However, determining thickness from absorbance measurements requires knowledge of both molar absorptivity and the concentration of the material comprising the film. It may be challenging to obtain the concentration of the molecules in such a film.

Film thickness can also be measured by mechanical methods such as dragging a stylus over the surface of a material. This method is the basis of the technique known as profilometry. A profilometer can be used to measure a film's thickness and roughness. In this method, a diamond stylus is moved vertically in contact with the surface of the nanofilm, then moved laterally across the surface for a specified distance and specified contact force. A profilometer can measure small surface variations in vertical stylus displacement as a function of position. A typical profilometer can measure small vertical features ranging in height from 10 nm to 50 µm. In fact, most of the world's surface finish standards are written for contact profilometers. Unfortunately, the technique is "invasive" in that the stylus makes contact with the surface and may well damage the film. Furthermore, profilometry requires a region of exposed substrate with an abrupt edge, such as a scratch in the film, to provide a reference height for calculating the thickness. There are noncontact profilometers that use light as a way of measuring the height of surface features. However, noncontact profilometry cannot measure the thickness of films on the order of 10^{-10} m (angstroms).

A technique known as ellipsometry has been used extensively to measure film thicknesses as low as a few angstroms. Ellipsometry is a noncontact and nondestructive technique used for measuring both the thickness and the refractive index of thin films on solid surfaces. Thin films ranging in thickness from a few angstroms to ~1 µm can be measured accurately and quickly using this method. Ellipsometry analyzes the state of polarization of light reflecting from a surface and uses the laws of electromagnetism (specifically, Maxwell's equations) to resolve the thickness and refractive index of the nanofilm.

8.3.1 Basic principles of electromagnetic theory and polarized light

In Chapters 5 and 6, we frequently discussed light in its classical representation as an electromagnetic wave. Ellipsometry measures the change in the polarization (direction of the electric field, not to be confused with polarization of electrons) of a light beam after its reflection from the solid surface of the sample being characterized. Although detailed coverage of the theory of ellipsometry is beyond the scope of this book, we include a brief overview of electromagnetic theory in order to better understand the basis of ellipsometry.

As previously discussed, light can be viewed as an oscillating electromagnetic field propagating through space. The oscillating field of light has two mutually perpendicular components; an electric field and a magnetic field. These are both perpendicular to the direction of light propagation, which we will arbitrarily define as the z-axis. Only the electric field component is considered here since the magnetic component does not interact appreciably with most molecules. The electric field can be represented mathematically as a complex exponential function

$$E(z, t) = E_0 e^{i\left(\frac{2\pi n z}{\lambda} - \omega t\right)} \tag{8.9}$$

Here, n is the refractive index, l is the wavelength of light, and ω is the angular frequency $(2\pi/\lambda)$ of the light. The quantity $2\pi n/\lambda$ is called the wavenumber, usually abbreviated as k (not to be confused with the imaginary component of the refractive index!), and refers to the number of wavelengths of light that fit within a certain length of material. The use of an imaginary exponential may seem counterintuitive, but by Euler's relationship,

$$e^{i\theta} = \cos\theta + i\sin\theta \tag{8.10}$$

a complex exponential is a compact method for representing periodic functions. You may have previously seen the Euler relationship written as $e^{i\pi} = 1$. Another convenient aspect of this functional form is that it can also represent exponential decay, as would be observed with a complex refractive index (absorption) and in other situations that we will discuss later in the chapter. However, a detailed discussion of the electrodynamics of light is beyond this text, and interested students are referred to texts in optics such as *Introduction to Modern Optics* by G. R. Fowles.

We can simplify the definition in Equation 8.9 by assuming that the light is passing through vacuum and that we are observing it as a single point as a function of time, such that

$$E(t) = |E| \cos(\omega t) \tag{8.11}$$

While Equation 8.11 is written as a scalar, it is actually a vector that depends on the orientation of the electric field. Thus, we consider light as an oscillating electric field whose amplitude and orientation can be represented by a line that we call the electric field vector.

The orientation of the electric field vector at a given moment in time is defined as the "polarization axis" of the light. This model of light is shown in Figure 8.5, where the length and direction of the solid arrows indicate the strength of the field and its orientation, respectively.

We now turn to the polarization of light. Usually, light emitted by most sources consists of photons whose electric fields are oriented in all directions that are perpendicular to the direction of propagation. This is unpolarized light. Conversely, **linearly polarized light** consists of photons whose electric fields are oriented in only one direction.

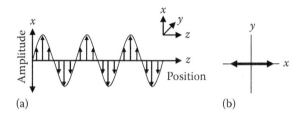

(a)   (b)

Figure 8.5   Classical representation of light as an electric field. The solid arrows' vectors represent the field's orientation and intensity. The magnitude of the vectors oscillates in time and their orientation defines the polarization axis. (a) A wave representation of the light. (b) A linear representation showing to which axis the electric field is confined.

In the simplest case of linearly polarized light, we can use a line to show the electric field oscillating along a single axis (arbitrarily defined as the z-axis), as shown in Figure 8.5(b). Linearly polarized light propagating along the z-axis can have its polarization axis oriented in any direction in the xy-plane. In other words, any source that produces linearly polarized light in the xy-plane can be thought of as being the linear combination of two vector components oriented along the x- and y-axes. The polarization axis of linearly polarized light is determined by the relative magnitudes of the two components. For linearly polarized light, the component light sources must have identical frequency and must also be in-phase with one another. By "in-phase" we mean that the minima and maxima of the electric field oscillations for each component must line up. A representation of linearly polarized light formed by a linear combination of two mutually perpendicular vector components is shown in Figure 8.6.

While we are on the topic of plane-polarized light, it is appropriate to introduce some common terminology when considering the reflection of light from a planar surface (Figure 8.7). For light incident on this surface,

Figure 8.6 (a) Linear and (b) wave representations of linearly polarized light of the same amplitude but various orientations. The polarizations are along (i) the x-axis and (ii) the y-axis. Vector addition of (i) and (ii) creates a linear polarization 45° from the x-axis (iii).

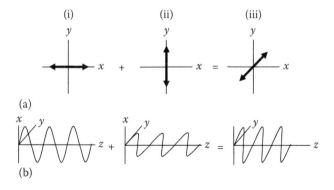

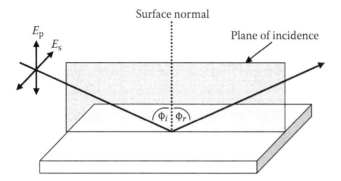

Figure 8.7 Light reflection off a planar surface. The plane of incidence contains the incident beam, the surface normal, and the reflected beam. If the polarization of the light is along the plane of incidence, then it is called p-polarized light. If the polarization vector is perpendicular to the plane of incidence, then the light is called s-polarized light.

we can define the plane of incidence as the plane that contains the surface normal and the light beam before and after the reflection (called the incident beam and reflected beam, respectively). If the electric field is polarized along the plane of incidence, we refer to the light as being p-polarized. Conversely, if the electric field is polarized along a plane perpendicular to the plane of incidence, we call the light s-polarized. A combination of in-phase s- and p-polarized electric fields can produce linearly polarized light with the electric field vector pointing at some angle within the xy plane. For example, if two s- and p-polarized light beams are completely in-phase and if they each have electric field vectors of the same magnitude, then upon addition the resulting beam would be linearly polarized with a polarization axis that would be 45° with respect to the plane of incidence.

Let's now consider what would happen if the two electric field vector components making up a polarized light beam were not in-phase. First, let's talk about a special case—when the two components are 90° out of phase or when one of the oscillations is a quarter of a wavelength ahead of the other. When viewed along the direction of propagation, the resulting polarization has constant magnitude (the length of the vector is the same), but its orientation changes in time such that the tip of the polarization vector traces a circular path in time. This kind of light is referred to as **circularly polarized light** and is depicted in Figure 8.8a and b. Depending on the relative phases of the electric field components, the tip of the polarization vector may trace out either a left- or right-handed screw and the corresponding light is described as right- or left-circularly polarized.

If two out-of-phase light waves combine (and not with a phase difference of 90°n where n is an integer), then **elliptically polarized light** is formed. The linear combination of the two vector components results in a polarization vector that traces an ellipse in time when viewed along the

Figure 8.8 (a) Two mutually perpendicular electric field components that are 90° out of phase will combine to generate circularly polarized light. (b) Vector addition of the two components leading to circularly polarized light. (c) Vector addition of the two components leading to elliptically polarized light.

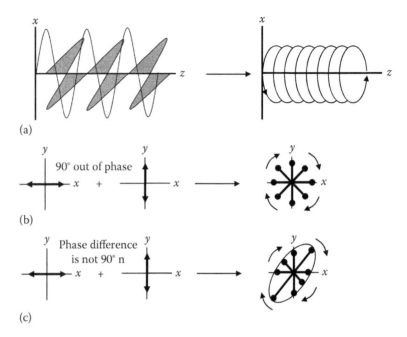

axis of propagation. Elliptically polarized light is depicted in Figure 8.8c. As with circularly polarized light, elliptically polarized light can be left- or right-handed depending on the relative phases of the components. The aspect ratio of the polarization ellipse is referred to as the ellipticity of the light. It should also be mentioned that elliptically polarized light can be produced by the linear combination of orthogonal components of circularly polarized light as well as linearly polarized light. In fact, linearly and circularly polarized light can be considered special cases of elliptically polarized light with aspect ratios of infinity and one, respectively.

8.3.2 Basic principles of ellipsometry

Now that we've discussed the electromagnetic nature of light and polarization, let's turn to our discussion of ellipsometry. Ellipsometry involves the reflection of light from a surface or the interface between two mediums with different refractive indices. The refractive index (n), previously discussed in Chapters 5 and 6, is the ratio of the speed of light in vacuum and the speed of light in a material. While, as discussed in Chapter 6, it is a complex number, where the imaginary component (k) is related to the extinction coefficient (absorption) of the material, we will limit our discussion of ellipsometry to situations in which absorption at the wavelength used for the measurement is insignificant; more complex models are required for absorbing films.

Consider a model reflective surface coated with a nanofilm of thickness d (Figure 8.9). For now we will limit our consideration to the interaction of an incoming light beam from a light source with the first surface (i.e., the interface between n_1 and n_2). In this model system, when light passes from one medium to a second medium, several phenomena occur at the interface. Some of the light is reflected from the surface and some enters the second medium.

When linearly polarized light reflects off a surface, there is a phase shift in both the parallel and perpendicular components (i.e., the s- and p-polarized components). As before, parallel and perpendicular are in relation to the plane of incidence of the incident light beam.

There may also be an amplitude difference between incident and reflected beams in both components. In fact, the phase shift and amplitude differences are usually not the same for both components. As a result, the reflected beam is elliptically polarized. The ellipticity depends on the optical properties of the substrate (reflecting surface) as well as the optical properties and thickness of any overlying films.

It is useful to obtain a parameter that describes how much each of the s- and p-components of the incident light are reflected or transmitted. This information is given by the Fresnel reflection coefficient, r, which is the ratio of the amplitude of the reflected wave to the amplitude of the incident wave. For a single interface, the Fresnel coefficients for s- and p-polarized light are equal and are given by

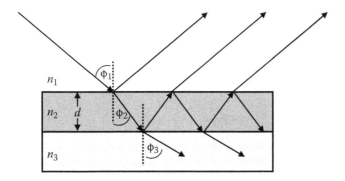

Figure 8.9 The reflection, refraction, and transmission of light through a model multiple interface system. In our case, the layer of thickness d can be a nanofilm assembled on a solid support. The n_i represent the refractive index of each phase.

$$r_{12}^s = \frac{n_1 \cos \phi_1 - n_2 \cos \phi_2}{n_1 \cos \phi_1 + n_2 \cos \phi_2} \qquad (8.12)$$

$$r_{12}^p = \frac{n_2 \cos \phi_1 - n_1 \cos \phi_2}{n_2 \cos \phi_1 + n_1 \cos \phi_2} \qquad (8.13)$$

In the two equations, the subscripts refer to medium 1 and medium 2 and the light transmitted through the interface is ignored.

So far, our discussion of ellipsometric principles has focused on the interaction of light with two media and a single interface. Now let's consider the rest of Figure 8.9, which shows a model of light interacting with multiple interfaces, in this case a reflective surface coated with a nanofilm of thickness d. The situation is only a bit more complex than for a single interface. The main source of complexity is that at each interface a light wave encounters, some of the light is reflected back from the interface and some is transmitted through it. Referring to Figure 8.9, we can realize that the result of this partial reflection/partial transmission is that some of the light that enters the thin nanofilm is internally reflected, "bouncing" between medium 1 and the reflective substrate surface. Furthermore, the intensity of the light inside the thin nanofilm eventually decays as it releases some light back into medium 1 at each "bounce." Each of these transmissions back into medium 1 is successively smaller and leads to a series of partial waves that combine to give a resultant total reflected wave. Therefore, our calculation of the amount of light reflected back into medium 1 from a system of multiple interfaces such as in Figure 8.9 must be a measurement of this total reflected wave and account for all of the small partial waves. This calculation is performed by modifying the Fresnel coefficients into total reflection coefficients R for multiple interfaces. For a three-layer system (e.g., a substrate, film, and air), these coefficients are given by

$$R^p = \frac{r_{12}^p + r_{23}^p e^{-i2\alpha}}{1 + r_{12}^p r_{23}^p e^{-i2\alpha}} \qquad (8.14)$$

$$R^s = \frac{r_{12}^s + r_{23}^s e^{-i2\alpha}}{1 + r_{12}^s r_{23}^s e^{-i2\alpha}} \qquad (8.15)$$

where

$$\alpha = 2\pi \left(\frac{d}{\lambda}\right) n_2 \cos \phi_2, \, i = \sqrt{-1}$$

and λ is the wavelength of the incident light in a vacuum. The total reflection coefficients for each component (p or s) are the ratios of the reflected wave amplitude to the incident wave amplitude. We see that the equations for these total reflection coefficients each incorporate an exponential decay term, $e^{-i2\alpha}$, which accounts for the increasingly smaller partial waves that are produced by the partial reflection/partial transmission at the nanofilm/medium 1 interface. We also notice that if $d = 0$ (or if there is no nanofilm on the surface) then Equations 8.14 and 8.15 reduce to Equations 8.13 and 8.14, as expected.

8.3.3 Obtaining the thickness of films: Optical parameters Del (Δ) and Psi (ψ)

Section 8.3.2 covered the necessary background needed to understand how ellipsometry can be used to yield thickness values for thin nanofilms. At this point, we need to define two parameters (δ_1 and δ_2) that describe the change in phase as light is reflected off a surface. Let δ_1 be the phase difference between the p-polarized component and the s-polarized component of the incident light. Let δ_2 be the phase difference between the p-polarized and the s-polarized component of the reflected light. We can now define one of the most important optical parameters used in ellipsometry, the parameter Del (Δ), as the phase difference between the p-polarized and s-polarized components of the incident light upon reflection. In other words, Δ is the resulting change in the phase difference between the s and p waves as the light is reflected from the sample (Equation 8.16).

$$\Delta = \delta_1 - \delta_2 \tag{8.16}$$

The two components (p and s) making up the incident light each have a given amplitude (length of the electric field vector), and those amplitudes may also change upon reflection. These amplitude changes are given by the total reflection coefficients in Equations 8.14 and 8.15. The second fundamental ellipsometric optical parameter psi (ψ) can be defined in terms of these coefficients and is given by the equation

$$\tan \psi = \frac{|R^p|}{|R^s|} \tag{8.17}$$

where ψ is defined as the angle whose tangent is the ratio of the magnitudes of the total reflection coefficients. We define the additional complex quantity ρ as the ratio of the total reflection coefficients, or

$$\rho = \frac{R^p}{R^s} \tag{8.18}$$

Using these three equations, we can now present the fundamental equation of ellipsometry:

$$\rho = \frac{R^p}{R^s} = \tan \psi e^{i\Delta} = \tan \psi (\cos \Delta + i \sin \Delta) \qquad (8.19)$$

Δ and ψ are the experimental quantities measured by the ellipsometer and fitted to a computer model to yield refractive index and film thickness (which are embedded in the equations for R^p and R^s).

Most ellipsometry experiments are performed on nanofilms at the solid-air interface. Silicon is a convenient substrate due to its availability, well-known optical constants, and ease of functionalization. However, many materials may be used, although materials that are transparent at the wavelength used for the experiment may require careful selection of observation angle and addition of a low-reflectivity opaque backing (which may be as simple as a piece of electrical tape) to limit reflection from the back of the substrate. The determination of refractive index and film thickness for these samples is often done as follows. Suppose we have a substrate covered by a film as in Figure 8.9. Generally, Δ and ψ are obtained for the bare substrate and the instrument determines the refractive index information of silicon and the native SiO_2 layer directly above it. This is done prior to any film deposition. A table of Δ and ψ values as a function of film thickness (called a del/psi trajectory) is determined. The del/psi trajectory is obtained using a computer program separate from the ellipsometer, although some instruments have programs to compute del/psi trajectories integrated with the instrumentation. The film is then deposited on the substrate and Δ and ψ are obtained for the substrate and film. The unknown thickness of the deposited film may then be obtained by comparison of the Δ and ψ values with the calculated del/psi trajectory for the bare substrate. The optical constants of the thin film, such as the refractive index, must be input in order for the program to produce the trajectory. The user makes an educated guess at these values.

Although the ellipsometer accurately determines Δ and ψ, these values are meaningless unless the program used to calculate the del/psi trajectory assumes the correct model. The model used is typically a two-layer model, such as that shown in Figure 8.9 for a silicon dioxide/silicon wafer substrate.

8.3.4 The ellipsometer

The layout of a typical ellipsometer is illustrated in Figure 8.10. In short, a laser beam is appropriately polarized and then reflected off the substrate at some angle; 70° is a common choice of observation angle. Some instruments allow selection of observation angle; the angle that gives the strongest and most reliable signal can depend on the system being measured. The polarization state of the reflected beam is measured using the appropriate optical equipment and its intensity is measured using a light-detecting photomultiplier tube.

Most common ellipsometers are usually equipped with a single-wavelength light source such as a helium–neon laser that emits a coherent beam of red light at a wavelength of 632.8 nm. However, because the simple models that we have discussed require that the sample does not absorb the incident light, some ellipsometers can make measurements at multiple wavelengths to allow selection of a nonabsorbed wavelength or if multiple wavelengths are not absorbed, more accurate modeling of the thickness and refractive index. Even more advanced spectroscopic ellipsometers have been developed that can rapidly scan over many wavelengths, producing a del/psi trace akin to a optical spectrum. These traces can be fit to more complex models to obtain the thickness and both real (n) and imaginary (k) components of the refractive index, allowing use on strongly absorbing materials. Even more accurate fits can be made by combining spectroscopic data from multiple observation angles in a technique called variable angle spectroscopic ellipsometry (VASE). These instruments can fit a series of psi/del traces to oscillators and then use Equation 6.7 to

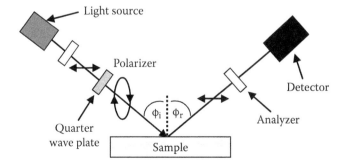

Figure 8.10 Layout of a typical ellipsometer. The angle of incidence (ϕ_i) is typically 70°. The light source is a helium–neon laser, and the detector is a photomultiplier tube.

simultaneously obtain the components of the refractive index. Figure 8.11 shows an example of data obtained for a thin film of the dye YLD-124 in poly(methylmethacrylate), illustrating the power of modern ellipsometric techniques.

However, to illustrate the utility of ellipsometry for characterizing film thicknesses, we will consider the simplest case of an instrument operating at a single wavelength. The polarizer linearly polarizes the coherent unpolarized light beam from the helium–neon laser. The polarizer is a special filter that transmits light only if the polarization axis of the light lines up with that of the analyzer. If the light does not line up with the polarization axis of the polarizer, then the light is separated into its parallel and perpendicular components relative to the polarization axis and only the parallel component is transmitted. The linearly polarized beam then passes through another optical element, the quarter-wave plate. The quarter-wave plate is also commonly referred to as a retarder. Its purpose

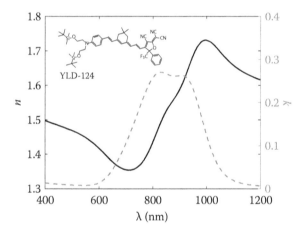

Figure 8.11 The real (n) and imaginary (k) components of a film containing YLD-124 dye embedded in poly(methyl methyacrylate), commonly known as PMMA or Plexiglas. The dye has an absorbance maximum near 800 nm. Note how n changes rapidly in the region where k is significant, representing a transition from non-resonant polarization to absorption. (Data provided by Dr. Delwin Elder, University of Washington. More information on YLD-124 can be found in Baer-Jones, T. et al. *Optics Express*, 2005, 13: 5216–5226.)

is to transform the linearly polarized beam into an elliptically polarized beam. The wave plate is an anisotropic material whose refractive index depends on the orientation of the propagating wave. P-polarized waves and s-polarized waves travel with different speeds through such a material. The thickness of the wave plate can be chosen to yield a beam whose components are exactly 90° out of phase with one another.

When linearly polarized light reflects off a surface, as previously explained, elliptically polarized light is typically produced. The rotating null ellipsometer produces an incident beam of varying ellipticities by varying the polarizer angle prior to the wave plate. When the beam reflects off the sample, the ellipticity of the beam changes. If the ellipticity is just right, then the change produced in the polarization of the beam by reflection from the sample produces a linearly polarized beam. The analyzer, which is identical to the polarizer, is then rotated until the polarization axis of the analyzer is perpendicular to the polarization axis of the reflected beam. The polarizer and analyzer angles are rotated sequentially until a null (angle at which the signal at the detector is zero) is located. These angles are used to determine the optical parameters Δ and ψ for the sample. In practice, the instrument completes this computation and the values for Δ and ψ are output.

A beautiful illustration of ellipsometry involves characterizing the sequential deposition of a polycation and a polyanion on a silicon surface. Polyelectrolyte nanoassemblies (the protocols and applications) are discussed in Chapter 10. Essentially, a clean silicon substrate is immersed into a solution of a polycation for about 5 minutes. It is then removed, washed, dried, and then immersed into a solution of a polyanion. The adsorption of the polyelectrolytes is driven by electrostatic attraction for each other and the resulting assembly can be described as a polyelectrolyte bilayer. The procedure can be repeated to form many layers.

After the construction of each layer, ellipsometry can be used to determine the film thickness. The thickness can be determined to about 0.2 nm. Usually some effort has to be made to ensure that the thickness is determined from the same area on the substrate. Figure 8.12 shows how the thickness of a layer-by-layer polyelectrolyte nanoassembly changes with each successive layer.

Figure 8.12 Thickness data as determined by ellipsometry for the layer-by-layer construction of a typical polyelectrolyte nanoassembly. The bilayer represents a layer of polycation complexed with a layer of polyanion.

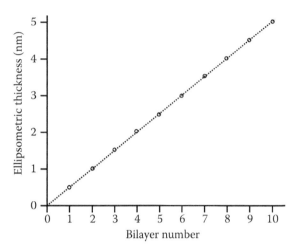

8.4 OTHER TECHNIQUES FOR MEASURING THICKNESS AND REFRACTIVE INDEX

Other techniques aside from ellipsometry exist to determine the refractive index and thickness of thin nanofilms based on the interaction of polarized light with a surface. We will discuss two such techniques, surface plasmon resonance (SPR) and dual-polarization interferometry, along with the optical principles necessary to understand these techniques.

8.4.1 Reflection phenomena at interfaces

The extent to which light is transmitted or reflected at interfaces depends on the refractive indices of the materials composing the interface as well as the angle of incidence of the light. We previously discussed reflection at interfaces in the context of ellipsometry, but two other key phenomena occur at such interfaces that can be exploited to characterize the properties of those interfaces or use those interfaces to perform a desired function such as sensing or data transmission. Those phenomena are total internal reflection (TIR) and evanescent waves.

8.4.1.1 Total internal reflection

When light is shone on the boundary between two materials with different refractive indexes ($n_1 \neq n_2$), the light passes through the boundary into the second medium and is refracted toward the surface normal if $n_2 > n_1$ and away from the surface normal if $n_1 > n_2$. This refraction obeys Snell's law:

$$n_1 \sin \theta_1 = n_2 \sin \theta_2 \qquad (8.20)$$

where n_1 and n_2 are the refractive indexes of the two substances, θ_1 is the angle of incidence, and θ_2 is the angle of refraction. In cases where the refractive of the second medium is less than that of the first (i.e., $n_1 > n_2$), an angle exists called the critical angle θ_{critical}, where the angle of refraction is 90°, or where the light in the second substance is refracted along the interface between the two substances. The critical angle can be calculated using Snell's law and basic algebra as

$$\theta_{\text{critical}} = \arcsin\left(\frac{n_2}{n_1}\right) \qquad (8.21)$$

When incident light strikes the interface between the substances at an angle that is greater than the critical angle, the light is reflected from the surface. This phenomenon is called **total internal reflection**.

8.4.1.2 Evanescent waves

A description of total internal reflection using classical physics says that the energy of the incident light is totally reflected by the interface between the two materials. However, some of the electric field from the light penetrates into the lower-index material. This "portion" of the light that enters the other medium is called an **evanescent wave**. While the penetration of the electric field into the lower-index material can be derived from Equation 8.9, such a derivation is beyond the scope of this text. However, recall that Equation 8.9 allows for the electric field to be either oscillating or exponentially decaying; the evanescent field decays exponentially as a function of distance,

$$E_x = E_0 e^{-x/d_p} \qquad (8.22)$$

where E_x is the electric field amplitude of the evanescent wave at a distance x from the interface, E_0 is the electric field at the interface, and d_p is the penetration depth defined as the distance at which E_0 is reduced to $1/e$ of its original value. If the conditions of TIR that generate an evanescent wave are known, then d_p can be calculated as

$$d_p = \frac{\lambda}{2\pi n_1 \sqrt{\sin^2\theta_{\text{incidence}} - \left(\frac{n_2}{n_1}\right)^2}} \qquad (8.23)$$

where $\theta_{incidence}$ is the angle of incidence and n_1 and n_2 are the refractive indexes of the two mediums. We see that the penetration depth d_p of the evanescent wave can be calculated for a given wavelength of light. In the visible region, the values of d_p typically range from 50 to 100 nm. Thus, the evanescent wave can be an excellent probe of the area near the boundary between the two substances (i.e., it is an excellent probe of surface modifications), and as such serves as the basis for a variety of nanomaterial characterization techniques such as SPR (Section 8.4.2), DPI (Section 8.4.3), and attenuated total reflection Fourier transform infrared spectroscopy (ATR-FTIR).

8.4.2 Surface plasmon resonance

SPR is another optical method that is employed to detect changes in thickness and refractive index of very thin organic films adsorbed to a metallic surface. SPR is often used to detect interactions between molecules, and it has emerged as perhaps the most widespread "surface method" for detecting and quantifying interactions between biological macromolecules at the nanoscale.

8.4.2.1 Principles of SPR

When incident light strikes the interface between a substance with a high index of refraction and another substance with a lower index of refraction, the light is completely reflected (total internal reflection) as long as the angle of the incident light is greater than that of the critical angle (Equation 8.21). Total internal reflection is normally observed when visible light is shone upon the interface between a glass prism ($n = \sim1.5$) and water ($n = \sim1.3$) at $\theta_{incidence} > \theta_{critical}$. However, if the surface of the prism facing the aqueous solution is coated with a thin layer of silver or gold as shown in Figure 8.13, then total internal reflection is not always observed. This loss of total internal reflection occurs because some of the incident light is "channeled" into the metal–water interface where it generates oscillating waves of surface charge density that move along that metal surface. These oscillating waves of surface charge density are called surface plasma waves or surface plasmons, and the phenomenon of their creation serves as the basis for SPR sensors.

The creation of these surface plasmons is angle-dependent, meaning that an angle of the incident light (greater than $\theta_{critical}$) exists at which the generation of the surface plasmons reaches a maximum. This angle is defined as the surface plasmon resonance angle θ_{spr} (or SPR angle).

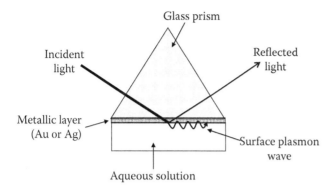

Figure 8.13 The generation of a surface plasmon wave. When polarized light is shone at the correct angle through a glass prism onto a metal–water interface, then a surface plasmon wave is generated that propagates along the interface. This results in a reduction of intensity of the reflected light.

The SPR angle therefore represents the greatest "channeling" of the incident light into the metal–water interface and consequently the greatest reduction in the intensity of the reflected light. So by measuring the reduction in reflectance as a function of the angle of incidence, the SPR angle (or angle at which a global minimum in reflectance is observed) can be determined.

Of more practical use, the electric field associated with the surface plasmons is not completely contained in the metal–water interface, but stretches slightly into the surrounding media, decaying exponentially as it extends away from the interface. Because of this penetration into each medium, any changes to the refractive index near the metallic surface alter the properties of the electric field, which in turn alters the SPR effect. Therefore, the SPR angle is sensitive to even very small changes in the index of refraction near the metal–water interface, such as might occur when a protein or polymer film is adsorbed to the metallic surface. In fact, for a given wavelength λ of light, the change in θ_{spr} is related to the change in refractive index at the surface $\Delta n_{surface}$ and the change in the thickness of the thin film Δd by

$$\Delta\theta_{spr}(\lambda) = c_1 \Delta n_{surface} + c_2 \Delta d \qquad (8.24)$$

where c_1 and c_2 are constants.

An SPR sensor typically operates by detecting changes in the SPR angle during the adsorption of molecules to the metal surface. Because $\Delta\theta_{spr}$ is a function of $\Delta n_{surface}$ and Δd, which in turn are a result of molecular interactions at the surface, an SPR sensor serves as a sensitive method to detect binding events of molecules to a surface. It should also be noted that $\Delta\theta_{spr}$ is typically reported in resonance units (RU), where 1000 RU corresponds to a change of 0.1°. For common analytes such as proteins, the correlation between $\Delta\theta_{spr}$ and the amount of substance adsorbed to the surface has been determined and a 1000-RU shift is approximately equivalent to an adsorption of 0.1 ng/mm^2 for most proteins.

It should be noted from Equation 8.24 that the SPR angle is a function of both the thickness d and the index of refraction n of the thin film, meaning that neither parameter can be extracted by itself without making certain assumptions (i.e., a change in thickness could be extracted if it were assumed that the refractive index remained constant). For this reason, the change in SPR angle is said to measure the *effective refractive index* or a *thickness- and refractive index-related* parameter analogous to the ψ and Δ parameters measured by an ellipsometer. In order to resolve thickness and absolute refractive index simultaneously, another method must be used, such as dual polarization interferometry or spectroscopic ellipsometry.

8.4.2.2 SPR instrument setup

A schematic diagram of a common SPR sensor (such as the prevalent Biacore instrument) is shown in Figure 8.14. The primary component is a prism that has been coated with a gold sensor surface so that surface

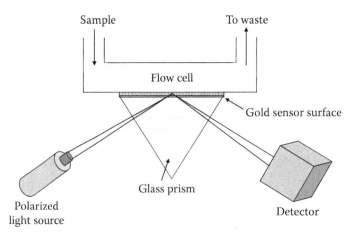

Figure 8.14 Typical setup of an SPR instrument. A polarized light source is used to excite the SPR effect at the gold–water interface. The detector monitors the changes in the intensity of the reflected light as a function of angle of incidence in order to detect changes in the SPR angle that may result from the adsorption of a thin nanofilm to the gold sensor surface. A flow cell allows for the easy exposure of the gold sensor surface to the desired sample solution.

plasmons are generated when polarized light of a particular wavelength is shone on it at the SPR angle. This gold sensor surface serves as the bottom of a flow cell through which solutions of various analytes can be passed. If the analytes interact with the sensor surface (or with molecules that have been pre-adsorbed to the surface), then a change in the SPR angle is observed by the detector and recorded by the computer software. The flow cell itself is typically very small (20–60 nL) and common flow rates are on the order of 1–100 μL/min.

While SPR sensors can be employed for a variety of purposes that involve the characterization of ultrathin films, they are primarily used as biosensors—to detect interactions between proteins and substrates, between strands of DNA, and between drug molecules and protein targets. SPR can also be used to monitor the formation of self-assembled monolayers on gold surfaces, particularly the formation of alkanethiol monolayers.

The SPR effect is not limited to planar surfaces, but can also be observed using gold or silver nanoparticles in solution. In this case, the nanoparticles have a wavelength-dependent absorbance that is a result of the excitation of surface plasmons within the nanoparticles, and this wavelength is measured rather than an angle-dependent response as with planar SPR (which would be difficult to measure for a nanoparticle). The wavelength at which the maximum absorbance is observed is a function of the particle size and particle shape. For example, spherical gold nanoparticles with a diameter of ~13 nm exhibit a maximum absorbance at 520 nm.

The wavelength at which the SPR effect is maximized also depends on the dielectric environment (i.e., the refractive index) near the nanoparticle surface and on the distance between neighboring nanoparticles. These two properties have allowed gold nanoparticles to be used as sensitive biosensors. For example, short pieces of DNA called oligonucleotides have been covalently attached to gold nanoparticles through thiol chemistry, and when complementary DNA is introduced into the solution, the oligonucleotide-coated nanoparticles aggregate as they hybridize to the strands in solution. This aggregation causes the distance between neighboring nanoparticles to decrease and a blueshift in the wavelength of maximum absorbance is observed. In this manner, gold nanoparticles can be used to determine the complementarity of two DNA sequences. Similar methods have also been developed to examine the changes in refractive index that occur on the surface of a gold nanoparticle when something adsorbs to its surface—for example, when an antigen binds to a surface-bound antibody.

8.4.3 Dual polarization interferometry

Unlike other surface characterization techniques such as QCM-D and SPR whose measurements offer only indirect estimations of the mass, density, or thickness of a thin film, dual polarization interferometry (DPI) is an optical technique that can provide accurate, simultaneous measurements of multiple film parameters. DPI uses two independent measurements of the effective refractive index of a thin nanofilm to simultaneously determine the mass, density, and thickness of that nanofilm. A variety of other waveguide spectroscopic techniques exist that are similar to DPI, such as optical waveguide lightmode spectroscopy (OWLS) and coupled plasmon waveguide resonance (CPWR) spectroscopy, but only DPI is discussed here.

8.4.3.1 Waveguide basics

As may be recalled from our discussion of the critical angle of reflection, if light is shone on the boundary between two substances, the light is totally reflected within one substance if it is shone at an angle greater than the critical angle ($\theta_{critical}$) and if the refractive index of the second substance is less than that of the first (Equation 8.21). If multiple interfaces are present, this phenomenon can be exploited to confine light within one layer of the system.

Consider a layer of a substance with one refractive index sandwiched between two layers of a substance with a lower refractive index. In such a setup, we would expect that light shone at an angle greater than $\theta_{critical}$ could undergo total internal reflection inside the middle layer, alternately reflecting off the top and bottom layers as it passed through. Indeed, such behavior is observed and the setup is called a waveguide, with the middle layer termed the waveguide core and the top and bottom layers called the cladding regions, as shown in Figure 8.15. When light is shone on the edge of a waveguide, total internal reflection can occur and the light passes through the core and emerges from the other side of the waveguide. A common application of total internal reflection is in fiber optic cables, which transmit light over long distance by confining it within a high refractive index fiber surrounded by a cladding layer with a lower refractive index.

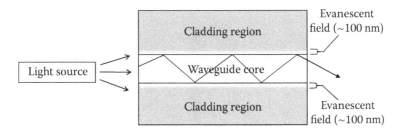

Figure 8.15   The setup of a basic waveguide. A waveguide is typically composed of a waveguide core sandwiched between two cladding regions. Light is transmitted through the waveguide core via total internal reflection. At each surface between the core and the cladding regions, an evanescent field is generated that extends ~100 nm into the cladding region.

When light undergoes total internal reflection within a waveguide, the light is not completely contained in the core, and some of the electric field penetrates into the cladding as an evanescent field (Equation 8.22), which decays exponentially with depth and is negligible beyond a distance of ~100 nm. Because of the interaction that occurs between the evanescent field and the cladding region, the speed at which the light propagates through the core of the waveguide depends slightly on the refractive index of the cladding region near the interface with the core. Thus, light traveling through a waveguide is sensitive to changes to the refractive index that occur within the first hundred nanometers of the core's surface, such as when a thin nanofilm is deposited on the surface of the waveguide core. Obtaining useful information on the refractive index of thin films deposited on the waveguide requires accurately measuring the speed of light through the waveguide; such measurements can be obtained using a technique called interferometry.

8.4.3.2 Waveguide interferometry and the effective refractive index

Interferometry is the study of the ways in which light waves interact or interfere with each other. Perhaps the most well-known example of early scientific interferometry is the diffraction pattern observed by Thomas Young in his double-slit experiments. Young noticed that if he shone a light beam on a screen that contained two closely spaced slits, then a characteristic interference pattern of alternating light and dark bands could be observed on a second screen located behind the first, as shown in Figure 8.16. This interference pattern could be explained by appealing to the wave nature of light—the light bands represented regions where the light waves emerging from each slit interfered with each other

Figure 8.16 The classic Young's double-slit experiments. Interference between the two wavefronts emerging from the double slit produce an interference pattern on a screen placed some distance away.

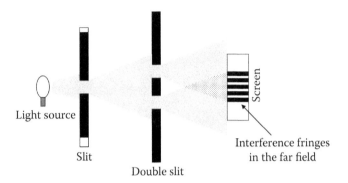

constructively, whereas the dark bands were the result of destructive interference of the light waves.

It is possible to construct a waveguide-based interferometer analogous to Young's interferometer using a setup as shown in Figure 8.17. In this type of waveguide, there are two waveguide cores rather than one, and the cores are separated by a thin cladding region. The top waveguide core is termed the sensing waveguide and the bottom core is termed the reference waveguide. If a broad beam of light is shone on the edge of this type of "waveguide stack," the light is totally internally reflected through both of the core regions and emerges from the other side to generate an interference pattern, as shown in Figure 8.17.

This waveguide interferometer can be used to determine the refractive index of a small nanofilm if the top cladding region is replaced by a sensing region (generally an aqueous solution maintained by a fluidic cell, as shown in Figure 8.17). In this setup, light travels through the sensing

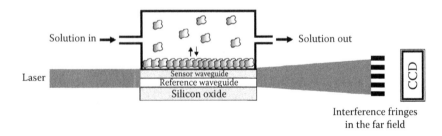

Figure 8.17 Typical architecture of a DPI flow cell. Light enters the stacked waveguide and upon exiting generates an interference pattern in the far field due to a phase shift. The phase shift occurs due to the adsorption of material onto the sensor waveguide.

waveguide at a speed that depends in part on the refractive index of the area within ~100 nm of the sensing waveguide surface (i.e., within the evanescent field). Hence, any thin films or molecules that are adsorbed to the sensing waveguide surface change the speed at which the light travels through the sensing waveguide (because the adsorbed molecules change the refractive index of the sensing region within the evanescent field). On the other hand, light always travels through the reference waveguide at the same speed because the cladding regions on either side are fixed. Since the relative speeds of the light emerging from the two waveguides affects the interference pattern produced, the interference pattern changes as a thin film is adsorbed to the sensor surface. More specifically, the relative positions of the light and dark bands (or fringes) shift as the film adsorbs to the surface. By using Maxwell's equations of electromagnetism and after applying some rather complex mathematics, the shifts in the fringe pattern can be used to calculate the refractive index of the film adsorbed to the surface of the sensing waveguide. Thus, a waveguide interferometer is able to characterize a thin nanofilm.

The refractive index measured by an interferometer such as described above is not the true refractive index of the film, but rather the *effective refractive index*, which is a complicated function of both the absolute refractive index and the thickness of the film. In this way, DPI is similar to SPR and ellipsometry in that it measures an optical parameter related to thickness or refractive index instead of directly measuring the thickness or refractive index. However, DPI has an advantage over single-wavelength ellipsometry or SPR in that it obtains the effective refractive index for two different polarizations of light. These two measurements can be used to mathematically obtain the thickness and refractive index in a manner similar to that used in spectroscopic ellipsometry.

8.4.3.3 Principles of dual polarization interferometry

DPI uses a waveguide interferometer setup as described above, only the light source is modified such that it alternately produces two different types of linearly polarized light via a polarizer switch (see Figure 8.17). One type of polarized light, named the transverse magnetic (TM) polarized mode, is composed of light waves with an electric field that oscillates perpendicular to the direction of the waveguide core, as shown in Figure 8.17. The other type, called the transverse electric (TE) mode, has light waves with an electric field that oscillates parallel to the direction of the waveguide. If we define the waveguide core as the plane of incidence, then the TM and TE modes directly correspond to the s- and p-polarized light modes described in our discussion of ellipsometry. Both of these polarization modes generate evanescent fields that extend into the cladding or sensing regions, but the fields produced by each mode are of different intensities and decay at different rates. Therefore, each polarized mode generates its own interference pattern on the detection screen and consequently each polarized mode provides a separate calculation of the effective refractive index.

It is important not to be misled by the direction of oscillation of the two polarized modes into thinking that no evanescent field would be generated by the TE mode, which oscillates parallel to the direction of the waveguide core. Regardless of polarization, the light waves still undergo total internal reflection (meaning the light beams still "bounce" off the cladding and/or sensing regions) and the two polarized modes still generate evanescent fields in the surrounding regions.

For the effective refractive index determined by each polarized mode, a large number of absolute refractive index and thickness values can be calculated that could possibly yield the observed effective refractive index, as shown in Figure 8.18. However, there is only one unique pair of absolute refractive index and thickness values that may generate the observed effective refractive index for both polarized modes. This pair represents the actual value of the absolute refractive index and the thickness of the film on the sensing waveguide surface. Therefore, by using two different polarized modes of light and by calculating the unique solution pair, DPI can be used to determine the actual refractive index and thickness of thin nanofilms. The technique is so sensitive that thickness changes of less than 1 Å are detectable. Furthermore, if one can assume that the refractive index of the thin film is a linear function of the density of its contents (a good assumption for many thin films), then the refractive index can be manipulated to yield density δ of the film according to

$$\delta = \frac{n_{\text{film}} - n_{\text{buffer}}}{dn_{\text{film}}/dc} \tag{8.25}$$

where dn_{film}/dc is the change in refractive index of the film per change in content density and n_{film} and n_{buffer} are the refractive indexes of the film and buffer (or solvent), respectively. In order to get the mass per unit area of the film, one need merely multiply the calculated density value by the average thickness of the film. Thus, DPI can be used to calculate the average density, mass, and thickness of the film simultaneously.

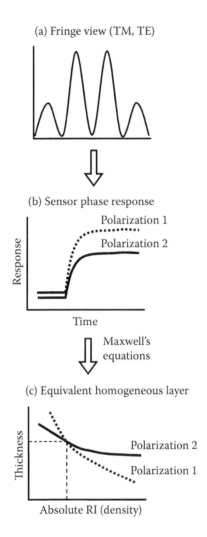

(a) Fringe view (TM, TE)

(b) Sensor phase response

Polarization 1

Polarization 2

Response

Time

Maxwell's equations

(c) Equivalent homogeneous layer

Thickness

Polarization 2

Polarization 1

Absolute RI (density)

Figure 8.18 (a) Typical representation of the fringe pattern observed in a DPI experiment. (b) The sensor phase response for the two polarization states. (c) Effective refractive index plots showing a unique solution for TE and TM modes. The point at which the curves cross gives the absolute refractive index and thickness of the film.

8.4.3.4 *Parameters of a DPI instrument and common applications*

DPI is a relatively uncommon technique for routine surface characterization; but a line of benchtop instruments was manufactured by Farfield Group Ltd, Manchester, UK. In their instrument, the DPI sensor surface is composed of silicon oxynitride, which allows for a wide range of covalent surface modifications through siloxane chemistry (discussed in Chapter 10). The flow cell that exists above the sensor surface has a volume of 2 µL, and typical flow rates are on the order of tens of µL/min. Their instrument boasts thickness detection limits of <1 Å and mass detection limits of 100 fg/mm^2. In addition, it can make multiple measurements per second, which allows it to monitor surface alterations in real time.

DPI has been marketed as a biosensor to measure the interactions between proteins and their substrates and to study the formation of membranes, supported lipid bilayers, and the interaction of surfaces with lipid vesicles. A nonbiological application of DPI is the characterization of the physisorption of nanospheres to hard surfaces. Because it is able to simultaneously determine both thickness and refractive index (and consequently density and mass), DPI is also used to monitor the changes in surface morphology of various thin films. For example, DPI can be used to estimate the shape of a protein adsorbed to a surface under conditions of differing pH.

8.5 SURFACE-SENSITIVE SPECTROSCOPIC METHODS

Several of the spectroscopic methods discussed in Chapter 6 can be adapted for studying surface phenomena by using experimental geometries that rely on principles discussed in the previous sections, such as total internal reflection or surface plasmons, to restrict the region of observation to within a few hundred nanometers of a surface. Several of these techniques will be discussed in the following sections.

8.5.1 Attenuated total reflection IR spectroscopy

ATR-FTIR spectroscopy is a powerful spectroscopic technique used to investigate the structure of adsorbates confined to the solid–air or solid–liquid interface. The technique is essentially IR spectroscopy of molecules present at the surface of a solid. It offers several advantages over its counterpart transmittance-mode IR absorption spectroscopy, which was discussed in Chapter 6. ATR-FTIR offers near-surface selectivity with only a minimal amount of sample and allows for the detection of samples of mass on the order of nanograms. Furthermore, using an infrared polarizer, it is easy to determine the orientation of anisotropic (well-ordered) samples such as self-assembled surfactant or lipid monolayers. The sample can also be exposed to external conditions (e.g., various solvents or different pH conditions), making in situ studies feasible.

ATR-FTIR makes use of evanescent waves in a manner that is similar to DPI, as both are able to measure the intensity of the waves that are formed from total internal reflection within the surface-sensor medium. One significant difference is that ATR-FTIR monitors the absorption of IR light of an adsorbate rather than changes in its thickness and refractive index like DPI. A system of mirrors is set up so that IR light is shone on a crystal with a high refractive index (n). This crystal is called an internal reflection element (IRE) and is structured in such a way that total internal reflection occurs, as shown in Figure 8.19. As with DPI, the total internal reflection of the IR beam results in the creation of an evanescent wave that penetrates the sample above the crystal and decays exponentially moving away from the crystal surface. (See Section 8.4.1.) For IR light, the penetration depth of the evanescent wave is typically on the order of a few micrometers (0.5–5 μm).

If the sample on the surface of the IRE happens to absorb the frequency of IR light that is being shone through the crystal, then the IR beam that emerges from the IRE and travels to the detector has a diminished intensity (or is *attenuated*). One can imagine that the adsorbate takes up the energy of the IR beam through the evanescent wave, but only at those IR frequencies that match the vibrational modes of the sample. Therefore, it is possible to obtain an IR absorbance spectrum for an adsorbate by scanning the entire IR spectrum and monitoring the frequencies at which the IR beam is attenuated on its emergence from the IRE.

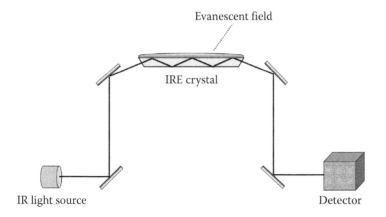

Figure 8.19 A schematic diagram of an ATR setup for an FTIR spectrometer. The evanescent wave that is generated at the surface of the IRE can penetrate the overlying region up to several microns.

There are several common ways to set up an ATR-FTIR system to monitor the IR spectrum of a thin nanofilm. Perhaps the most straightforward method is to build the thin nanofilm on an external substrate and then clamp the substrate on top of the IRE, sandwiching the thin nanofilm between the substrate and the IRE. An IR absorbance spectrum of the nanofilm can then be obtained as described above. While this method is easy to implement, it does not allow for continuous monitoring of the thin nanofilm during its creation. To acquire that type of data, it is necessary to construct a flow cell on top of the IRE and build the thin nanofilm on the surface of the IRE itself. Using this setup, the IR spectrum can be monitored continuously and changes that occur in the spectrum during the creation of the thin film can be recorded.

For the purposes of determining the IR spectrum of a thin nanofilm as described above, a horizontal ATR (HATR) setup is commonly employed. In this setup, the IRE is a parallel-sided crystal plate with dimensions on the order of 1 cm by 5 cm. For an IRE of this size, the IR beam is typically reflected between 5 to 10 times at each surface, depending on the exact dimensions of the crystal and the angle of incidence. The upper face of the crystal is exposed to the thin film through a clamping method or a flow cell setup, as described. The most common crystal materials used for HATR setups are zinc selenide (ZnSe) and germanium. Diamond is a more robust material that can be used as an ATR crystal, although its higher cost is a drawback.

8.5.2 Reflection absorption IR spectroscopy

Another surface-sensitive IR spectroscopy method is **reflection absorption infrared spectroscopy**, or RAIRS. In RAIRS, infrared light is shone onto a metal surface at a grazing angle of incidence. The vibrational spectrum of molecules adsorbed onto the surface is obtained by comparing the intensity of the reflected light from a clean surface to one covered by a thin film. The vibrational modes that are observable in RAIRS are governed by a certain metal surface selection rule.

The selection rule states that only modes that have a component of their transition dipole moment perpendicular to the metal surface appear in a RAIRS spectrum. Hence, the appearance of a RAIRS spectrum indicates that the transition dipole moments of specific functional groups in the film molecules have a component parallel to the surface normal, and combined with other structural data can provide information on the ordering of the adsorbed layer.

8.5.3 Surface-enhanced Raman spectroscopy

Raman scattering is normally a very weak effect. Under normal conditions, the intensities of the Raman scattered peaks in a spectrum are ~0.001% of the incident light. Therefore, if a Raman spectrum is desired for an analyte with a relatively small concentration (such as for a thin film deposited on a surface), some method of signal enhancement must be employed in order to gather any useful information.

One method that is commonly used to enhance the Raman signal is adsorption of the analyte onto a rough metal surface or colloidal particle to obtain the Raman spectrum from the adsorbed species. Most often silver is used, but gold and copper have also been observed to produce similar effects. This type of signal enhancement is called **surface-enhanced Raman spectroscopy** (SERS). SERS is a convenient method that allows amplification of the Raman signal while simultaneously probing the region near a metallic surface, thus allowing Raman spectra to be obtained for low quantities of a substance that are adsorbed to a metallic surface.

The exact mechanism by which adsorption to a metallic surface produces enhancement of the Raman signal is not fully understood. Two predominant theories have emerged to explain the effect: the electromagnetic enhancement theory and the chemical enhancement theory. Each theory is able to explain some observations, but not others.

The electromagnetic theory claims that the generation of surface plasmons by the incident light is the means by which Raman scattering is enhanced. As may be recalled from our discussion of SPR (Section 8.4), a surface plasmon is a collective oscillation of the electrons at the surface of a conducting metal (typically gold or silver) that is induced when light is shone on its surface. The generation of surface plasmons greatly enhances the electromagnetic field near the metallic surface and therefore would be expected to enhance the strength of the Raman scattered light. This, then, is the fundamental idea behind the electromagnetic enhancement theory. The incident light produces surface plasmons, which enhance the electromagnetic field near the surface, which in turn produces a stronger Raman signal.

The chemical enhancement theory appeals to a charge-transfer mechanism to explain the enhancement of Raman signal upon adsorption to a metallic surface. The charge-transfer mechanism essentially states that new electronic states are made available to the molecule due to its interaction with the metallic surface. These new allowable electronic states can be of intermediate energy between the highest occupied molecular orbital of the molecule and its lowest unoccupied molecular orbital of the molecule. Hence, the new electronic states can serve as resonant intermediaries in the process of Raman scattering. Therefore, excitations of the molecule that use the charge-transfer mechanism occur at much lower energies than are normally required for the molecule to be excited. Thus, Raman scattering can happen more readily and the signal is enhanced.

SERS is typically carried out on a "rough" metallic surface, with roughnesses on the order of 10–100 nm. Such surfaces are commonly manufactured by sputtering or evaporation of the metal onto a substrate or by roughening of a metallic electrode surface during oxidation–reduction cycles. Colloidal particles of the metal are also used to produce SERS. These colloids may either be suspended in solution or adsorbed onto a substrate. On these types of "rough" metallic surfaces, the SERS effect extends up to tens of nanometers from the surface, allowing for the effective probing of ultra-thin nanofilms.

8.6 NONLINEAR SPECTROSCOPIC METHODS

8.6.1 An introduction to nonlinear optics

Thus far, the optical techniques that we have discussed rely on *linear* optical processes, describable in terms of a single electric field of a specific frequency propagating through materials that can be further described by a single refractive index or as excitation, emission, or scattering processes involving single photons. However, not all optical processes are linear—linear optics provides a good description of most materials over typical intensities of light, analogous to how Newtonian mechanics provides a good description of massive objects moving at speeds well below the speed of light. At sufficiently high electric field strengths and/or in certain, highly ordered materials, nonlinear effects, which involve the interaction of multiple photons, can become important. Unlike linear optical processes, nonlinear optical processes can change the frequency of the light moving through a material or change the instantaneous refractive index of a material without any change in composition. Exploiting nonlinear optical processes in the laboratory typically requires powerful pulsed lasers that can generate brief (femtosecond to nanosecond) bursts of extremely strong electric fields. Although the theory of **nonlinear optics** is beyond the scope of this book, a cursory treatment is provided to facilitate understanding of how nonlinear optical effects can be used to probe the properties of molecules in nanoassemblies.

Linear optical effects such as reflection, refraction, absorption, and inter-ference are observed from all light sources regardless of intensity. Linear optical effects are based on a linear relationship between an oscillating electric field and an induced dipole moment in a molecule. Thus, when the oscillating electric field of light interacts with a molecule, the elec-tron cloud in the molecule also begins to oscillate. This electron den-sity oscillation sets up an oscillating dipole moment in the molecule. The strength of this dipole (μ) depends linearly on the strength of the incoming electric field (E) according to Equation 8.26:

$$\mu_{\text{ind}}(\omega) = \alpha(\omega)E(\omega) \tag{8.26}$$

We previously discussed the molecular polarizability α in Chapter 5, but like the refractive index, it is a frequency-dependent complex number. Note that we have written frequency here as the angular frequency $\omega = 2\pi\nu$, discussed earlier in the chapter and commonly used in electrody-namics for reasons of mathematical convenience. In the context of this section, we will only consider the real component of the polarizability. The real component of the polarizability can be related to the refractive index of a material through the Clausius–Mossotti equation, which we introduced in Chapter 5. For a multicomponent system,

$$\frac{n^2 - 1}{n^2 + 2} = \sum_i \frac{N_i \alpha_i}{3\varepsilon_0} \tag{8.27}$$

where N_i is the number density (concentration) of each species in the material, α_i is the polarizability of that species, and ε_0 is the permittivity of vacuum.

A key characteristic of light/matter interactions described by linear molecular polarizability is that when light of frequency ω interacts with a material, the frequency of the light is unchanged. For example, when incident light of frequency ω bounces off a surface, the reflected light will also be of frequency ω. Now, Equation 8.26 is actually the linear term in a power series equation with quadratic and higher-order terms:

$$\mu = \mu_0 + \alpha E + \beta E^2 + \gamma E^3 + \dots \tag{8.28}$$

In essence, representations of polarization in response are a Taylor series where each term represents a derivative of the polarization versus electric field at that order (i.e., $\alpha = d\mu/dE$). The higher-order terms become important only if E is large or for specific, highly polarizable structures in certain orientations. The constants β and γ are known as the first and second **hyperpolarizability**, respectively. These constants are essentially zero for most systems when ordinary low-intensity light interacts with a molecule. However, when light from an intense pulsed laser is used, these nonlinear terms become significant. Equation 8.28 is the general equation describing the effect of light on a molecule. It is important to realize that all materials are nonlinear; it is a matter of the magnitude of the perturbing electric field that is needed to set in the anharmonicity and to drive it to nonlinear behavior. Extending this treatment to a bulk material, a similar equation is obtained, where the dipole (μ) is now replaced by the average polarization (P) of the bulk material:

$$P = P_0 + \chi_1 E + \chi_2 E^2 + \chi_3 E^3 + \ldots \tag{8.29}$$

The constants χ_1, χ_2, and χ_3 are known as first-, second-, and third-order susceptibilities. They are related to the sum components of the corresponding hyperpolarizabilities averaged over orientations of the molecules in the bulk material. Like with the refractive index, each term in the expansion has real and imaginary components, where the real components represent refraction/scattering processes and the imaginary terms represent absorption/emission processes. The constant term P_0 is typically zero, but a few materials such as lithium niobate ($LiNbO_3$) and barium titanate ($BaTiO_3$) retain electric polarization at zero field, acting in an analogous manner to permanent magnets. These are **ferroelectric** materials that are crucial for many instruments that rely on nonlinear optical effects. The complete mathematical descriptions of the various χ terms are ignored here, but it is should noted that at optical frequencies, χ_1 is proportional to the square of the refractive index. We can separate each polarization term in Equation 8.29, such that $P_1 = \chi_1 E$ represents the linear polarization, $P_2 = \chi_2 E^2$ represents the second-order nonlinear polarization, and so on. In fact, many of the nonlinear optical effects used to study nanomaterials are based on the second-order nonlinear polarization. We focus on this term.

Let's consider two incident intense laser sources with frequencies ω_1 and ω_2 incident on a material. These two frequencies have two oscillatory electric fields (with fields E_1 and E_2) simultaneously acting on the material, given by the equation

$$E = E_1 \cos \omega_1 t + E_2 \cos \omega_2 t \tag{8.30}$$

Substituting this equation into the expression for the second-order nonlinear polarization, $P_2 = \chi_2 E^2$, yields the following interesting result:

$$P_2 = \frac{\chi_2}{2} \left\{ \left(E_1^2 + E_2^2\right) + \left(E_1^2 \cos 2\omega_1 t\right) + \left(E_2^2 \cos 2\omega_2 t\right) \right.$$
$$\left. + 2E_1 E_2 \cos(\omega_1 + \omega_2)t + 2E_1 E_2 \cos(\omega_1 - \omega_2)t \right\} \tag{8.31}$$

This is an interesting result because of the terms $2\omega_1$, $2\omega_2$, $\omega_1 + \omega_2$, and $\omega_1 - \omega_2$, which tell us that when two intense light frequencies (ω_1 and ω_2) interact with a material, we can produce resulting light of doubled frequencies ($2\omega_1$, $2\omega_2$), and even resulting light of combined ($\omega_1 + \omega_2$) and subtracted ($\omega_1 - \omega_2$) frequencies. The doubled frequency production is known as **second-harmonic generation** (SHG), and the latter two products are usually referred to as **sum-frequency generation** and **difference-frequency generation** (SFG and DFG). Two photons can pass through a material and combine in energy to produce a single photon. If we had only one frequency of light, then two photons would combine to produce a single photon of twice the energy (and hence frequency) of the incident photon. In this case, there is no distinction between SHG and SFG. We are combining two photons to produce one, so the conversion efficiency of this process has a theoretical upper limit of 50%.

Not all materials generate a second-order polarization, and those that produce this effect do so at varying efficiency. The key determining factor is χ_2, the second-order susceptibility. The mathematical properties of χ_2 depend on factors such as molecular orientation, and these properties ultimately determine the nonlinear optical conversion efficiency. The properties of χ_2 are discussed in the next section.

Example 8.3   Combining Photon Energies

A pulsed laser light source composed of two wavelengths, (a) green light (500 nm) and (b) infrared light (900 nm), is passed through a nonlinear optically active material. Determine the color of the second harmonic and sum-frequency light emitted from the material.

Solution   Convert all wavelengths to frequencies:

$$\text{(a) } v = \frac{c}{\lambda} = \frac{2.998 \times 10^8 ms^{-1}}{500 \times 10^{-9}m} = 5.996 \times 10^{14} \text{ Hz}$$

(b) $v = \dfrac{c}{\lambda} = \dfrac{2.998 \times 10^8 ms^{-1}}{900 \times 10^{-9}m} = 3.331 \times 10^{14}$ Hz

Combine the frequencies:

SHG:

(a) $2(5.996 \times 10^{14}) = 1.199 \times 10^{15}$ Hz, or ~250 nm (deep UV light)

(b) $2(3.331 \times 10^{14}) = 6.662 \times 10^{15}$ Hz, or ~450 nm (blue light)

SFG: $5.996 \times 10^{14} + 3.331 \times 10^{14} = 9.327 \times 10^{14}$ Hz, or ~320 nm (UV light)

Second-order susceptibility in a molecular material is a function of three major factors: (a) The hyperpolarizabity of individual chromophores in the material, (b) the number density (concentration) of chromophores, and (c) the extent to which the molecules are ordered such that their dipole moments are parallel to each other:

$$\chi_2(\omega_1, \omega_2) = gN\beta(\omega_1, \omega_2)\langle\cos^3\theta\rangle \qquad (8.32)$$

where N is the number density of chromophores, and $\langle\cos^3\theta\rangle$ is the cubic average of the angle between the dipole moments of all of the chromohores in the system and the axis of the electric field of the light propagating through the system. The local field factor g depends on the linear susceptibility of the system at the frequencies of the light propagating through the system.

Bulk second-order susceptibility requires that the components of the system have nonzero molecular hyperpolarizability. This response can be due to either the structure of the molecules or induced by their surrounding environment. Generally speaking, conjugated molecules are easier to polarize than nonconjugated molecules because the electrons are delocalized along the carbon backbone. However, high polarizability is not sufficient to obtain a second-order response. The molecules also just lack **inversion symmetry**. Inversion involves passing each atom in a molecule through the center of the molecule and placing it on the opposite side of the molecule. Electron donating and/or withdrawing groups can break symmetry and induce a permanent dipole moment. Typical nonlinear optically active (NLO) chromophores contain an electron acceptor (A) and an electron donor (D) bridged by a π-conjugated structure. The donor and acceptor produce a charge separation (dipole moment) in the molecule, which makes it more energetically favorable for the molecule to polarize parallel to its dipole moment than to polarize antiparallel to it. The conjugated, polarizable bridge allows electrons to move from donor to acceptor more easily. NLO response can be increased by increasing the strength of the donor or acceptor or length of the conjugated structure (moving the partial charges of the dipole farther away from each other).

The hyperpolarizability of such a system can be estimated via the two-state model, which is derived from quantum mechanical perturbation theory and approximates the response of a material in terms of the transition dipole for the lowest energy excited state, μ_{ge}, the change in dipole moment between the ground state and that state, $\Delta\mu$, and the energy difference ΔE_{ge} between the HOMO and the orbital that forms the lowest energy excited state,

$$\beta \propto \frac{\mu_{ge}^2 \Delta\mu}{\Delta E_{ge}^2} \qquad (8.33)$$

Note the similarities between the two-state model and the oscillator strength discussed in Chapter 6. However, further note that the hyperpolarizability is a ground-state property, and—at least at wavelengths away from absorbance maxima—the excited state properties are used to represent the extent to which electrons can polarize (e.g., the real component of the hyperpolarizability) instead of an absorbance process. While highly simplified, it is useful for qualitative understanding of molecular properties, similar to how the highly simplified Onsager model can be used to understand dielectrics.

The presence of molecules with a high hyperpolarizability is not sufficient for a bulk second-order response. χ_2 is zero in a centrosymmetric environment ($\langle \cos^3\theta \rangle = 0$). Figure 8.20 shows molecules with and without inversion symmetry and examples of layers of molecules that are centrosymmetrically and noncentrosymmetrically ordered. If a molecule possesses inversion symmetry, or if a system of molecules is ordered in a manner that has inversion symmetry, χ_2 is reduced to zero. Structural changes do not have to be as dramatic as adding strong electron donating and accepting groups. Let's consider the simple hydrocarbons ethylyne (or acetylene) and propadiene (or allene). Ethylyne has a linear structure, but the two pairs of hydrogen atoms in propadiene are rotated by 90° from each other. After inversion, the propadiene molecule looks different, but the position of the hydrogens is the same for ethylyne after inversion. Therefore, propadiene has inversion symmetry and propadiene does not. Interestingly, a molecule at an interface always lacks inversion symmetry,

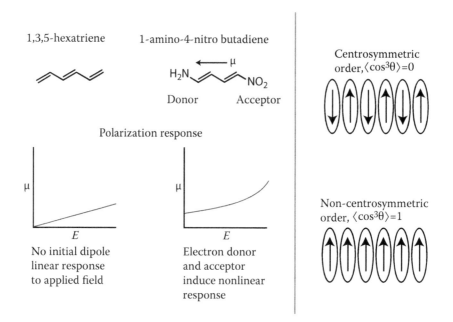

Figure 8.20 Left: The polarization response of two simple conjugated molecules to an electric field applied along the axis of the conjugated chain. 1,3,5-hexatriene has no dipole moment in the absence of a field, and applying a field induces a weak dipole that increases linearly with field strength as per Equation 8.20. It has zero first hyperpolarizability (β). In contrast, 1-amino-4-nitrobutadiene has a large dipole moment and polarization is enhanced when the applied field is parallel to its dipole moment, generating a nonlinear response. Right: For a molecule to have nonzero χ_2, its constitutent molecules must be aligned noncentrosymetrically such that their individual NLO responses add together (bottom) instead of canceling (top).

even ethylyne (Figure 8.21). The symmetry breaking for otherwise centrosymmetric molecules at a surface is due to differences in the interactions of the molecule with the interface versus the bulk. Therefore, surfaces and interfaces are inherently noncentrosymmetric regions, and molecules confined to such regions produce SHG and SFG light. This fact allows SFG and SHG to be used as surface-specific probes to investigate molecular structure at interfaces. Many nanomaterials are supported on surfaces and can be probed using such nonlinear optical techniques.

8.6.2 Second-harmonic generation

SHG involves the conversion of frequency ω to 2ω. Like any second-order nonlinear optical effect, it requires that the material exhibiting SHG is noncentrosymmetric and is strongest for regions that are strongly

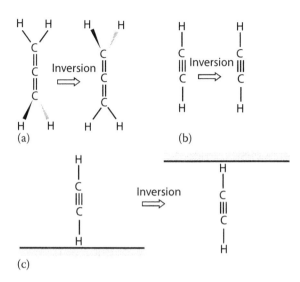

Figure 8.21 An inversion operation on simple molecules. (a) Propadiene does not have inversion symmetry because the positions of all the hydrogen atoms are changed after the operation. (b) All atoms in ethylyne are unchanged after an inversion. This molecule has inversion symmetry. If ethylyne is on a surface, as in (c), the center of symmetry is broken. Surfaces represent noncentrosymmetric regions where there is no inversion symmetry.

acentrically (noncentrosymmetrically) ordered and for molecules that exhibit a large first hyperpolarizability (β). To first order, hyperpolarizability is proportional to dipole moment and to polarizability. Consider an amphiphile containing a high-β head group (Figure 8.22). The polarizable head group can be considered as an NLO chromophore, in which lone pairs on the alkoxy group function as an electron donor and the carboxylic acid group functions as an electron acceptor, with a conjugated azobenzene structure acting as the polarizable bridge between donor and acceptor. A monolayer comprised of this molecule produces a highly oriented assembly with a large net dipole. These conditions meet the requirement of a strong SHG signal. An additional layer can be added to the assembly either with the same orientation (asymmetric assembly) or with the opposite orientation (symmetric assembly). The latter reduces the SHG signal because of the centrosymmetric nature of the symmetric bilayer.

Constructing nanoassemblies that have a large NLO response is important to a number of technologies, including optoelectronics (using photons instead of electrons in electronic devices), communications, and data storage. Materials with efficient frequency-doubling properties

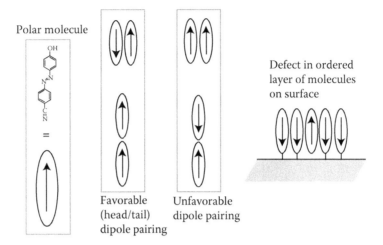

Figure 8.22   An azobenzene dye capped with an electron donating group (such as OH) and an electron accepting group (such as CN) has a large hyperpolarizability (β) and strong dipole moment (left). However, in solution, the dipole moments of the dye interact with each other (see Chapter 5), favoring head-to-tail alignment, with centrosymmetric pairing of adjacent molecules (top left) being the most energetically favorable, leading to a lack of bulk second-order response. While noncentrosymmetric alignment can be induced by anchoring molecules to a surface, amphiphilic structure, or external factors such as applied electric fields, even well-aligned systems have centrosymmetric defects due to repulsion between parallel dipoles (right).

are also used in lasers to generate light of different wavelengths. Self-assembly methods can be used to guide molecular building blocks into preferred orientations that yield new materials with NLO-active properties. Even though this field is leading the development of next-generation electronic and computational breakthroughs, fundamental limitations exist in fabricating functional nanoassemblies for NLO applications. These limitations arise from the inherent difficulty in obtaining asymmetrically oriented NLO chromophores in layered thin-film assemblies. As a result, the NLO response does not increase quadratically with the film thickness, rendering the assembly useless. Layer-by-layer methods, covered in Chapter 10, seem to be a promising approach to creating multilayered films containing NLO blocks throughout the 3D assembly. An ideal NLO-active film would contain hundreds of discrete layers of highly oriented molecules self-organized on a solid support. In the case of NLO materials, if electron donor–acceptor chromophores were organized in such a way that the dipole moments were oriented in the same direction, then a quadratic increase of the optical second-harmonic signal would be observed with increasing film thickness (number of bilayers) as predicted by Equation 8.34, where L is the film thickness, I_ω is the intensity of incident light, n_ω is the film's refractive index at the incident light frequency, and $n_{2\omega}$ is the refractive index at the doubled frequency.

$$I_{2\omega} = \frac{2\omega^2 d_{eff}^2 L^2}{c^3 \varepsilon_o n_\omega^2 n_{2\omega}} (I_\omega)^2 \frac{\sin^2(\Delta kL/2)}{(\Delta kL/2)^2} \tag{8.34}$$

The Langmuir–Blodgett (LB) method is a commonly used technique to organize amphiphilic molecules into densely packed 2D structures at the air–water interface. The 2D structure can then be conveniently transferred to a substrate. This method is discussed further in Chapter 10. By repeating the transfer process, ordered 3D nanostructures of desired thickness can be obtained. If asymmetric structural order is retained throughout the assembly, one can then achieve a sum effect of the functionality represented by an individual layer. However, the idea of relying on the sum effect of individual LB layers to obtain an enhanced NLO functionality in 3D materials is seriously challenged by studies that demonstrate a certain level of intermixing between layers. Intermixing randomizes the chromophore orientation, resulting in a reduction of the SHG intensity. We are interested in an asymmetric structure with a large optical nonlinearity. This type of deposition is usually accomplished using two different species (discussed further in Chapter 10). Disorder in LB assembled materials is due to the presence of kinked alkyl tails within the assembly, resulting in a decreasing packing density. Furthermore, disorder may also arise from a lack of in-plane ordering due to unfavorable dipole–dipole interaction. As discussed in Chapter 5, dipole–dipole interactions are most energetically favorable when the dipoles are arranged head-tail (i.e., positive charge adjacent to negative charge). Favorable and unfavorable dipole configurations are shown in Figure 8.23. In a typical LB assembly involving two different species, both materials from a given deposition cycle may occupy defects in underlying layers. Materials of either species have a tendency to spread in a direction perpendicular to the substrate, leading to enhanced interpenetration, despite the maintenance of equivalent bilayer thicknesses over many layers. In other words, equivalence in deposition amount and thickness over many layers may indicate similar net surface conditions in each layer, but care must be taken in using the notion of a similar surface to infer consistency in molecular order within layers. Even disordered systems can display seemingly regular deposition patterns—a film with disorder on the molecular level may appear markedly ordered at the macroscale. Such considerations are far-reaching in a field largely focused not only on spatial

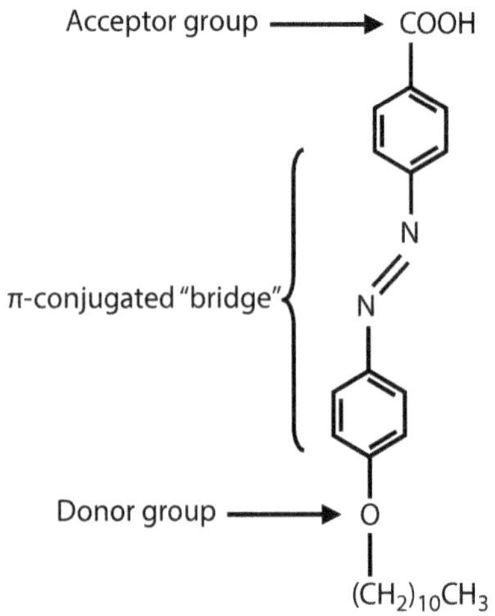

Figure 8.23 An amphiphile containing an NLO-active head group. This moiety is comprised of an electron donor (the oxygen atom) connected to an electron acceptor (the carboxylic acid group). The azobenzene group serves as a π-conjugated bridge between the electron donor and acceptor groups. This head group is highly polarizable.

arrangement of molecules, but also on the chromophore orientation at the molecular level.

In addition to measuring the NLO response from a nanomaterial, SHG can also be used to follow the assembly process in real time, obtain the density of NLO-active chromophores within the assembly, and even image ordered domains in nanofilms. By obtaining the relative amount of SHG generated from s- and p-polarized light, the orientation of the NLO-active chromophore with respect to the surface normal can also be calculated.

8.6.3 Sum-frequency generation spectroscopy

Photons of two different frequencies (ω_1 and ω_2) can combine to generate light at the sum frequency ($\omega_{SFG} = \omega_1 + \omega_2$). Similar to SHG, SFG is produced from noncentrosymmetric environments. SFG is routinely used as an IR spectroscopic method to obtain molecular structure information of nanofilms confined at interfaces. In this technique one of the two frequencies is fixed at a visible wavelength, typically 532 nm (ω_{vis}). The other frequency is variable in the IR region (ω_{IR}). The two light beams are overlapped on a surface and the intensity of the SFG beam is measured. It turns out that this intensity is greatly enhanced when ω_{IR} is in resonance with a vibrational mode of the molecule at the surface. Thus, a plot of the SHG intensity versus ω_{IR} provides a vibrational spectrum of molecules on the surface. It is worth noting that in contrast to ATR-FTIR spectroscopy, **SFG spectroscopy** is surface-specific to within a few nanometers, whereas ATR methods probe regions within as much as a few microns from the surface. A theoretical expression for the magnitude of the SFG signal is given by Equation 8.35:

$$I_{SFG} = 128\pi^3 \left(\frac{\omega_{SFG}}{hc^3}\right) |K_{SFG}K_{vis}K_{IR}|^2 |\chi_2|^2 \left(\frac{I_{vis}I_{IR}}{AT}\right) \qquad (8.35)$$

The signal intensity, I_{SFG}, is the number of SFG photons produced per laser pulse and depends on the beam intensities (I_{vis} and I_{IR}), the laser pulse length (T), the area of overlap of the beams (A), the various geometric Fresnel factors (K), and the second-order susceptibility.

In practice, in order to generate SFG light, precise incidence angles of the fixed frequency visible and tunable IR light have to be used. Furthermore, the emitted SFG light is observed at a precise angle. This is known as phase matching, and the appropriate angles can be calculated using Equation 8.30. Figure 8.24 shows the typical geometry of an SFG experiment, indicating the various angles shown in Equation 8.36.

$$\omega_{SFG} \sin\theta_{SFG} = \omega_{vis} \sin\theta_{vis} - \omega_{IR} \sin\theta_{IR} \qquad (8.36)$$

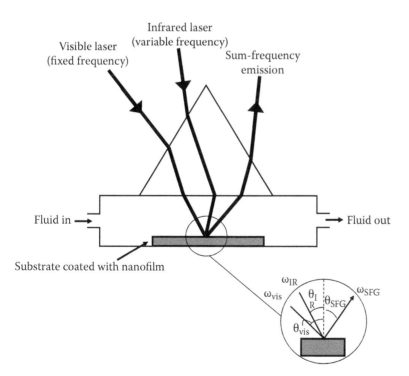

Figure 8.24 Typical geometry of an SFG experiment in which a prism is used to couple the various laser beams onto a surface. The various angles of the IR, visible, and SFG beams are indicated in the circle.

SFG spectroscopy has been used extensively to study the conformation of alkyl chains in monolayers assembled at interfaces. If the alkyl chain is in an all-trans conformation, then all the CH_2 groups are in a locally centrosymmetric environment, as illustrated in Figure 8.25. The presence of a gauche defect breaks this centrosymmetry and renders the CH_2 sum-frequency active. The power of SFG spectroscopy is illustrated by the following example. Consider a monolayer of dodecane thiol (DDT) chemisorbed to a gold surface (Figure 8.26) and under water. Details of this type of monolayer are discussed further in Chapter 10. This monolayer is tightly packed with all the CH_2 groups in an all-trans conformation. Only the terminal CH_3 groups are noncentrosymmetric. The SFG spectrum of DDT shows only features that are attributed to the vibrational modes of the terminal methyl group. In comparison, Figure 8.27 shows the SFG spectrum of a mercaptododecanoic acid (MDA). This molecule is structurally similar to DDT, except that it contains a large carboxylic head group instead of a CH_3 group. The SFG spectrum of MDA shows no CH_3 features, but does show prominent vibrational features that arise from CH_2 groups. The presence of the methylene features indicates a significant amount of gauche defects in the alkyl chains of the MDA monolayer, which is not surprising since, compared to DDT, MDA has a bulky, charged, head group that prevents the monolayer from packing tightly. This characteristic leads to a greater amount of space for the alkyl chain to occupy, resulting in the formation of kinks in the hydrocarbon chain.

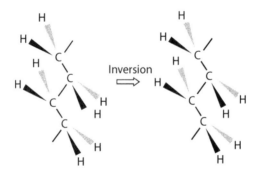

Figure 8.25 Portion of an all-trans alkyl chain. The CH_2 are locally centrosymmetric because after an inversion operation, the positions of all atoms remain unchanged. Therefore, these groups in this conformation will not produce SHG or SFG light. A "kink" in the chain or a gauche defect will create a noncentrosymmetric environment of CH_2 groups, rendering these groups SFG active.

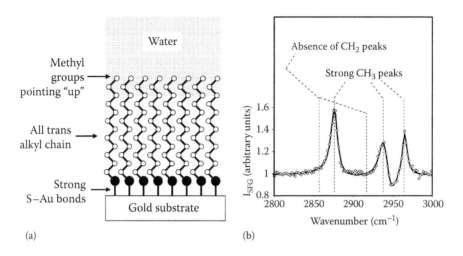

(a) (b)

Figure 8.26 (a) A self-assembled dodecane thiol monolayer at the gold–water interface. An all-trans conformation due to strong interalkyl chain hydrophobic interactions creates a close-packed structure with the terminal CH_3 groups pointing toward the water phase. The SFG spectrum in (b) shows only features due to the noncentrosymmetric CH_3 groups. The CH_2 groups are centrosymmetric and so do not appear in the SFG spectrum.

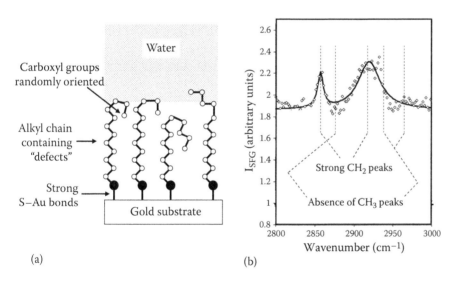

(a) (b)

Figure 8.27 (a) A self-assembled mercaptododecanoic acid monolayer at the gold–water interface. The large anionic head group prevents the monolayer from packing tightly and creates kinks and gauche defects in the chain. This randomized the head groups. The SFG spectrum in (b) shows only features due to the noncentrosymmetric CH_2 groups.

SFG spectroscopy has established itself as a powerful nanocharacterization method. The technique can probe any interfacial medium that is optically accessible, and has been used to probe many interfacial processes such as corrosion, surface phase transitions, detergency, and the structure of cell membranes.

8.7 IMAGING NANOSTRUCTURES

One of the more intuitive methods that we have not yet discussed is taking a "picture" of the nanostructure of interest. Indeed, so-called nano-imaging methods are often among the first used to study a specific nanomaterial. There are a variety of nanoimaging methods that exist. In the next few sections, we discuss a sampling of these methods, ranging from subdiffraction limit optical methods, such as near-field scanning microscopy, to nonoptical methods that avoid the diffraction limit by imaging using electrons or highly precise mechanical sensors.

8.7.1 Imaging ellipsometry

8.7.1.1 Imaging using conventional ellipsometry

We have already encountered ellipsometry as a powerful method for determining the thickness and optical constants of films as small as a few angstroms (for a review of ellipsometry, see Section 8.3). This method involves analyzing the change in the polarization state of a light beam on reflection off a planar surface. The ellipsometric parameters recorded from monitoring this change in the polarization state of the light beam can be translated into thickness values for the surface after using an appropriate mathematical model. It should also be highlighted that a thickness value obtained in this manner actually represents the average thickness value within the beam spot on the surface. In order to achieve a better lateral (or horizontal) resolution, the light beam must be focused on a smaller spot on the surface. Thus, perhaps the most straightforward approach to image a surface using ellipsometry would be to maximize the lateral resolution by using a tightly focused light beam then determining the film thickness as the beam scans the surface. We could then obtain a three-dimensional image of the surface topography. However, using conventional ellipsometry to scan the sample in this way can be time-consuming, so a quicker method called imaging ellipsometry has been developed.

8.7.1.2 Principles of modern imaging ellipsometry

Modern **imaging ellipsometry** can be thought of as combining an ellipsometer and a microscope. Rather than using a tightly focused light beam and scanning it across the sample, modern imaging ellipsometers generally employ a light beam with a large diameter (often on the order of millimeters) such that the entire sample is illuminated. The reflected image of the entire object is then focused onto a high-resolution charge-coupled device (CCD, a type of digital sensor) camera using an objective lens. From this image, ellipsometric data of the sample (including thickness) can be extracted pixel by pixel, generating a surface topography map of the sample. Therefore, the ellipsometric data for the entire sample can be determined quickly, resulting in a 3D image of the sample.

8.7.1.3 Methods for extracting ellipsometric data in imaging ellipsometry

The "pixel-by-pixel" ellipsometric data from an imaging ellipsometer setup can be determined using one of several methods. A popular method is to employ "off-null" mode analysis. We recall from our discussion of ellipsometry in Section 8.3 that conventional ellipsometers are often operated in null mode. In null mode, the change in the polarization state of the reflected light is determined by changing the polarizer, compensator, and analyzer so that a null condition is achieved (no light passes through the analyzer). Depending on the values of the polarizer, compensator, and analyzer used to achieve this null condition, different ellipsometric parameters may be calculated. In "off-null" mode, the imaging ellipsometer is "zeroed" by determining the null conditions for the bare substrate. These conditions are then kept constant as the entire sample is imaged. Since the sample is generally not of the same thickness and refractive index as the substrate, light of greater and greater intensity passes through the analyzer for thicker regions of the sample. Therefore, the intensity of the light in each pixel can be related to the thickness of the sample at that point. It should be noted, however, that this type of off-null mode analysis is substrate-, sample-, and thickness-dependent, so comparisons with a reference material of known thickness and refractive index are often recommended. For example, for a pure silicon substrate and a biological sample, the intensity I under an off-null condition has been reported to be related to the film thickness d by $I = kd^2$ where k is a proportionality constant (*not* the imaginary component of the refractive index), and this equation is valid with a deviation of approximately 2% up to $d \sim 5$ nm.

An alternative and perhaps more straightforward method for obtaining the "pixel-by-pixel" ellipsometric parameters is to continually adjust the polarizer, compensator, and analyzer in order to determine the null condition for each pixel (or group of pixels) and then extract the parameters from those null conditions as would be done in conventional ellipsometry. While this approach is slightly more time-consuming, it does not require the use of any reference samples or the intensity-thickness assumptions employed in "off-null" mode analysis.

8.7.1.4 Image focusing

One of the major obstacles in imaging ellipsometry is that only the center of the image is in focus due to the large incident angle of the light beam. To compensate for this limited focus area, a scanner can be incorporated into imaging ellipsometer instruments. The image may then be taken at several different positions, and the computer software combines these images into one focused image of the entire sample. An alternative approach has been to use the Scheimpflug method, which is an optical technique that allows the entire sample to be placed in focus by tilting the lens to a certain degree, called the Scheimpflug line.

8.7.1.5 Resolution of an imaging ellipsometer

The lateral (x,y) resolution for an imaging ellipsometer as described above is generally limited by the resolving power of the CCD camera and is usually on the order of a few microns. The depth (z) resolution, on the other hand, is comparable to conventional ellipsometry at a few tenths of a nanometer. Because of its nondestructive nature, compatibility with a wide variety of sample types, and excellent temporal and depth resolution, imaging ellipsometry is a powerful technique to image nanostructures and is suited for thin-film analysis. Figure 8.28 shows an ellipsometric thickness image of a single phospholipid bilayer assembled on a glass surface. The bilayer thickness is around 5 nm. The figure shows how imaging ellipsometry can be used to follow the slow hydration-induced spreading of the bilayer over the uncovered portion of the substrate. Thus, imaging

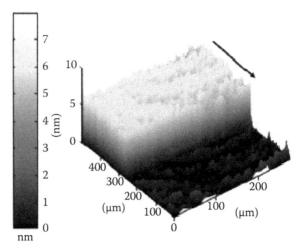

Figure 8.28 An ellipsometric thickness image showing the height of a phospholipid bilayer assembled on a solid support. (Image provided by Professor Atul Parikh, University of California, Davis.)

ellipsometry provides topographical data with excellent vertical resolution. However, because the lateral resolution is on the order of microns, nanostructures cannot be resolved in the xy plane. In order to image nanostructures at atomic-level resolutions in all three dimensions, methods such as the scanning probe microscopies must be used.

8.7.2 Scanning probe methods

The scanning probe microscopies are powerful imaging techniques in which a very sharp tip is scanned across a surface, producing an image with near atomic-level resolution. **Scanning tunneling microscopy** (STM) and **atomic force microscopy** (AFM) are the most common examples of scanning probe methods. Both STM and AFM produce three-dimensional images of surfaces that approach atomic or molecular resolution, making them ideal for the study of nanostructures on surfaces.

8.7.2.1 Scanning tunneling microscopy

STM operates by monitoring the "tunneling" current that is produced when a sharp tip is brought extremely close to a surface that is able to conduct electricity. In order to bring and maintain the tip so close to the surface, a piezoelectric transducer is used. As a refresher, we recall that piezoelectric transducers possess the ability to physically expand or contract in response to an applied voltage (for a more complete discussion, see the discussion of QCM techniques in Section 8.3). Therefore, a sharp metal STM tip, usually made of platinum, is attached to a piezoelectric scanner that can cause the tip to move small distances in the x, y, and z directions. Depending on the type of piezoelectric material used to make the scanner (usually some form of ceramic) and its dimensions, the contraction or expansion of the material can be as small as 1 nm per volt applied, allowing for the metal tip to be brought close to the surface being imaged. In early models of STMs, the piezoelectric scanners consisted of transducers arranged in the x, y, and z directions, as shown in Figure 8.29, but more recent models have made use of a tubelike piezoelectric transducer to achieve better resolution.

Using a piezoelectric scanner, the STM tip can be brought close (within 1 nm) to a conducting surface. A small voltage, generally between 2 mV and 2 V, is then applied between the conducting substrate and the metal tip, causing electrons to tunnel between the tip and the surface, creating a current. The magnitude of this tunneling current depends exponentially on the distance (h) between the conducting surface and the STM tip. For a

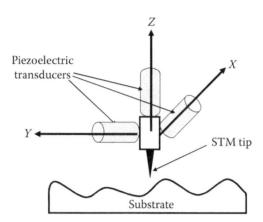

Figure 8.29 Schematic diagram of the piezoelectric transducers that control the position of an STM tip near the sample surface. For many piezoelectric transducers used, a distance of as little as 1 nm can be affected with a single applied volt. When the STM tip is positioned sufficiently close to the sample surface and a potential is applied across them both, a tunneling current will be induced between the tip and the sample. This tunneling current is the basis of the STM measurement.

constant applied voltage, the tunneling current, I, is given approximately by

$$I \approx e^{-2\kappa h} \tag{8.37}$$

Equation 8.37 tells us that the tunneling current decreases exponentially as we move the tip away from the surface. The constant κ is called the electronic decay length of the electron and is a measure of how the probability density of a confined electron decays with distance, or more precisely, it is the decay length of the electronic wavefunction. Not surprisingly, κ depends on how tightly the electron is bound to the conducting surface, called the **work function** of the surface. The ability of electrons to travel or "tunnel" through a nonconducting medium between two conducting materials that are close together is strictly a quantum mechanical effect, which was previously discussed in Chapter 4 as a consequence of finite potential energy barriers and involves the wavefunction of electrons penetrating into the classically forbidden region (in this case, the space between the conductors). As the wavefunction decays exponentially in the forbidden region, tunneling exhibits an exponential decay as a function of the distance between the two materials.

As the STM tip is scanned across a rough surface, the tunneling current changes as the tip encounters bumps or dips in the surface (or as h

changes). There are two obvious methods for imaging the surface. In the first method, known as "constant-height" STM, the tip is held at a constant vertical position and the tunneling current is plotted as a function of the x–y region scanned. Because the tunneling current is dependent on the distance between the tip and the surface (h), the surface can be imaged. The second method, known as "constant-current" STM, is to maintain a constant tunneling current as the tip is scanned across the surface by varying the tip's vertical position. The piezoelectric transducer moves the tip up or down as the surface is scanned, ensuring that the tunneling current is maintained at the predetermined value. Essentially, h remains constant throughout the scanning process, so the instrumentation has to monitor only the changes in the tip's vertical position to provide a topographical image of the surface. In practice, this second method is generally used to generate a surface image.

As with most of the scanning probe microscopies, STM surface images are typically generated by scanning the tip across the surface in a *raster pattern*, or line by line, to create a rectangular grid. In some STM models, the scanning process is controlled by moving the substrate and not the tip, which is moved only vertically to maintain a constant tunneling current.

Example 8.4   Current Changes in STM

Consider a metal surface with an electronic decay length κ of 10 nm^{-1}. By how much does the tunneling current change when the height h between the surface and the STM tip increases from 1.0 nm to 1.1 nm?

Solution   From Equation 8.31, we have $I \approx \exp(-2\kappa h)$. We can determine the approximate factor by which the tunneling current changes by taking the following ratio:

$$\frac{I_{1.0 \text{ nm}}}{I_{1.1 \text{ nm}}} \approx \frac{\exp\left(-2 \times 10 \text{ nm}^{-1} \times 1.0 \text{ nm}\right)}{\exp\left(-2 \times 10 \text{ nm}^{-1} \times 1.1 \text{ nm}\right)} \approx 8$$

Thus, we see that the current changes by almost an order of magnitude when the distance h varies by 0.1 nm.

Example 8.4 emphasizes the sensitivity of STM to changes in height. In fact, if the tunneling current is kept constant then the height remains constant to within 10^{-3} nm, meaning that atomic-level resolution is achieved.

While such resolutions constitute a major achievement, STM suffers from its inability to image surfaces that do not conduct electricity. Nonconducting surfaces are unable to be imaged unless they are modified or coated with a conducting substance, which may damage the sample. Atomic force microscopy, discussed below, does not suffer from this disadvantage, although STM provides superior resolution.

For a typical STM, the lateral range of the scanner is generally from tens of angstroms to approximately 100 μm. The allowable height is from the sub-angstrom to ~10 μm range. The tip is generally constructed by manually cutting a platinum wire or by electrochemical etching of tungsten metal.

One example of an application of STM is the atomic resolution STM image of a graphite surface, shown in Figure 8.30. In this image, the hexagonal arrangement of the carbon atoms in graphite is clearly discernible, as well as the valleys in between adjacent carbon atoms, thus demonstrating the utility of STM in studying conducting surfaces.

Figure 8.30 STM image of a graphite surface. Notice the angstrom-level resolution that is characteristic of STM. (Image from F. Atamny et al., *Phys. Chem. Chem. Phys.*, 1999, 1: 4113–4118. Reproduced by permission of the PCCP Owner Societies.)

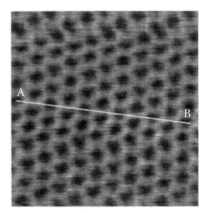

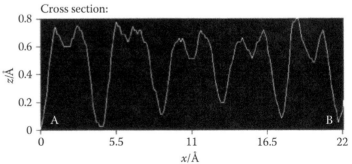

8.7.2.2 Atomic force microscopy

Atomic force microscopy, or AFM, operates on principles similar to those of STM. As with STM, AFM functions by scanning a sharp tip across a surface to generate an image. However, rather than monitoring the tunneling current between the tip and the surface, AFM monitors the height of the tip as it physically interacts with the surface at a constant force.

The setup of a typical AFM is shown in Figure 8.31. A sharp tip, often made of diamond or silicon nitride, is attached to a cantilever spring and placed in physical contact with the substrate at constant force. The position of the tip is monitored by reflecting a laser beam off the back of the cantilever and monitoring the deflection of this beam as the tip is scanned across the surface. If the tip encounters a bump or a dip, the reflection of the laser beam is slightly altered and monitored by the detector. Rather than moving the tip to scan the surface, the substrate is typically mounted on a piezoelectric tube scanner, allowing for the surface to be moved in a raster pattern under the tip. The cantilever itself is tens of microns in length, <10 microns in width, and ~1 micron in thickness. The tip is usually cone or pyramid shaped, with a height of several microns and a base width of several microns.

AFM is able to image both nonconducting and conducting surfaces, unlike STM. However, in AFM, the physical contact between the tip and the surface can damage the surface, resulting in a distorted image. This drawback is particularly problematic if one wishes to image a "soft" surface such as a biological membrane or a surfactant film. This problem of surface destruction can be somewhat overcome by scanning the

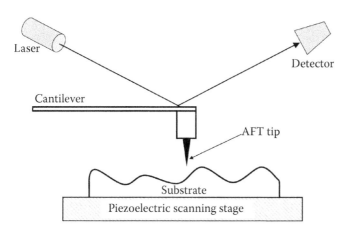

Figure 8.31 Schematic diagram of an atomic force microscope. The AFM tip is held in constant force against the sample surface by the cantilever. A laser beam is reflected off the back of the cantilever to monitor the tip height. The sample is scanned underneath the AFM tip by a piezoelectric stage.

substrate in *tapping mode*. In tapping-mode AFM, the tip is "tapped" on the surface, placing it in contact with the surface only for a short amount of time. This "tapping" is usually accomplished by oscillating the cantilever with constant driving force and ensuring that the cantilever is positioned so that the tip touches the surface only at the bottom of the oscillation. Typical frequencies of oscillation are on the order of a few hundred kilohertz. By using tapping-mode AFM, surfaces that would normally be destroyed are able to be imaged properly.

AFM can be used under water or on other liquids as well as in air, allowing for imaging at solid–liquid interfaces. This ability is particularly important for biological samples, which might be distorted at the surface-air boundary. One application of AFM in surface imaging is the detection of different functional groups on a surface, called **chemical force microscopy** (CFM). In CFM, the AFM tip is chemically functionalized, often by coating the tip with gold and then functionalizing it with a self-assembled monolayer using thiol-gold chemistry (Figure 8.32). For example, a gold-coated AFM tip might be functionalized with 11-mercaptoundecanoic acid, effectively resulting in a –COOH coated AFM tip [R is a $-(CH_2)_nCOOH$ group in Figure 8.32]. This type of chemisorption is discussed in Chapter 10. This chemically functionalized AFM tip can then be scanned across a surface composed of different functional groups. Depending on the interaction between the surface-bound functional group and the functional group on the AFM tip, a different "height" is reported. A –COOH coated tip would be expected to experience more frictional force from surface-bound –COOH groups as compared to $-CH_3$ groups, and therefore the –COOH groups would appear to be "taller" than the $-CH_3$ groups. For example, a –COOH-coated AFM tip can be used to scan across a surface containing both $-CH_3$ and -COOH groups. A different image is obtained when the same surface is scanned with a $-CH_3$-coated

Figure 8.32 A method for the creation of a CFM tip. An AFM tip is coated with Au and then is functionalized using thiol-gold chemistry. The functional group R can then be used to probe the surface.

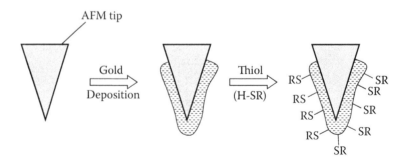

AFM tip. The results are striking and highlight the utility of CFM to detect different functional groups on a surface.

Aside from being used to image surfaces and surface-bound material, AFM has other applications. One unorthodox application of AFM in a non-imaging setting is in the recent development of a cell nanoinjector. In the past, in order to introduce a substance into a living cell, the cell had to be permeabilized with an electric current or chemical agent, or a bulky micropipette had to be used as an injector. This process could result in damage to the cell membrane, inducing undesirable side effects. However, scientists have been able to attach delivery moieties such as a carbon nanotubes to the end of an AFM tip for use as "nanoinjectors." Because of the tip's small dimensions, the cell membrane is not irreversibly perturbed.

The substance (or cargo) to be injected into a cell is chemically attached to the delivery moiety (through a disulfide linkage, for instance). A carbon nanotube as a delivery moiety acts like a "nanoneedle" when it is attached to the AFM tip, which is then used to inject the carbon nanotube (with the molecular cargo) through the cell membrane. Inside the reducing environment of the cell interior, the disulfide linkage between the cargo and the carbon nanotube is severed and the cargo is released into the cell. The AFM tip then withdraws the nanoneedle from the cell interior. Because the nanoneedle is of such small dimensions, the cell membrane is not perturbed greatly and the cell remains unharmed. Furthermore, by using the AFM machinery, excellent spatial precision in the placement of the cargo can be achieved. Figure 8.33 shows a schematic diagram of a nanoinjector and Figure 8.34 shows electron microscope images of the nanoneedle before and after the attachment of molecular cargo. Although the AFM-operated nanoinjector is admittedly limited to delivering cargo to one cell at a time, it still promises to be a useful tool in a wide variety of biological studies.

8.7.3 Transmission electron microscopy

Unlike the scanning probe microscopies, which utilize physical interactions between a sharp tip and a surface to create an image, electron microscopy techniques use a beam of electrons to visualize a sample. Because electrons have much smaller wavelengths than visible light, electron microscopies are able to achieve resolutions far greater than optical microscopies.

Figure 8.33 Schematic steps of the AFM-operated nanoinjector. The nanoinjector is linked to the molecular cargo through a disulfide bond and then poised above the cell membrane. The nanoinjector is then inserted through the cell membrane and the molecular cargo is released inside the cell as the disulfide linkages are reduced. Finally, the nanoinjector is withdrawn from the cell, ideally causing little to no membrane damage.

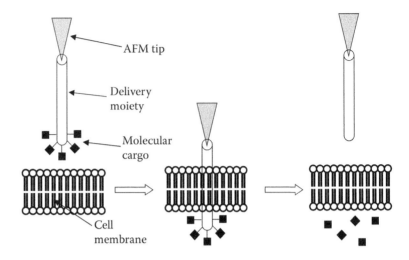

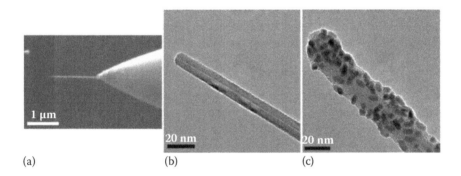

(a)   (b)   (c)

Figure 8.34 Imaging of the nanoneedle before and after the molecular cargo is attached. (a) SEM image of the carbon nanotube affixed to the AFM tip. (b) TEM image of the carbon nanotube shown in (a). (c) TEM image of the nanoneedle after the molecular cargo has been attached (a quantum dot in this case). (Image taken from X. Chen et al., 2007, *PNAS*, 104: 8218–8222. Copyright 2007 National Academy of Sciences, USA. With permission.)

Scanning electron microscopy (SEM) is a commonly used electron microscopy method that functions by monitoring the electrons that are backscattered after interacting with a sample and uses those backscattered electrons to reconstruct an image of the sample. On the other hand, **transmission electron microscopy** (TEM) makes use of electrons that are transmitted through a thin sample to produce an image. For our current discussion, we focus on TEM as a tool to image nanomaterials.

8.7.3.1 Principles of TEM

The resolution of an image produced by an optical microscope is limited by the wavelength of the radiation being used. If we use the Abbe equation applied to the resolution of a light microscope, then the maximum resolution δ is approximately given as

$$\delta \approx 0.61 \left(\frac{\lambda}{n \sin (\beta)} \right) \tag{8.38}$$

where λ is the wavelength of the light being used, n is the refractive index of the viewing medium, and β is a property of the magnifying lens called the semiangle of collection. Together, $n \sin(\beta)$ is often called the numerical aperture (NA) of the objective. To provide a rough estimate of δ, let's assume $n \sin(\beta) = 1{-}1.5$. We see that the resolution of a light microscope is approximately 50%–60% of the wavelength of the light being used. Visible light has wavelengths in the range of ~350–750 nm, so conventional optical microscopies are unable to resolve objects that are smaller than a few hundred nanometers. Since most nanomaterials of interest possess structures that are much smaller than several hundred nanometers, optical microscopies are only moderately useful in the imaging of nanomaterials. In order to visualize these nanomaterials, something with a much smaller wavelength than visible light must be used.

Electrons, as all small particles, can be thought of as being both particles and waves, with wavelengths that depend on their momentum (see Chapter 4). Recall that de Broglie's famous equation relates the wavelength λ of a particle to its momentum p by

$$\lambda = \frac{h}{p} \tag{8.39}$$

where h is Planck's constant, equal to 6.626×10^{-34} J s. Therefore, in order to calculate the wavelength of an electron (which gives us an idea of the maximum possible resolution of an electron microscope), we must be able to calculate its momentum. This goal can be achieved by understanding the basic physics of a TEM.

In an electron microscope, an electron produced in the electron gun is accelerated by an electric potential V toward the sample to be studied. This acceleration imparts a kinetic energy K to the electron that is equal to the accelerating potential or voltage. Therefore, using classical physics, we can write

$$V = K = \frac{1}{2} m_e v^2 \tag{8.40}$$

where m_e is the rest mass of the electron (9.11×10^{-31} kg) and v is the electron's velocity. From classical physics we also know that $p = mv$, so using Equation 4.40 we can write

$$p = m_e v = \sqrt{2m_e \left(\frac{1}{2} m_e v^2\right)} = \sqrt{2 m_e V} \tag{8.41}$$

Therefore, we can calculate the momentum of an electron if we know the accelerating voltage used in the TEM. Plugging this result into de Broglie's relationship, we get

$$\lambda = \frac{h}{p} = \frac{h}{\sqrt{2 m_e V}} \tag{8.42}$$

Therefore, we can calculate the wavelength of an electron from its accelerating voltage in an electron microscope. Furthermore, we see that as we increase the accelerating voltage, we decrease the wavelength of the electron. Ignoring other effects, an electron microscope can achieve better resolution by accelerating the electrons to higher energies. However, as the energy of the electrons increases, so does the likelihood that it can destroy or damage the sample being studied. This is one limitation to the maximum resolution that can be achieved in a TEM.

The magnitude of accelerating voltages typically used in most TEMs (on the order of hundreds of keV) accelerates the electron so much (near the speed of light) that relativistic effects must be accounted for. To account for these relativistic effects, Equation 8.42 becomes

$$\lambda = \frac{h}{\left[2m_e V \left(1 + \dfrac{V}{2m_e c^2}\right)\right]^{1/2}} \tag{8.43}$$

where c is the speed of light in a vacuum. Using this equation we can calculate that an electron that has been accelerated to 100 keV has a relativistic wavelength of ~3.7 pm. For a TEM, we can approximate the maximum resolution by modifying Equation 8.38 to

$$\delta \approx 0.61 \left(\frac{\lambda}{\beta} \right) \qquad (8.44)$$

where β in this case is a property of the electron "lens" used to focus the electron beam. Therefore, using Equation 8.44 we see that with a 100-keV electron beam ($\lambda \sim 4$ pm), we can achieve a theoretical maximum resolution ($\delta \sim$ several picometers) that is smaller than the diameter of an atom. While such maximum resolution is not feasible because we are unable to build perfect electron lenses, it highlights the fact that TEM is able to achieve a high level of resolution.

8.7.3.2 TEM instrumentation

In a simplified way, a transmission electron microscope operates like a slide projector. In a slide projector, a beam of light is transmitted through a slide. Some of the light is reflected or absorbed by the slide, so when the transmitted light is projected onto a screen, an image is produced. A TEM functions in essentially the same manner, with the only difference being that a beam of electrons is transmitted through the sample rather than a beam of visible light.

As seen in Figure 8.35, an electron gun produces a beam of electrons, often by heating a metal (most commonly tungsten) to such high temperatures

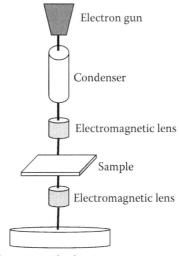

Figure 8.35 Schematic diagram of a transmission electron microscope. Electrons are produced by an electron gun and condensed into an electron beam in the condenser. This electron beam is focused onto the very thin sample by electromagnetic lenses. The transmitted electrons are collected by another electromagnetic lens and projected onto a fluorescent screen where they produce a visible image, which can be viewed directly or monitored on a computer.

that electrons are ejected from its surface. These ejected electrons are condensed and then focused by electromagnetic lenses, which are analogous to optical lenses, but function on different principles. The focused beam of electrons is shone on the sample, and those electrons that have been transmitted are gathered by another lens and then shone onto a fluorescent screen or other detector to produce an image of the sample. The setup of a TEM can be modified so that it is able to scan an entire sample, and this technique is called **scanning transmission electron microscopy** (STEM). As with AFM and STM, STEM typically scans a given sample in a raster pattern to produce an image.

In order to prevent deflection of the beam of electrons due to interactions with gas molecules inside the machine, the interior of a TEM must be operated under high vacuum conditions. This requirement is one of the principal disadvantages of TEM, although recently there have been rapid developments in the field of environmental TEM that allow operation under lesser vacuums. We will focus on the more common, high vacuum TEM. For this technique, the sample being examined must be able to withstand the high vacuum; otherwise, the image won't be an accurate representation of the sample under normal conditions. For nonsolid samples, two principal methods of sample preparation have emerged. These methods allow for the imaging of biological samples or other soft nanomaterials that otherwise could not be imaged using TEM. The first method is to dehydrate the sample, then stain the sample or coat it with metal to produce the necessary contrast. The second is to cryogenically freeze the sample and image the frozen specimen. One requirement of this second method is that the sample must be frozen so quickly that it does not have time to rearrange into a crystalline state. Otherwise, the ordered ice crystals produce a diffraction pattern of the electron beam that obscures the image of the sample. If the sample is frozen quickly enough, amorphous ice is produced and one can obtain an accurate image of the sample in its "natural" state.

The second major difficulty with TEM is that it is limited to observing very thin samples. In order to be effectively imaged, the sample must be thin enough to be transparent to the incoming beam of electrons. As a general rule, the sample should be less than 100 nm thick, although the exact suitable thickness depends on the material being examined and the energy of the electrons used. Thicker samples can be examined if electrons with higher energies are employed, but at higher energies the electrons may begin to destroy the sample. In terms of sample mounting, the sample can sometimes be placed on a sample holder that is very thin and transparent (or mostly transparent) to the electron beam. For extremely small samples, a sheet of thin, amorphous carbon films have been used as sample holders. These thin carbon films are suitable for such applications because they are electron-transparent nanomaterials that possess considerable durability even at a thickness of only one atom.

There are difficulties associated with using TEM to image nanomaterials, but it is a useful technique for a variety of applications and boasts impressive resolutions. Normally TEM is recommended for use in conjunction with other imaging methods to obtain an accurate understanding of the material being studied.

Figure 8.34 shows a TEM image of a carbon nanotube that has been attached to an AFM tip to serve as a nanoinjector (the operation of a nanoinjector is discussed in the previous section). The molecular cargo attached to the surface of the nanotube is clearly visible, thus demonstrating the resolving power of TEM.

8.7.4 Near-field scanning optical microscopy

In our introduction to TEM, we discussed the limited utility of conventional light microscopies in imaging nanostructures. In general, the resolution of an optical microscope is limited by the size of the spot to which the light beam can be focused using magnifying lenses. This limitation is often referred to as the **diffraction limit**. The diffraction limit on resolution is wavelength-dependent, as given by Equation 8.31 For modern objectives examining samples in an aqueous medium, *NA* is usually in the range of 1.3 to 1.5. Therefore, the resolving power of a conventional optical microscope is approximately half the wavelength of the incident light, typically ~200 nm for visible light.

The diffraction limit can be avoided using the nonoptical techniques discussed in the previous sections; however, all of these techniques have limitations with regard to sample preparation, sample type, or sample damage. Furthermore, none of these alternative methods offer the types of information that are available to optical methods—such as spectroscopic information, excellent time resolution, fluorescence detection capabilities, information about refractive index and reflectance of the sample, and contrasting power using different staining agents.

Near-field scanning optical microscopy (NSOM or SNOM) is an optical microscopy that can operate with resolutions below the diffraction limit, and therefore is able to offer the advantages of optical microscopies with a resolution that is actually useful for the study of nanomaterials.

8.7.4.1 History and principles of NSOM

Although the fundamental idea of NSOM is relatively simple, its practical implementation proved to be rather difficult. The original idea was developed by Edward Synge and was published in a series of papers beginning in 1928. Synge realized that if the diffraction limit was imposed by the practical limit to which a beam of light could be focused, then the limit might be overcome by shining light through a very small hole (or aperture) that was smaller than the wavelength of the light itself. If this hole were placed close to the sample, the light would not have time to diffract outward and destroy the resolution of the image. Thus, a sample could be imaged at a resolution below the diffraction limit.

Despite Synge's development of the theory of NSOM in the early twentieth century, it was not implemented until 1972 when Ash and Nicholls used an NSOM setup with microwave radiation ($\lambda \sim 3$ cm) to image a metal grating sample. They demonstrated that a resolution of 1/60th of the wavelength of the incident radiation was achievable using their method. Their results validated Synge's theory, but practical considerations prevented the development of an NSOM using the much smaller wavelengths of visible light until the mid-1980s. During this period, scientists overcame the technical difficulties of implementing NSOM with visible light. Their practical setup serves as the basis for modern NSOM instruments.

8.7.4.2 Modern NSOM instrumentation and different NSOM operating modes

The major component of any NSOM microscope is the aperture tip or NSOM probe. A variety of NSOM probes exist. One common NSOM probe is manufactured by heating and pulling a fiber-optic cable into a very fine point, then coating the tapered end with reflective metal, except for a very small aperture at the point. SEM images of this type of NSOM probe are shown in Figure 8.36. Laser light shone through the fiber-optic cable emerges from the aperture as a beam with a diameter that is smaller than the wavelength of light. For this type of fiber-optic NSOM probe, the fundamental maximum resolution is ~12 nm, but the practical limit is typically ~50 nm. Another common NSOM probe can be made by using electron beam lithography to create a nanometer-scaled aperture through

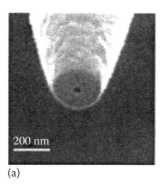
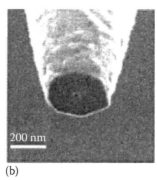

(a)   (b)

Figure 8.36 SEM images of two NSOM probes that have been constructed by tapering a fiber-optic cable into a very fine point and then coating that tip with aluminum, leaving only a very small aperture at the point. The ends have also been flattened using a focused ion beam. In these NSOM probes the apertures are approximately (a) 120 nm and (b) 35 nm. (Image reprinted with permission from J. A. Veerman et al., *Appl. Phys. Lett.*, 1998, 72: 3115–3117 as shown in R. C. Dunn, *Chem. Rev.*, 1999, 99: 2891–2927. Copyright 1998 American Institute of Physics.)

silicon or another metal and then shining the laser through this nano-metric aperture. Both of these probes can be manufactured with aperture diameters of ~50–100 nm. Recent developments have also been made in so-called apertureless NSOM probes. These probes exhibit great promise in achieving even better subdiffraction limit resolutions.

An NSOM setup can be operated in at least five different modes, as listed below (see Figure 8.37).

1. *Transmission mode.* Light is shone through the NSOM probe and then detected on the other side of the sample.

2. *Reflection mode.* Light is passed through the NSOM probe and reflected from the sample. This reflected light is detected and compiled to produce an image.

3. *Transmission-collection mode.* Light is shone on the sample from underneath and the transmitted light is collected in the NSOM probe and passed to a detector.

4. *Reflection-collection mode.* Light is shone on the sample from above and reflected light is collected by the NSOM probe and sent to a detector.

5. *Illumination-collection mode.* The NSOM probe is used as both the light source and the light collector.

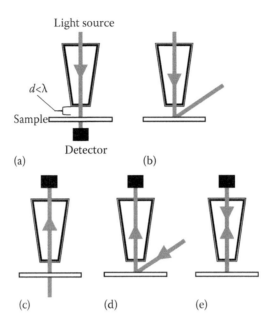

Figure 8.37   The different operating modes of NSOM. (a) In Transmission mode, the light travels from the probe through the sample to a detector on the other side. (b) In Reflection mode the light from the probe is reflected off the sample surface and captured by the detector. (c) In Transmission-Collection mode, the sample is illuminated from underneath and captured by the probe, through which the light travels to a detector. (d) In Reflection-Collection mode, the sample is illuminated externally to the probe and the reflected light is captured by the probe and channeled to the detector. (e) In Illumination-Collection mode, the probe is responsible for both illuminating the sample and collecting the light that reflects from its surface.

Depending on the type of sample examined and the kind of study undertaken, different NSOM operating modes may be employed.

Aside from issues related to manufacturing a suitable NSOM probe, the second major difficulty in the development of NSOM was probe positioning. In order to function correctly, the NSOM probe must be placed close to the surface of the sample being studied, usually less than a few nanometers. Such precise positioning is no easy task, especially when the probe is scanned across a "rough" sample surface. Furthermore, if the tip comes into forceful contact with the surface, either the probe or the sample could be damaged. Therefore, scientists have developed different feedback mechanisms to ensure that the probe remains at the correct distance above the sample, similar to those used to maintain sample-tip distance in STM or AFM. The shear-force feedback mechanism and the tapping-mode feedback mechanism are among the most common feedback mechanisms employed. While the details of each feedback mechanism are beyond the scope of this text, the essential idea behind each mechanism is that the probe is oscillated at a certain frequency and the force that is placed on the probe as it approaches the surface is monitored. By using a continuous feedback loop, the probe can be oscillated at the correct height above the sample. Furthermore, if the tip position is monitored as the sample is scanned in the xy plane, then topographical information about the sample can be obtained (much like with AFM or STM) in addition to any optical information. Finally, to ensure accurate probe positioning, the entire NSOM instrument is usually operated on a vibration isolation table.

A schematic of a typical modern NSOM microscope is shown in Figure 8.38. The setup depicted is an NSOM microscope with a fiber-optic probe operating in transmission mode. The laser light is initially passed through a bandpass filter and a combination of half-wave and quarter-wave plates to control the wavelength and polarization of the light, respectively. The light then enters the fiber-optic cable and is transmitted to the probe tip through which it is shone on the sample. The transmitted light is collected by the detector and processed into an image. The height of the probe tip is controlled by the feedback mechanism, and the probe's vertical position can be monitored to produce topographical information. The sample itself is placed on a piezoelectric stage that can move in the x–y direction, allowing for the sample to be scanned in a raster pattern, much like with AFM or STM.

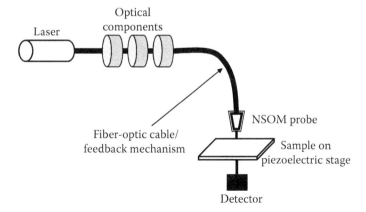

Figure 8.38   Schematic of a near-field scanning optical microscope operating in Transmission mode. Laser light is channeled into a fiber-optic cable that ends in an NSOM probe tip positioned very close to the sample surface. The transmitted light is captured by the detector to produce an image. The sample is scanned by the piezoelectric scanning stage, and the NSOM probe height is monitored and controlled by the feedback mechanism.

An example of an NSOM image is shown in Figure 8.39, which shows the topographical and cross-polarized optical NSOM images of spherical and toroidal (donut-shaped) liquid crystal droplets suspended in a polymer. Liquid crystals are materials that have properties between those of a liquid and those of a solid. There are many types of liquid crystals based on the orientation of molecules, but for our present purposes we merely note that the general shape of the droplets is clearly seen in the topographical images and that the orientation of the crystals can be seen in the cross-polarized images.

Figure 8.39 NSOM images of spherical and toroidal liquid crystals suspended in a polymer. (Image from E. Mei et al., *Langmuir*, 1998, 14: 1945–1950 as shown in R. C. Dunn, *Chem. Rev.*, 1999, 99: 2891–2927. With permission.)

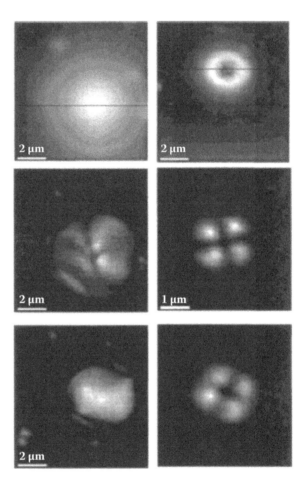

NSOM setups are often used in conjunction with other surface imaging methods to provide complementary information. For example, NSOM microscopes are commonly incorporated into conventional fluorescence microscopes, and surface-enhanced Raman NSOM setups have also been employed.

Overall, NSOM is a useful imaging method that can offer a wealth of information about a nanomaterial being studied. While its maximum resolution (often ~50 nm) is not quite as impressive as AFM, STM, or electron microscopy, its versatility in being able to image a wide variety of samples under "normal" conditions and its richness in the types of information it provides usually compensate for this lower resolution.

End of chapter questions

1. Equation 8.1 can be used to determine the surface tension of a liquid by measuring how far the liquid travels up a narrow capillary tube. How far would water rise up a narrow capillary of radius 1 mm? Assume that water makes a zero contact angle against the surface of the tube.

2. A paper plate of width 10 mm and thickness 1 mm is withdrawn from a surfactant solution. A force of 0.77 mN is measured just before the plate detaches from the surface of the solution. Estimate the surface tension of the surfactant solution.

3. In a QCM experiment, the following values of frequency (at the third overtone, $n = 3$) were measured in real time as a nanofilm of mammalian DNA was adsorbed on a quartz crystal with Sauerbrey constant $C = 17.7$ ng Hz^{-1} cm^{-2}.

Use the Saurbrey equation to generate a plot of time versus adsorbed mass. Assume the adsorption follows first-order kinetics; that is, [Adsorbed DNA mass] $= (1 - Ae^{-kt})$ where A is a constant and k is the first-order rate constant. Use this equation to obtain a value for k.

4. Consider solutions of bovine serum albumin (BSA), a common blood protein, made at pH 3, 5, and 7 flowed over a silicon waveguide in a DPI experiment. pH 5 is close to the isoelectric point of BSA (or the pH at which it has no net charge), where it is expected that each BSA protein experiences little electrostatic repulsion with neighboring proteins at pH 5.

   (a) Predict at what pH you would expect the maximum adsorption of BSA. Explain your answer.

Time (min)	0	5	10	15	20	25	30	35	40	45	50	55	60	65
ΔF (Hz)	0	0	−2	−5	−7	−15	−25	−42	−55	−60	−63	−64	−65	−65

(b) In bulk solution, BSA adopts an ellipsoid shape with dimensions of 140 Å × 40 Å and it has a molecular weight of 66.43 kDa. Using this information in conjunction with the DPI mass values, how would you calculate the area occupied by each protein molecule on the waveguide surface at each pH?

(c) If we were to assume that the BSA proteins adsorbed parallel to the waveguide surface to form a saturated monolayer and that they were not distorted on adsorption, then we could use the bulk phase dimensions of BSA to calculate the nanofilm thickness (it's ~40 Å and each molecule occupies ~5600 Å² on the surface at pH 5). The nanofilm layer thickness values are much less for the BSA proteins at pH 3 and 7, and the area-per-molecule values are much greater than predicted. Thus, it can be inferred that the BSA molecules at pH 3 and 7 are spread much more thinly on the waveguide surface and that they assume a much more distorted form. The relative rates of adsorption of BSA at different pH to the silicon surface are pH 3 > pH 5 > pH 7. What forces are responsible for the adsorption of BSA to the waveguide surface at each pH? What explanation can be offered for the observed order of the relative rates of adsorption of BSA at differing pH?

5. The following table shows the refractive index of polyethylenimine (PEI) as a function of concentration. The refractive index of a film of PEI on the surface of a DPI waveguide is 1.55. Assuming this film is under a pure water solution, estimate the density of the film. In a separate experiment, ellipsometry determined this film to have a thickness of 1 nm. What is the mass of the film?

Concentration (mM)	1.0	5.0	10	15	20
Refractive Index	1.44	1.45	1.46	1.47	1.48

6. Common IRE materials used in ATR-FTIR are ZnSe and germanium (Ge). Using tabulated values in the scientific literature, we can find the refractive index values for ZnSe and Ge at a variety of incident light wavelengths. Using other known experimental parameters allows us to calculate the penetration depth of the evanescent wave into the sample being studied. For example, at IR frequencies of ~1700 cm^{-1} (the typical stretching frequency of a C = O bond), ZnSe has an refractive index of ~2.35, whereas the refractive index of Ge is ~4.0. Assuming that the angle of incidence used in an IRE of an ATR-FTIR is 45°, compare the evanescent wave penetration depths in an aqueous sample (n ~1.5) for an IRE made of either ZnSe or Ge when IR radiation of 1700 cm^{-1} is used. Hint: See the Section 8.4.1 for a useful equation to calculate the penetration depths.

7. Consider a nanofilm comprised of discrete layers of a chromophore, such as a dye molecule. Ellipsometry can measure the thickness of such a film. If the molar absorptivity of the molecule is known, show how Equation 6.9 can be used to determine the mass of the film.

8. Consider the ellipsometric refractive index data for YLD-124 shown in Figure 8.11. Assuming that you had this spectrum and a number of YLD-124 samples of unknown thickness on silicon substrates, could you accurately measure the thickness of the films using a single-wavelength, single-angle ellipsometer with a helium–neon laser (632.8 nm)? Explain your reasoning. If not, what changes could you make to your instrumental setup to improve your accuracy that do not involve purchasing an entire new instrument?

9. The Sauerbrey equation underestimates the mass adsorbed to the crystal surface in liquids because it was originally developed for oscillation in air and does not account for the decrease in resonant frequency due to the change in viscosity. It is possible to amend the Sauerbrey equation for an accurate measurement of mass in liquid by correcting for these viscosity effects according to the following equation:

$$\Delta f = f_o^{3/2} \sqrt{\left(\frac{\eta_l/\rho_l}{\pi \rho_q \mu_q} \right)}$$

where f_0 is the resonant frequency. ρ_l is the density of the liquid, η_l is the viscosity of the liquid, ρ_q is the density of quartz (2.648 g/cm³), and μ_q is the shear modulus of quartz for AT-cut crystal (2.947×10^{11} g/cm s²). In which liquid, water or deuterium oxide, is the detection limit of a typical QCM-D instrument at its fundamental resonant frequency more sensitive? What is the ratio of their sensitivities? (The interested student should see the following reference for more details: Kanazawa, K. K. and Gordon, J. G. II, The Oscillation Frequency of a Quartz Resonator in Contact with a Liquid, *Anal. Chimica Acta*, , 1985, 175: 99–105.

10. (a) The text mentions that many techniques cannot by themselves provide absolute measurements of properties such as refractive index, thickness of a film, or density. Choose two nonspectroscopic techniques discussed in the chapter and describe how their combined use provides more information than either would alone. Outline a set of experiments that you would undertake to learn as much as possible about a particular film-forming substance such as SDS (or provide a compound).

(b) Choose a spectroscopic method, such as IR, and describe how this technique could add to your understanding of the material and its adsorption properties.

11. Scanning tunneling microscopy is used to image a metal surface with an electronic decay length of 5 nm⁻¹. If the tunneling current increases by a factor of 5, then how much has the height changed? Describe what this height refers to in your own words. Hint: See Example 8.4.

12. Rank from tallest to shortest the height of an AFM tip that has been functionally coated with –COOH as it traverses a surface of the following makeup:

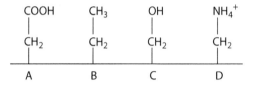

13. I'm using NSOM to detect the incorporation of malaria-causing *Plasmodium* proteins in the membrane of malaria-infected human erythrocytes. I place my sample 300 nm from the NSOM probe but see that the resulting pictures are not resolved. I then realize that the wavelength of the light source from my laser is 150 nm. What must I do to increase the resolution?

14. This question concerns SFG spectroscopy. Consider the overlap of two light beams (one visible of wavelength 532 nm and the other IR of wavelength 2.2 μm) at a solid surface. The angle of incidence of the visible beam is 30° and that of the IR beam is 40°.

(a) Calculate the angle of the emitted SFG beam.

(b) What is the wavelength of the emitted SFG beam?

(c) Describe qualitatively the difference between the SFG sprectra of SDS, CTAB, hexanol, and decanol in terms of the CH and CH bands. Assume that monolayers of these molecules are present between a hydrophobic surface and an aqueous phase.

15. In a study conducted at the Tokyo Institute of Technology, QCM was used to directly monitor the reactions of the Klenow fragment of *Escherichia coli* DNA polymerase I. The reactions take place on DNA oligonucleotides, containing either a $(TTTTC)_3$ or $(TTTTC)_{10}$ template and the primer necessary for polymerase binding, which have been immobilized on the QCM. This enzyme acts as a catalyst for complementary base pairing of dATP and dGTP monomers from solution into a second strand. The results of two trials with different initial conditions for the $(TTTTC)_3$ template are shown below:

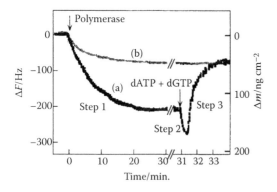

In (a), polymerase was added at first, and then excess monomers were added after 30 minutes. In (b), polymerase was added in the presence of excess monomers.

(a) What is physically occurring at steps 1, 2, and 3 of (a) in terms of the biological components of the reaction? What is physically occurring in (b) that causes a different curve? Why does it make sense that the final changes in frequencies are the same?
Hint: The frequency change in step 1 of (a) is of roughly equivalent magnitude to the frequency change in step 3 of (a), and the frequency change at step 2 of (a) is roughly equivalent to the total frequency change in (b).

(b) What can be said about the relative magnitudes of the rate constants for each of the three steps in the DNA polymerase reaction mechanism based off the data above?

(c) What relative changes in frequency, coupled mass, and rate constants would be expected if using the $(TTTTC)_{10}$ template instead of the $(TTTTC)_3$ template?

References and recommended reading

Evans, D. F. and Wennerström, H. *The Colloidal Domain*, 2nd ed. 1999. Wiley-VCH, New York. Chapter 2 provides a light read on AFM and STM.

Purcell, E. M. and Morin, D. J. *Electricity and Magnetism*, 3rd Ed. 2013. Cambridge University Press, Cambridge, UK. An excellent introduction to electrostatics and electrodynamics that provides background relevant to both coulombic interactions in matter and the interaction of electric fields with matter.

Fowles, G. R. *Introduction to Modern Optics*, 2nd ed. 1975. Dover Publications, New York. This undergraduate textbook provides a good introduction to wave and ray optics, including elementary nonlinear optics.

Tompkins, H. G. *A User's Guide to Ellipsometry*. 2006. Dover Publications, Mineola, NY. This is a graduate-level book that covers the theory of ellipsometry in detail and provides many interesting applications of the technique.

Vickerman, J. C. *Surface Analysis—The Principal Techniques*. 2003. John Wiley & Sons, Chichester, West Sussex, UK. This book provides an excellent coverage of surface science techniques, many of which are not covered in this chapter. The book is recommended for students who wish to gain a better understanding of vibrational spectroscopy at surfaces and scanning probe methods.

Boyd, R. W. *Nonlinear Optics*. 1992. Academic Press, San Diego, CA. This book is recommended for the advanced student interested in nonlinear methods such as second-harmonic generation and sum-frequency generation. The book provides an excellent fundamental treatment of nonlinear optics.

Bonnel, D. *Scanning Probe Microscopy and Spectroscopy: Theory, Techniques, and Applications*, 2nd ed. 2000. Wiley-VCH, New York. This book focuses mainly on STM and is only recommended for the student who is seriously interested in this method.

CHAPTER 9

Introduction to Functional Nanomaterials

CHAPTER OVERVIEW

This chapter consists of a survey of several common nanomaterials and their applications, combined with the fundamental physics required to understand the function of those materials, including transfer of energy and/or charge between components of a device. The second half of the chapter is written more like a scientific review of nanomaterials, with the goal of helping the student to transition from the fundamental textbook-style material to the primary scientific literature. There are many other potential applications of nanomaterials, and both the range of applications and technologies implemented is constantly evolving; the examples discussed represent only a small subset of emerging materials and technologies. Thin films and other 2D systems will not be heavily covered in this chapter, as they will be discussed in detail in Chapter 10.

9.1 NANOSCALE MACHINES

Nanoassemblies consist of many components held together by a series of specific interactions, whether covalent or noncovalent. **Functional nanoassemblies**, which are capable of responding to the presence of a target compound or energy source and responding to it by changing their own structure or performing work on their surroundings, are common in nature. Living organisms contain a wide variety of proteins that consist of many different chemical functional groups connected to one or more polymer chains and cofactors (small molecules or inorganic clusters) that allow them to harness energy from a chemical reaction (such as the hydrolysis of ATP to ADP and phosphate) or light (e.g., photosynthesis) and perform some task. An example of a complex natural nanoscale machine is nitrogenase, a protein found in certain soil bacteria such as *azotobacter* vinelandii, which uses energy from hydrolyzing ATP to break the strong triple bond in N_2, transports electrons produced from other reactions to the active sites on the proteins, and then reduces the N_2 to ammonia (NH_3). The protein complex consists of two types of proteins, each composed of multiple chains and with multiple iron–sulfur clusters, for a total of nearly 50,000 atoms and with the longest dimension in excess of 15 nm. A simplified view of the structure showing the different chains and metal centers is shown in Figure 9.1. The protein undergoes a complex series of motions (Danyal et al., 2016) that coordinates catalysis of the different steps of nitrogen reduction, essentially functioning as a tiny machine. The resulting natural nanosystem can perform the nitrogen reduction process at ambient temperature and pressure, in contrast with the Haber–Bosch process, which has been used for industrial ammonia production since the early twentieth century and requires high temperatures and pressure to overcome the activation energy of the reaction, such that producing ammonia requires approximately 2% of the world's primary energy production. Developing new catalysts that could mimic the functionality of nitrogenase and similar enzymes with engineered nanosystems could lead to far more efficient industrial processes. However; functional materials are not limited to biological systems or catalyzing reactions; molecular components can be combined to form materials for a wide range of functional systems, including sensors, medical treatments, electronic devices, and structural materials.

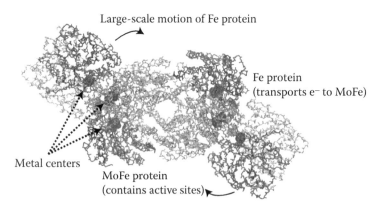

Figure 9.1 The structure of the nitrogenase protein (PDB ID: 2afk) showing only the peptide backbone and metal centers (no side chains). Electrons are transferred from the iron–sulfur center in the Fe protein to the active site at a metal center called FeMo-co, where they are used to reduce N_2 to NH_3. The protein undergoes large-scale motions that likely control when electron transport can occur (Danyal et al., 2016).

Substantial research effort has been undertaken in how to combine molecules, materials, or components in to a wide variety of functional nanosystems. **Supramolecular chemistry** is concerned with the study of the basic features of interactions between molecular components and with their implementation in specially designed nonnatural systems. As multiple components are combined into a nanoassembly, the range of possible functions for that material increases. Assemblies consisting of two or more components can transfer energy and/or electric charges between the components, allowing the assemblies to convert energy from an external source such as light or an electric circuit into useful work. The broad category of all such nanoassemblies has been dubbed molecular machines (if components are covalently attached) or supramolecular machines (if components are held together by noncovalent interactions). These machines are comprised of organized systems of cooperating molecules, components of a macromolecule, or a combination of solid materials and molecules. (Supra)molecular machines are deliberately and specifically engineered with the ability to function as sensors, processors, and custom catalysts. Other examples of nanosystems that involve energy or charge transfer between materials, dyes, and polymers to accomplish a desired function include solar cells, light-emitting diodes, and transistors.

One, more fanciful but illustrative example of nanoscale machines is "nanocars," or molecules that have mobile, functional components that are able to convert an external source of energy in to linear motion. An initial design reported in *Nano Letters* (Shirai et al., 2005) relied on thermally driven rotation of fullerenes to move a large (2 nm × 3 nm) planar organic molecule across a surface, as observed by STM. A later variant (Kudernac et al., 2011) used isomerization of double bonds driven by the tunnel current of a STM combined with vibrational motion of an alkyl side chain to produce linear motion in an intended direction. Figure 9.2 shows the structure and motion of such a nanocar. The same molecular engineering techniques and physical processes used to make nanocars have also recently been applied to make light-driven molecular machines capable of killing cancer cells by breaching their lipid membranes, illustrating some of the potential that this technology holds (García-López et al., 2017).

In the following sections, we will discuss key processes for functional nanomaterials, such as charge and excitation transfer. We will also survey organic and inorganic materials of different dimensionalities that can be used to construct functional nanomaterials. Functional biological

Figure 9.2 An electrically driven nanocar. Panel (a) shows the molecular structure of the nanocar and panel (d) shows a STM image of the nanocar and a schematic of the excitation process. Panels (b), (c), and (e) show motion of the nanocar driven by rotation of the fluorene "wheels," which are rotated by isomerization of a double bond. (Reproduced from T. Kudernac et al., *Nature* 2011, 479: 208–211. With permission from Springer Nature.)

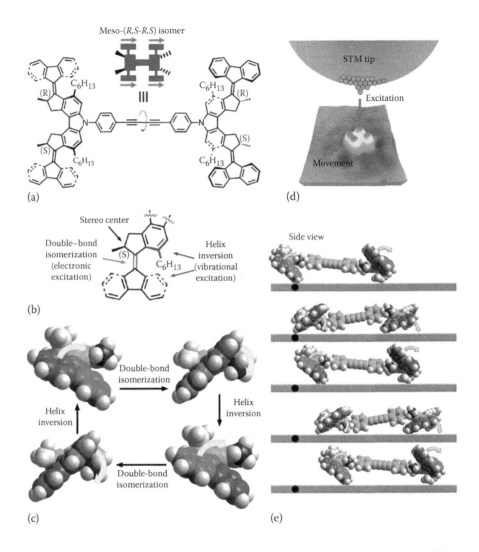

materials are beyond the scope of this chapter, although many excellent review articles exist, such as Huang et al., 2016.

9.2 CHARGE TRANSFER

Charge transfer is the movement of an electron from one molecule to another, or one portion of a molecule to another. The source of the electron is called an electron donor (D) and the component receiving an electron is called an electron acceptor (A). The general form of such a process can be written as

$$D + A \xrightarrow{k_{ET}} D^+ + A^-$$ (9.1)

where k_{ET} is the rate constant of the electron transfer process. Note that the oxidization states in Equation 9.1 are generic; an electron transfer process does not need to start with two neutral molecules, but each ET step does involve a species being reduced and another species being oxidized. The reaction does not even need to change the composition of the system. For example, a reaction such as

$$Fe^{+3}_{(aq)} + Fe^{+2}_{(aq)} \rightarrow Fe^{+2}_{(aq)} + Fe^{+3}_{(aq)}$$ (9.2)

is a valid electron transfer process analogous to the exchange of H^+ between water molecules in aqueous solution. Just like other reactions, electron transfer processes can be broken into multiple elementary steps and approximations can be applied to allow analysis of these processes. One of the most common models for charge transfer is the **Marcus theory** (Marcus, 1956, Marcus and Sutin, 1985).

9.2.1 The Marcus model of charge transfer

The Marcus model of charge transfer was originally developed for simple ionic exchange reactions such as Equation 9.2, but has since been generalized for a wide variety of electron transfer processes between molecules or structures such as nanoparticles. The Marcus model treats the electron transfer process as a tunneling process (Chapter 4) between two overlapping states. It assumes that electron transfer kinetics can be modeled using an Arrhenius-like approach (Chapter 3), in which the activation free energy ($\Delta G^{\ddagger}$), analogous to E_a, is a function of the free-energy change ΔG of moving an electron from donor to acceptor and the energy change for the environment around the donor and acceptor to respond to the new charge distribution. This second energy change is known as the solvent reorganization energy, λ. A reaction energy diagram showing these parameters is shown in Figure 9.3.

Under these assumptions, the rate constant of an electron transfer reaction can be written as

$$k_{ET} \propto e^{-\frac{\Delta G^{\ddagger}}{RT}} = \frac{2\pi}{h} \frac{e^{\frac{(-\Delta G + \lambda)^2}{4\lambda k_B T}}}{\sqrt{4\pi\lambda k_B T}} |T_{DA}|^2$$ (9.3)

where T_{DA} is the overlap between the wavefunctions of the donor and acceptor analogous to the transition density for exciting an electron

Figure 9.3 A schematic of the Marcus model of electron transfer as a function of an arbitrary reaction coordinate showing the transition state, ΔG, of the electron transport process and solvent reorganization energy λ.

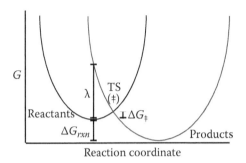

within a molecule, h is Planck's constant, and k_B is Boltzmann's constant (R/N_A). While Equation 9.3 may appear complex, it shows a systematic relationship between the structure and energetics of a donor and acceptor and a crucial process for materials. The solvent reorganization energy λ consists of two components. The first component is the change in bond energies known as the inner-sphere component, which can be calculated from a change in vibrational frequencies. The second component represents the difference in electrostatic interactions between the donor and acceptor and is called the outer-sphere component. It can be represented as

$$\lambda_{\text{outer}} = e^2 \left(\frac{1}{2r_D} + \frac{1}{2r_A} - \frac{1}{r_{DA}} \right) \left(\frac{1}{n^2} - \frac{1}{\varepsilon} \right) \tag{9.4}$$

where e is the fundamental unit of electric charge, r_D is the radius of the electron donor, r_A is the radius of the electron acceptor, r_{DA} is the distance between the donor and acceptor, n is the refractive index of the medium, and ε is the static dielectric constant of the medium. For most systems, the outer-sphere (electrostatic) component has a much larger contribution than the inner-sphere component.

Marcus theory provides a useful framework for predicting rates of electron transfer between molecules and/or nanostructures when the energetics of the states involved are understood. Understanding the energetics of these states in extended systems requires generalizing our understanding of molecular electronic structure from molecules to larger systems such as nanoparticles, nanowires, or surfaces. These systems have many states that are close in energy to each other, forming continuous bands separated by gaps, which can be described using **band theory**.

9.2.2 Band theory

Band theory is the generalization of molecular orbitals to larger systems. While often used in the context of bulk inorganic solids, it is also useful for understanding the electronic structure of large molecules and of nanomaterials. Consider the three molecules shown in Figure 9.4. As previously discussed in Chapter 4, ethylene (left) has one filled π bonding orbital and one unfilled π^* antibonding orbital, to which electrons could be excited by light with an energy equal to the energy difference between orbitals. Hexatriene (center) has three filled π bonding orbitals, each with a corresponding antibonding orbital. Since the electrons are delocalized over a longer region, the energy difference between the HOMO and LUMO is smaller and an electron could be excited by longer-wavelength light. Now, what happens if the size (and number of electrons) in the molecule is increased even more? Polyacetylene (right) is a very long chain of CH units connected by alternating single and double bonds with a large number of π electrons, such that the energy difference between individual filled or unfilled orbitals becomes negligible compared to the thermal energy $k_B T$ of the environment that the molecule is in. In this situation, the HOMO represents the top orbital of a continuous range of filled states called the **valence band** (VB), and the LUMO represents the lowest-energy orbital of a continuous range of empty states called with the **conduction band** (CB). As with the molecular systems that we have previously discussed, electrons can only be excited to a different energy level by light with an energy equal to the energy difference between orbitals. For a system containing many electrons, this means that an electron could be excited from any orbital in the VB to any orbital in the CB, but not into the region between bands. The energy difference between the top of the VB and the bottom of the CB is called the band gap (E_g), and the

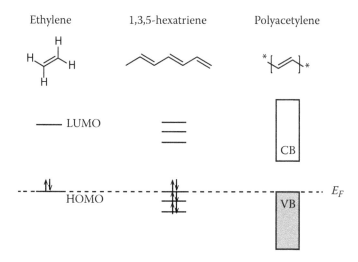

Figure 9.4 Formation of bands of electronic states. Ethylene (left) has only one occupied π orbital with a corresponding antibonding orbital. Extending the chain of carbon atoms increases the number of contributing orbitals and decreases the spacing between them until they become effectively continuous (right). Note how the band gap also decreases as the chain is extended, as electrons are delocalized over a longer distance.

energy of the highest-energy state containing an electron is called the Fermi level (E_F).

However, orbitals within bands are not necessarily evenly spaced in energy. The number of orbitals at each energy is known as the **density of states**, or $g(E)$. For example, for electrons confined in a 3D box, the density of states is given by

$$g(E)_{3D} = 4\pi V \left(\frac{2m_e}{h^2}\right)^{3/2} \sqrt{E} \qquad (9.5)$$

where V is the volume in which the electrons are confined. Densities of states for complex systems can also be determined using spectroscopic techniques or computer simulations. As another example, the density of states for the nonlinear optical dye YLD-124 in the solid phase, as determined using a computational technique known as density functional theory, is shown in Figure 9.5.

The concept of band theory and density of states can be used to classify how well a material conducts electricity as well as its ability to be excited by light of different wavelengths. If a material has a band gap of more than about 3–4 eV (where 1 eV is the energy of an electron in a 1 V potential, or about 96.5 kJ/mol), which corresponds to a wavelength range of 310–413 nm, it is an **insulator**, and does not appreciably conduct electricity. It also does not absorb light at visible wavelengths and appears colorless. If a material has a smaller but nonzero band gap, it is a **semiconductor** and absorbs light at visible wavelengths. However, if the Fermi level is within a band of orbitals, then electrons can freely move into unoccupied orbitals

Figure 9.5 Computed density of states for a crystal of the dye YLD-124, calculated using density functional theory. Occupied states are indicated by a black line and unoccupied states by a gray line.

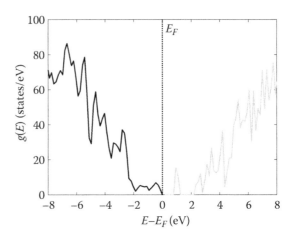

and the material is a **conductor**. Conductors can absorb (or, alternately, reflect, depending on the relationship between the real and imaginary components of the refractive index) light at all wavelengths. Simplified densities of states are shown for these types of materials in Figure 9.6.

In the limit of a large number of states, the probability of a state with a given energy being occupied can be written as a continuous function of temperature,

$$f(E) = \frac{1}{e^{(E-E_F)/k_B T} + 1} \tag{9.6}$$

This relation is known as the Fermi–Dirac distribution. At the Fermi level, a state in a material has a 50% chance of being occupied and the probability of an electron being in a state with higher energy decays exponentially as a function of temperature. Since the thermal energy $k_B T$ at room temperature is only about 0.025 eV, the probability of charge carriers being in the conduction band is incredibly small in a semiconductor with a band gap of even 1 eV.

In order to improve the number of charge carriers in a semiconductor to allow it to transport charge better, impurities can be added to the material via a process called **doping**. For a molecule or organic material, this could mean oxidizing or reducing some of the molecules in the system, and for an inorganic material, this typically means substituting atoms in the material with atoms that have either one fewer (p-doping) valence electron or one more (n-doping) valence electron than the atom they are

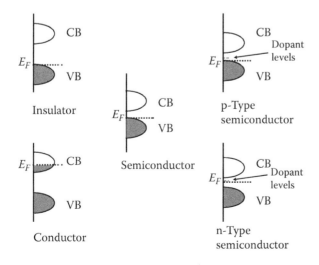

Figure 9.6 Schematic densities of states for insulators, conductors, and semiconductors. p-type semiconductors have extra unoccupied states above their valence band that allow them to more easily conduct holes (empty orbitals) through their valence band, while n-type semiconductors have additional occupied states below their conduction band that allow them to more easily conduct electrons through their conduction band.

replacing. For example, silicon—presently the backbone of the semiconductor industry—has four valence electrons. Silicon can be n-doped with phosphorous (five valence electrons) to form an n-type semiconductor or p-doped with boron (three valence electrons) to form a p-type semiconductor. These impurities either add occupied orbitals in the band gap (n-doping) or empty orbitals in the band gap (p-doping). Schematic densities of states for n-type and p-type semiconductors are shown in Figure 9.6. For either type of doping, the probability that an electron can be excited is greatly increased compared to an undoped semiconductor.

Note that molecular materials and nanomaterials may have densities of states that resemble a doped system and may be intrinsically n-type or p-type without further modification; the ability control electronic structure via molecular design or particle size is greatly beneficial for nanomaterials in contrast with conventional bulk semiconductors.

In an n-type material, electrons in the conduction band are the mobile charge carrier. However, in a p-type system, the mobile charge carriers are unoccupied orbitals in the valence band, which are called holes (h^+) and are treated as having a positive charge since they represent the absence of an electron. The difference in electronic structure charge carrier type between different types of semiconductors can be exploited to build functional devices such as solar cells, light emitting diodes, and transistors.

9.2.3 Solar cells and light-emitting diodes

One illustrative example of a class of semiconductor device that uses charge transfer to convert energy from an external source to useful electrical work is solar cells, also known as **photovoltaic cells**. Solar cells consist of a layer of a p-type material in contact with a layer of an n-type material (Figure 9.7). The difference in the band energies creates an electric field in the region between the materials. If light at a wavelength sufficient to excite an electron in the p-type material from the valence band to the conduction band is shone on the device, electrons are excited to the conduction band, then are pulled into the conduction band of the n-type material by the electric field induced by difference in band energies between the two materials. A hole is left behind in the valence band of the p-type material. The electron and hole each migrate to electrodes and pass through an electric circuit where they recombine and deliver their energy (Figure 9.7).

The maximum voltage (called the open-circuit voltage, or V_{OC}) of an individual photovoltaic cell is limited by the energy difference of the

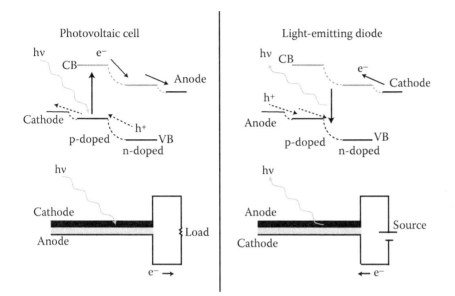

Figure 9.7 Schematic of the operation of both a photovoltaic cell (left) and a light-emitting diode (right). The relevant energy levels (top of valence band and bottom of conduction band) are shown in the upper portion of the figure; the physical structure of the devices are shown in the bottom half.

Fermi levels of the anode (positive electrode) and cathode (negative electrode). More fundamentally, it is limited by the longest wavelength of light absorbed; for example, a solar cell with a band gap of 2 eV (620 nm) could produce no more than 2V. The current produced by a solar cell is proportional to the number of photons absorbed; maximum current is produced when the solar cell is connected to itself without powering any load (short-circuit current). The power generated is equal to the product of voltage and current, or proportional to V_{oc} and I_{sc},

$$P = IV = FF \cdot V_{OC}I_{SC} \qquad (9.7)$$

where *FF* is a constant of proportionality known as the fill factor. The efficiency of photovoltaic cells is far less than 100% as not all photons are absorbed and some charge carriers recombine in the device, losing energy by reemitting light or nonradiative processes (Chapter 6). At present, the best efficiency available in conventional crystalline silicon solar cells with the most advanced cells, which use multiple junctions and concentrate light with lenses or mirrors, is about 40% efficient (National Renewable Energy Laboratory, 2017).

What happens if a P–N junction with a band gap in the visible range is connected such that a power source injects electrons into the n-type material? In this situation, the electrons will migrate to the junction and drop down to the valence band, recombining with holes in the p-type material and emitting light at a wavelength corresponding to the energy of the band gap, forming a **light-emitting diode**, or LED. LEDs are essentially solar cells running in reverse. Of course, for practical devices, the materials are optimized for their intended application (e.g., preventing recombination of charge carriers in solar cells while encouraging it in LEDs, or tuning a solar cell to absorb a wavelength range that generates maximum power vs. tuning a LED to generate a desired color). These optimizations can be performed by selecting from a wide variety of materials, dopants, and additional layers of materials outside the P–N junction.

Both photovoltaic cells and light-emitting diodes have been active areas of nanotechnology research. Blends of organic polymers and fullerenes can be used to make thin and flexible photovoltaic cells (Mazzio and Luscombe, 2015). Examples of two common materials found in organic photovoltaic cells are shown in Figure 9.8. The efficiencies of organic photovoltaic cells can be further enhanced by blending with metal nanoparticles, using plasmons (discussed in the context of SPR in Chapter 8) to concentrate light in the device (Gan et al., 2013). While organic photovoltaic systems are still rare compared to conventional silicon systems, organic LEDs have become ubiquitous in consumer electronics, and a wide variety of materials have been developed to generate a broad spectrum of colors in thin and long-lasting devices (Thejokalyani and Dhoble, 2013) Alternately, devices can be constructed using dyes adsorbed to nanoparticles and embedded in a liquid or solid electrolyte (Hagfeldt et al., 2010), which enables simple device fabrication

Figure 9.8 Two common organic semiconductors used in organic photovoltaic cells, P3HT (left) and PCBM (right).

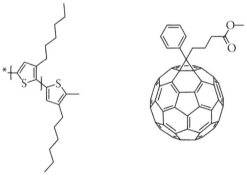

P3HT
(poly(3-hexyl thiophene))
p-Type semiconductor

PCBM
(phenyl-C61-butyric acid methyl ester)
n-Type semiconductor

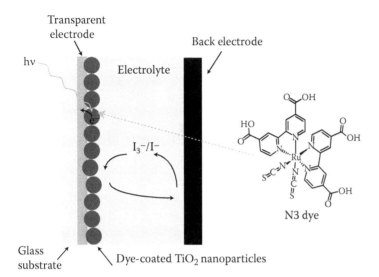

Transparent electrode

Back electrode

hν

Electrolyte

I_3^-/I^-

N3 dye

Glass substrate

Dye-coated TiO_2 nanoparticles

Figure 9.9 A schematic of the operation of a dye-sensitized solar cell. Light excites a dye, which transfers an electron to a nanoparticle, which in turn injects the electron in to an electrode where it can do work in an electrical circuit. The dye is reduced back to its initial state by an electrolyte such as I_2/I_3^-, which transfers an electron and is in turn reduced by electrons reentering the device at the opposite electrode.

compared to conventional semiconductor fabrication processes. A schematic of a dye-sensitized solar cell is shown in Figure 9.9. Similarly, quantum dots (a type of nanoparticle) are being explored for use in light-emitting diodes in conjunction with organic materials (Anikeeva et al., 2009). Nanoscale technologies are likely to expand the capabilities of both photovoltaic devices and light-emitting diodes over the next few years.

9.2.4 Field-effect transistors

Another type of semiconductor device that is important to mention due to its importance in computing, is the **field-effect transistor (FET)**. Field-effect transistors consist of a layer of a semiconductor material (n-type or p-type) in which two channels of a semiconductor of the opposite type (p-type or n-type) are embedded. The region between the two channels is covered with a thin layer of an insulator, and both the channels and the insulator are attached to metal electrodes. A schematic of a FET is shown in Figure 9.10. If no voltage is applied to the electrode above the insulator (the gate electrode), there is a large energy barrier for charge carriers to pass between the channels. In the NPN configuration shown, electrons in the n-type material cannot reach the conduction band of the p-type material and are unable to pass through it. However, if a sufficient voltage is applied to the gate, the electric field from the electrode shifts the band energies of the material under the gate, allowing charge carriers to pass through.

Figure 9.10 A simplified diagram of a field-effect transistor (FET).

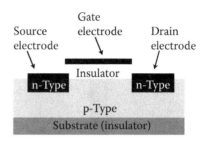

Among many other applications, field effect transistors form the backbone of modern integrated circuits, acting as nanoscale switches for processing data. Present technologies for conventional semiconductors have allowed industrial-scale fabrication of FETs with components as small as 14 nm, and a variety of other materials such as organic polymers, have been explored for flexible and printable materials.

9.3 EXCITATION TRANSFER

In the previous section, we discussed devices that involve the transfer of electrons from one material, molecule, or component of a molecule to another. However, quanta of energy can also be transported by transferring excitation from a donor to an acceptor; essentially, an electron in the donor relaxes to the ground state and an electron in the acceptor is promoted to an excited state, in a process with the following general form

$$D^* + A \xrightarrow{k_{XT}} D + A^* \tag{9.8}$$

here, k_{XT} is the rate constant for the **excitation transfer** process and the * refers to an excited state. While we will discuss excitation transfer in the context of assemblies of organic dyes, the same principles apply to nanoparticles and other nanoscale materials.

9.3.1 Functional dye assemblies

Photon energy is a highly favorable way to deliver energy to a molecular or supramolecular machine for a variety of reasons, including the ability to tune the wavelength (color) of the source, minimal invasiveness, and the ability to selectively excite various sites of an assembly depending on the wavelength applied. Processes that harness light energy for later use as chemical energy are abundant in nature, and as we previously discussed, for solar cells, and thus make the understanding and fabrication of such devices an important task.

The most basic method by which photon energy can be harnessed by molecular systems is through the exploitation of the specific absorption-fluorescence characteristics of dye molecules. Let us examine the concept of energy transport between dyes using a model system of two interacting dye molecules, a donor (D) and an acceptor (A), as shown in Figure 9.11. In our example system, the donor can be irradiated with light of wavelength 450 nm (blue light), resulting in excitation of the dye and fluorescence at 550 nm (green light). A different dye, acceptor A, is excited upon exposure to 550-nm wavelength light and fluoresces red light at 620 nm. We can exploit the overlap of D's fluorescence and A's absorbance in order to fabricate a simple molecular machine. Specifically, we can monitor the intensity I of fluorescence from dye monolayers on fatty acid precursors and spacers to determine their distance of separation, d. These films can be assembled easily via the hydrophobic/hydrophilic interactions in a Langmuir–Blodgett (discussed later) or electrostatic self-assembly (ESA) deposition process. As shown in Figure 9.12, the hydrophilic fatty acid head groups associate with a glass substrate, exposing the molecule's hydrophobic tail. Subsequent depositions of dye molecules D, a certain number of fatty acid spacers, and molecules of A result in a simple molecular machine. Using photoquenching equations derived by Drexhege et al. (1963) and later applied to this system by Mobius (1969), the intensity I of the green (D) and red fluorescence (A) can be calculated at any separation distance d and verified by monitoring I in spectroscopic experiments.

Figure 9.13a expresses the quantum yield as a ratio of green fluorescence intensity from D when A is located at a distance d ($I_{d,D}$) to the intensity of green fluorescence when d is infinitely large or A is absent ($I_{\infty,D}$). This value is a function of the separation distance of the two dyes, where d_0 is the distance where half of the fluorescence from D is quenched by A. Conversely, Figure 9.13b expresses the ratio of red fluorescence intensity from A when precursor D is located at a distance d ($I_{d,A}$) to the intensity of red fluorescence when d is infinitely large ($I_{\infty,A}$) as a function of d. On the

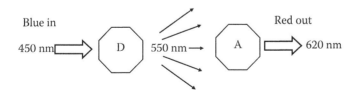

Figure 9.11   Excitation of donor molecule (D) by blue light (450 nm) and its subsequent fluorescence of green light (550 nm). The emitted green light is absorbed by acceptor molecule (A) which finally emits red light (620 nm).

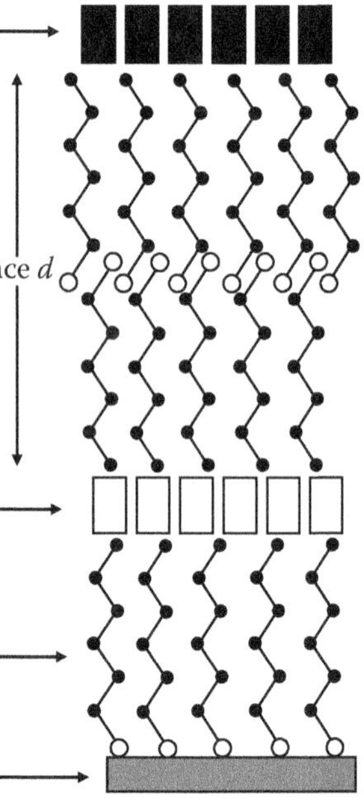

Figure 9.12 A layer-by-layer assembly in which a mono-layer of donor molecules is separated from a monolayer of acceptor molecules by "inert" fatty acid multilayers. The distance d is determined by the number of discrete fatty acid monolayers between the A and D monolayer.

Acceptor molecule (A) ⟶

Distance d

Donor molecule (D) ⟶

Fatty acid $CH_3(CH_2)5COOH$ ⟶

Glass substrate ⟶

Figure 9.13 Quantum yield profiles illustrating energy transfer from D to A. (a) Green fluorescence yield from D as a function of distance d. (b) Red fluorescence yield from A as a function of distance d. As d increases, energy transfer to A is reduced, resulting in more fluorescence from D and less from A.

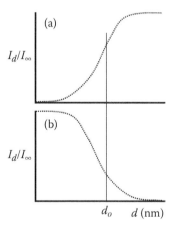

(a)

I_d/I_∞

(b)

I_d/I_∞

d_o d (nm)

left, I_d/I_∞ for donor fluorescence is plotted versus d. As expected, green fluorescence *increases* with d since more green fluorescence is transmitted as acceptor dye A becomes less accessible by d increasing. In Figure 9.13b, I_d/I_0 for acceptor fluorescence is plotted against d. As expected, red fluorescence *decreases* with d since A is excited with less efficiency as it becomes more distant from the excitons (the excited regions of the assembly) released from D, which are needed to create fluorescence in A.

These results themselves do not furnish us with much information beyond a confirmation that our theoretical model for energy transfer between the dyes is valid. However, Lehn (in *Supramolecular Chemistry*, VCH Weinheim, 1995) presents a compelling application for such behavior in dye assemblies. Consider two antibodies that both bind specifically and at a different site to an antigen. By coupling dyes A and D to the two different antibodies, the dyes become sufficiently close to allow energy transfer. By introducing an excess of the modified dyes, spectroscopic analysis could be used to perform a simple immunoassay. After accounting for background fluorescence from modified antibodies unbound to the antigen interacting with one another directly, fluorescence from acceptor A could be monitored and would indicate the specific concentration of the antigen in question.

9.3.2 Photorelaxation

The functionality of the aforementioned dye assemblies hinges on the differences in absorbance and emission spectra, or Stokes shift, of the two entities. We have already discussed the Stokes shift in the context of Raman spectroscopy; it is a shift in energy of a photon due to a transition between vibrational levels when a molecule is excited to, relaxes from, or scatters photons via an excited electronic state. As we discussed in Chapter 6, when a molecule is excited to a higher electronic state, it can relax via fluorescence, but often undergoes nonradiative vibrational relaxation first, losing energy to its environment. The typical lifetimes of excited electronic states are very short, often on the order of 10 ns, after which the molecule reverts to its ground state. When the molecule relaxes to the ground state, it emits energy corresponding the energy difference between the excited state and ground state in the form of a photon of the corresponding wavelength, as per Einstein's relation

$$E_{photon} = \frac{hc}{\lambda} \qquad (9.9)$$

where $h = 6.62606896 \times 10^{-34}$ J s and $c = 2.998 \times 10^8$ m/s. The difference between the excitation photon wavelength necessary to induce an electronically excited state and the wavelength of the subsequently emitted photon is known as the Stokes shift.

$$\Delta E_{stokes} = hc \left(\frac{1}{\lambda_{excitation}} - \frac{1}{\lambda_{emission}} \right) \qquad (9.10)$$

The Stokes shift is most often positive; in other words, the energy of the exciting photon is partially diminished via other nonradiation energy loss processes (see Figure 9.14) during excitation, and thus the emitted photon is of lower energy (longer wavelength).

Example 9.1 Stokes Shift

Estimate the Stokes shift in kJ/mol for the spectra shown in Figure 9.14.

Solution The absorbance maximum is about 350 nm and the emission maximum is about 450 nm, such that $\Delta\lambda$ is about 100 nm. However, we wish to calculate the energy of the Stokes shift, not just its wavelength. According to Equation 9.10,

$$\Delta E = hc\left(\frac{1}{350 \text{ nm}} - \frac{1}{450 \text{ nm}}\right)$$

$$= \frac{(6.626 \times 10^{-34}\text{Js})(2.998 \times 10^{8}\text{m/s})}{1575 \times 10^{-9}\text{m}}\left(\frac{1 \text{ kJ}}{1000 \text{ J}}\right)\left(\frac{6.02 \times 10^{23}}{1 \text{ mol}}\right)$$

$$= 75 \text{ kJ/mol}$$

This large Stokes shift represents a substantial loss of energy via nonradiative processes, over 1/5 the amount of energy required to break a C–C covalent bond.

In certain materials, however, anti-Stokes shifts occur and are interesting phenomena. At first glance, an anti-Stokes shift seems to violate conservation of energy, as a lower-energy photon is transformed into a higher-energy one. However, we have seen similar phenomena before in the context of nonlinear optics, where multiple photons can combine their energies to produce a higher-energy photon. Where is the extra energy coming from in this case? Instead of another electronic process, the extra energy comes from vibrational relaxation, or in essence, the

Figure 9.14 A typical excitation (absorption) and emission (fluorescence) spectrum of a dye molecule. The difference $\Delta\lambda$ is a measure of the Stokes shift.

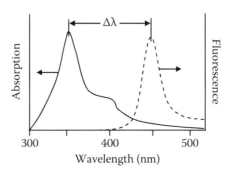

system losing heat. In materials exhibiting anti-Stokes shift behavior, photon energy is absorbed; however, before reverting to ground state, the exciton—an electrostatically coupled donor and acceptor—is also vibrationally excited via an intermolecular vibrational mode known as a phonon. This extra energy absorption creates a higher-energy photon emission. However, loss of this vibration energy creates a loss of heat in the material; it cools as it photorelaxes.

9.3.3 Resonant energy transfer

If a dye or nanoparticle emits light at a wavelength that can be absorbed by another dye or nanoparticle, energy can be transferred between the two systems via the emitted photon. This is called the trivial mechanism of excitation transfer, as it is unrelated to the specifics of the system. However, other methods of energy transfer exist that are related to electrostatic coupling between the donor and acceptor molecules. One common energy transfer mechanism that is important for several fluorescence spectroscopy techniques is **Förster transfer**, which involves coupling between the transition dipoles of two molecules or nanosystems. In Förster transfer, the oscillating electrons in the donor produces a transition dipole that induces an oscillating dipole in the acceptor, transferring energy between the donor and acceptor. The rate constant for this transfer process is proportional to the square of the transition dipoles,

$$k_{XT} \propto \frac{\mu_D^2 \mu_A^2}{r_{DA}^6} \tag{9.11}$$

where μ_D and μ_A are the transition dipoles of the donor and acceptor and r_{DA} is the distance between them. The excitation transfer rate falls off with the sixth power of distance, which makes Förster transfer very sensitive to the location of donor and acceptor and enables its use in measuring donor–acceptor distances.

9.3.4 Formation and properties of excitons

Thus far, the energy transfer mechanism that we have discussed involve an isolated donor molecule and acceptor molecule. However, when we examine materials that involve different dyes or semiconductor materials in interaction with one another, for example, a layer of an acceptor dye A deposited on top of a donor dye D, we find that the absorption spectrum is different from the sum of their individual spectra. Specifically, the donor, absorbing at shorter wavelengths, is shifted to even shorter wavelengths (higher energy), and the acceptor dye, absorbing light at longer wavelengths, is shifted to longer wavelengths (lower energy). In order to explain the cause of this difference, we can approximate the behavior of the dye molecules by oscillating charges exposed to the oscillating electric field of the incident light. When considering our system as a whole, we must account for the interactions among the electrons to determine the energy required to excite the system. As incident light reaches the system and excites electrons, each π-electron system can either oscillate in phase or out of phase with its neighbors. In-phase oscillation produces a strong absorption band because both oscillating charges are accelerated by the electric field of the excitation, increasing the magnitude of the transition dipole. In out-of-phase oscillation, a weak absorption band is realized because one charge is accelerated while its neighbor feels an opposing electric field and is decelerated, reducing the magnitude of the transition dipole.

In an aggregate layer such as our example donor–acceptor system, the oscillator that we use to represent a dye molecule will be strongly coupled to all the other dye molecules (oscillators) in the layer due to the tight packing in the system. Namely, the π-electrons of each molecule to be excited are in close contact with all the other electrons and their oscillations are in phase. This means that as incident light is exposed to the dye interface, an in-phase oscillation among all the layers' oscillators occurs and an excited domain forms, having absorbed the energy from the incident photons. We refer to this excited domain as an **exciton**. Essentially, excitons consist of excited electrons that are electrostatically bound to vacant orbitals (holes) but can diffuse through a system and can transfer energy between each other when their oscillations couple. The size of an exciton depends on the dielectric constant of its environment, and the binding energy between each electron and its corresponding hole is given by Coulomb's law (Chapter 5), as

$$\Delta U_{\text{binding}} = -\frac{1}{4\pi\varepsilon_0\varepsilon} \frac{e^2}{r_{\text{exciton}}} \tag{9.12}$$

The binding energy of an exciton decreases the energy of an excited state, shifting absorbance to longer wavelengths. It is counteracted by the thermal energy $k_B T$ of the system, such that the radius at which the binding energy of an exciton consisting of a single electron and a single hole becomes equal to the thermal energy can be calculated as

$$r_{\text{exciton}} = \frac{1}{4\pi\varepsilon_0\varepsilon} \frac{e^2}{k_B T} \qquad (9.13)$$

therefore, large dielectric constants and high temperatures favor larger excitons. In many organic materials, excitons are localized to one or a few molecules due to weak dielectric screening, and are known as Frenkel excitons. In high-dielectric environments such as many crystalline semi-conductors or dyes in solvents with large dielectric constants, excitons can become much larger than the distance between molecules and are known as Wannier–Mott excitons.

Excitons have a finite lifetime. We can quantify the approximate lifetime of the exciton in our model system. It has been shown that the fluorescence lifetime of a monomer τ_0 in terms of our oscillator model is

$$\tau_0 = \frac{3m\varepsilon_0 c_0^3}{2Q^2 \pi v_0^2 n} \qquad (9.14)$$

where m, Q, v_0 is the mass, charge, and frequency of the oscillator, ε_0 is the permittivity of the vacuum, c_0 is the speed of light in vacuum, and n is the refractive index of the material. The fluorescence lifetime is the inverse of the fluorescence rate constant,

$$k_0 = \frac{1}{\tau_0} \qquad (9.15)$$

We will now consider excitons involving multiple molecules, such that N oscillators in an aggregate are oscillating in phase, contributing to the excited domain. Therefore, we should replace m by Nm and Q by NQ in Equation 9.14, obtaining

$$\tau_{\text{agg}} = \frac{3(Nm)\varepsilon_0 c_0^3}{2(NQ)^2 \pi v_0^2 n} = \frac{1}{N}\tau_0 \qquad (9.16)$$

We can exploit the fact that the excited domain must exist through a balance between attractive forces of the in-phase oscillating dipoles (excited portion) and the thermal motion tending to disrupt such a domain in order to find the number N of molecules contributing to the excited domain. The binding energy ($\Delta U_{\text{binding}}$) in a system of oscillators decreases the energy of the excited state, making it more favorable. Thermal energy ($k_B T$) works against the former quantity and makes it less favorable, thereby pushing the oscillator out of phase. We can estimate the number of molecules, N, involved in an exciton by setting the binding energy equal to $k_B T$, as we did in Equation 9.13. Therefore, for an aggregate of dyes,

$$N = -\frac{\Delta U_{binding}}{k_B T} \tag{9.17}$$

Example 9.2 Exciton Size

The binding energy of an aggregate has been measured to be $\Delta U_{binding} = -0.24$ eV. Determine the number of molecules involved in an exciton at 300K and the lifetime of that exciton, given $m = m_e$, $Q = e$, $n = 1.5$, and $v_0 = 0.75 \times 10^{15}$ s^{-1}.

The number of molecules in the exciton can be found using Equation 9.17:

$$N = -\frac{\Delta U_{binding}}{k_B T} = \frac{0.24 \text{ eV}}{(8.61 \times 10^{-5} \text{ eV} \cdot \text{K}^{-1})(300 \text{ K})} = 10$$

We can now use the number of molecules N contributing to the excited domain to obtain the fluorescence lifetime of the domain. First, we must find the fluorescence lifetime of an isolated molecule,

$$\tau_0 = \frac{3 m_e \varepsilon_0 c_0^3}{2 \pi e^2 v_0^2 n}$$

$$= \frac{3(9.109 \times 10^{-31} \text{ kg})(8.854 \times 10^{-12} \text{ C}^2 \cdot \text{J}^{-1})(2.997 \times 10^8 \text{ m} \cdot \text{s}^{-1})^3}{2 \pi (1.602 \times 10^{-19} \text{ C})^2 (0.75 \times 10^{15} \text{ s}^{-1})^2 (1.5)}$$

$$= 5 \times 10^{-9} \text{ s}$$

The fluorescence lifetime of the aggregate is therefore

$$\tau_{agg} = \frac{\tau_0}{N} = \frac{5 \times 10^{-9} \text{ s}}{10} = 5 \times 10^{-10} \text{ s}$$

Excitons are capable of moving through a material; we will consider the two-dimensional example of an exciton propagating through a monolayer of dyes on a surface. The area covered by an exciton of width L during τ_{agg} can be be calculated as

$$A = L_{\text{exciton}} \cdot v \cdot \tau_{\text{agg}} \qquad (9.18)$$

where v is the speed at which the exciton propagates.

We will consider a simplified case in which a donor photon has a 50% probability of reaching an acceptor molecule. Under this condition, A must be equal to the area in which just one acceptor molecule is present. Thus, we redefine A as Za where a is the area covered by a donor molecule and Z is the number of donor molecules per acceptor.

$$v = \frac{A}{L_{\text{exciton}} \cdot \tau_{\text{agg}}} = \frac{Za}{L_{\text{exciton}} \cdot \tau_{\text{agg}}} \qquad (9.19)$$

Excitation propagation is very important for devices such as solar cells in which an exciton needs to move to an interfacial region where it can split into free-charge carriers before it recombines. We will consider an example of exciton propagation on a surface.

Example 9.3 Exciton Propagation

Consider a dye containing system where the donor is in great excess, with $Z = 10,000$, where the size of an individual dye molecule is 1.5 nm $\cdot$ 0.4 nm, where the length of an exciton at 300 K is 5 nm, and where $\tau_{\text{agg}} = 5 \cdot 10^{-10}$ s (as in Example 9.2). Determine the speed of the exciton.

$$V = \frac{(10000)(1.5 \times 10^{-9} \text{ m})(0.4 \times 10^{-9} \text{ m})}{(5 \times 10^{-9} \text{ m})(5 \times 10^{-10} \text{ s})} = 2400 \frac{\text{m}}{\text{s}}$$

For more information on excitation transport in aggregates, see Chapter 23 in Kuhn and Försterling's textbook, *Principles of Physical Chemistry: Understanding Molecules, Molecular Assemblies, Supramolecular Machines*.

Certain experimental observations support a calculation of such a fast-moving exciton. Most importantly, this theoretical number helps explain how the excited domain fluoresces or excites another dye molecule. In this oscillation model, charges in the excited domain change simultaneously and the nearest-neighbor charges are always opposite; thus an attraction of nearest-neighbor charges occurs, leading to a compression of the excited domain. This compression creates a wave that carries the exciton, which does not change shape or lose energy until its destruction through absorption by an acceptor or through fluorescence. This is another possible mechanism for how energy is able to be transported through a medium.

9.4 QUANTUM DOTS

In the previous sections in this chapter, we have considered basic physics needed to understand charge and excitation transfer in functional nanomaterials with several examples of systems and devices. We will now begin a brief tour through a wide variety of functional nanomaterials, starting with zero-dimensional systems and working our way up in dimensionality.

Quantum dots are inorganic semiconductor nanoparticles, typically 2–10 nm in size. The excitons (electron-hole pairs) within quantum dots are confined along all three dimensions, such that they are 0-dimensional nanomaterials. In contrast, nanowires are confined in two dimensions (1D nanomaterials), and quantum wells are confined along one dimension (2D nanomaterials). As the confining dimensions of a nanomaterial decrease toward a limit, the energy gap starts to change with it. This change in the energy gap can be quantified using models such as the particle-in-a-sphere model, which we discussed in Chapter 4. The small size of quantum dots confines electrons to a region that gives them optical and electronic properties in between those of bulk semiconductors and discrete molecules (Nirmal, 1999). Each dot is a mesoscopic entity with individual properties, differentiating it from a colloidal particle.

Quantum dots typically consist of a core, shell, and final coat, as shown in Figure 9.15. They are characterized by the nature of each of these layers, their size and aspect ratio, their quantum efficiency in optical materials,

Figure 9.15 Structure of a quantum dot. The size and nature of the inorganic core and the shell material can be varied. Quantum dots are often capped with an organic thin film.

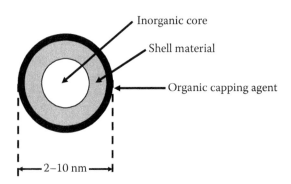

Inorganic core

Shell material

Organic capping agent

2–10 nm

and their coercivity in magnetic materials. Properties of an ensemble of quantum dots are additionally determined by particle size distribution and differences in morphology within the ensemble.

9.4.1 Optical properties of quantum dots

Quantum confinement gives quantum dots distinct optical properties compared to bulk semiconductors. Quantum confinement occurs when the radius of the quantum dot is smaller than the exciton radius in the corresponding bulk semiconductor. For CdSe, this transition occurs at a radius of 5.6 nm (Nirmal, 1999). Below this size, the quantum dot has clearly distinguishable discrete energy levels, and its band gap is shifted by an energy difference that can be calculated from the particle-in-a-sphere model as

$$E \frac{h^2}{8mr^2} \qquad (9.20)$$

The effects of quantum confinement give quantum dots very high molar extinction coefficients (of the order of $0.5\text{-}5 \cdot 10^6 \text{ M}^{-1} \text{ cm}^{-1}$) making them particularly bright probes when light intensities are severely attenuated by scattering and absorption—their absorption rates are faster than those of organic dyes at the same excitation photon flux. Their large extinction coefficients and short fluorescence lifetimes allow quantum dots be 10–20 times brighter than organic dyes for fluorescence analyses, yielding much better signal-to-background ratios.

Additionally, the broad absorption profiles of quantum dots allow simultaneous excitation of multiple wavelengths; thus molecular and cell targets can be tagged with different colors. The fluorescence emission wavelengths can be continuously tuned from 400 to 2000 nm by varying particle size and chemical composition. For example, if one compares the emission spectra of rodent skin and that of quantum dots obtained under the same excitation conditions, one finds that the quantum dot signals can be shifted to a spectral region where autofluorescence is reduced in order to improve signal-to-background ratios. Moreover, quantum dots are remarkably resistant to photobleaching compared to organic dyes and are thus well suited for continuous tracking studies over a long period of time. Signal-to-background ratios can be further improved by time-domain imaging, which is used to separate quantum dot fluorescence from background fluorescence due to the longer excited states of quantum dots (Figure 9.16). Finally, quantum dots often exhibit large Stokes shifts. The difference between the excitation and emission peaks can be as large as

Figure 9.16 Comparison of the excited state decay curves (monoexponential models) between quantum dots and common organic dyes. Line (a) indicates excitation. Lines (b) and (c) indicate the decay curves for the organic dye and quantum dot, respectively. A measure of the fluorescence lifetime is indicated by time t_1 (for the dye) and t_2 (for the quantum dot).

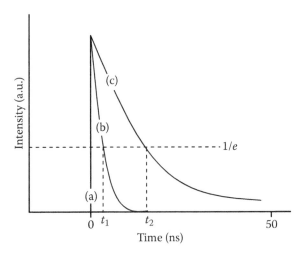

300–400 nm, making the quantum dot signal easily recognizable above the background.

9.4.2 Synthesis of quantum dots

9.4.2.1 Precipitative methods

Precipitative synthesis of quantum dots is similar to colloidal synthesis in that it involves supersaturation of a solution followed by crystallization. Like traditional chemical processes, colloidal semiconductor nanocrystals are synthesized from precursor compounds dissolved in solutions. The precursors are heated to temperatures high enough to convert them to monomers. Once the monomers reach a high enough supersaturation level, the nanocrystal growth starts with a nucleation process. The temperature is a critical factor in determining optimal conditions for crystal growth—it must be high enough to allow for the rearrangement and annealing of atoms during the synthesis process while being low enough to promote crystal growth. Furthermore, the size of the crystal depends on temperature (i.e., it is larger at higher temperatures). Another critical factor is the monomer concentration. At high monomer concentrations, the critical size of the crystal is relatively small, resulting in the growth of nearly all particles and thus facilitating monodispersity (uniformity of the particle size in the system). When the monomer concentration is depleted during growth, the critical size becomes larger than the average size present.

The preparation of CdS quantum dots requires the controlled nucleation of CdS in dilute aqueous solutions of cadmium sulfate and ammonium sulfide. The dynamic equilibrium between solvated ions and solid CdS in acetonitrile as a solvent in the presence of a styrene or maleic anhydride copolymer allows the preparation of stable CdS nanoparticles (average size 3.4–4.3 nm). To obtain highly monodispersed nanoparticles, postpreparative separation techniques like size exclusion chromatography or gel electrophoresis are used.

Weller et al. (1985) injected phosphine (PH_3) into solutions containing metal salts to make particles of Zn_3P_2 and Cd_3P_2. The particle size can be controlled by varying temperature and phosphine concentration. Even though this method is cheap, it has several disadvantages. It lacks reaction control, which is problematic in large-scale syntheses. Furthermore, important semiconductors like GaAs or InSb are not available for use in this method as they are air and/or moisture sensitive.

9.4.2.2 Reactive methods in high-boiling-point solvents

Another approach for making quantum dots involves reaction of a precursor compound and a high-boiling, hydrophobic solvent. Solutions of dimethylcadmium, $(CH_3)_2Cd$ (in tri-n-octylphosphine, TOP), and tri-n-octylphosphine selenide (TOPSe) in hot tri-n-octylphosphine oxide (TOPO) at 120°C–300°C produce TOPO-capped nanocrystals of CdSe. The solvent molecules (e.g., TOPO) both serve as a reaction medium and also coordinate with unsaturated metal atoms on the QD surface to prevent the formation of bulk semiconductors.

The TOPO method is one of the most widely used methods of quantum dot synthesis as it has several advantages over other methods. The size of the dots can be controlled by temperature (i.e., they are larger at higher temperatures) and monodispersity can easily be achieved. Additionally, the yields of the reactions are usually high, producing hundreds of milligrams in a single experiment. Moreover, this method is readily adapted to the production of core-shell structures with materials having high quantum efficiencies.

9.4.2.3 Gas-Phase synthesis of semiconductor nanoparticles

Quantum dots can also be synthesized in the gas phase. This method involves the atmospheric or low-pressure evaporation of either powders of a preformed semiconductor, or coevaporation of two elemental components (like Zn metal and sulfur). It is not a very favorable method due to the large size distribution of the particles (10–200 nm) and the tendency of particles to aggregate due to the absence of a surface-capping agent.

9.4.2.4 Synthesis in a structured medium

Quantum dots can also be synthesized in a confined medium. A matrix is used to define a reaction space—it provides a mesoscopic reaction chamber in which the crystal can grow only to a certain size. The properties of the system control the properties as well as the size of nanoparticles. Substances typically used as matrices are layered solids, micelles/microemulsions, gels, polymers, and gasses. This technique is often used in biologically related processes. An example of this is the use of empty polypeptide cages found in the iron storage protein ferritin in the synthesis of bioinorganic nanocomposites of CdS-ferritin.

9.4.3 In vivo imaging with quantum dots

Due to their optical properties as discussed earlier, quantum dots are better than organic dyes for the purpose of in vivo molecular and cell imaging. Researchers have achieved real-time visualization of single-molecule movement in single living cells using quantum dots. Their high electron density allows correlated optical and electron microscopy studies of cellular structures. Quantum dot bioconjugation can be achieved by

Figure 9.17   The structure of a multifunctional quantum dot probe. Schematic illustration showing the capping ligand TOPO, an encapsulating copolymer layer, tumor-targeting ligands (such as peptides, antibodies, or small-molecule inhibitors), and polyethylene glycol. (Reprinted from Gao, X., Yang, L, Petros, J. A., Marshall, F. F., Simons, J. W. and Nie, S. *In vivo Molecular and Cellular Imaging with Quantum Dots. Current Opinion in Biotechnology* 2005, 16: 63–72. With permission from Elsevier.)

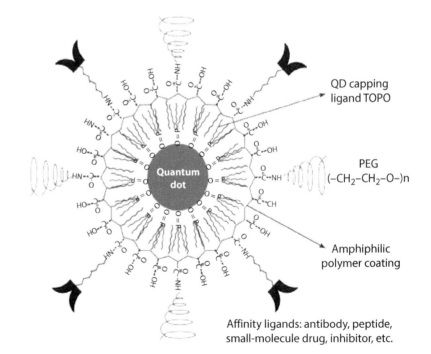

QD capping ligand TOPO

PEG ($-CH_2-CH_2-O-$)n

Amphiphilic polymer coating

Affinity ligands: antibody, peptide, small-molecule drug, inhibitor, etc.

Quantum dot

passive absorption, multivalent chelation, or covalent bond formation as shown in Figures 9.17 and 9.18 (Gao et al., 2005).

Streptavidin-coated quantum dots preferentially bind to biotin with a K_D of approximately 10^{-14} M, one of the strongest covalent bonds known in nature. Such bioconjugation is widely used for molecular and cellular imagining. An application of this is seen in an experiment conducted to identify β-sheet aggregate-binding ligands using one-bead one-compound screening. The ligands under study were each attached to a bead. When treated with the biotinylated peptide, only one of these three ligands binds to it. The streptavidin-coated quantum dots preferentially bind to this peptide-ligand combination, and thus the ligand is easily detected due to

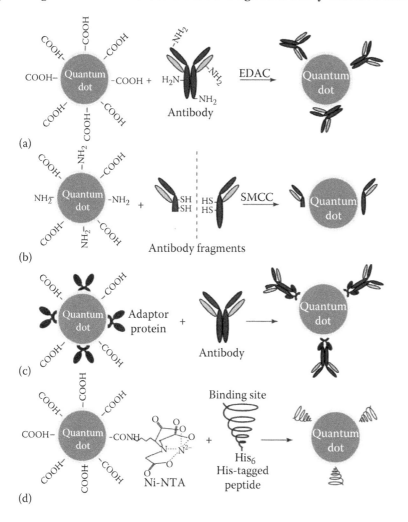

Figure 9.18 Methods for conjugating QDs to biomolecules. (a) Traditional covalent cross-linking chemistry using EDAC (ethyl-3-dimethyl amino propyl carbodiimide) as a catalyst. (b) Conjugation of antibody fragments to QDs via reduced sulfhydryl-amine coupling. SMCC, succinimidyl-4-N-maleimidomethyl-cyclohexane carboxylate. (c) Conjugation of antibodies to QDs via an adaptor protein. (d) Conjugation of histidine-tagged peptides and proteins to Ni-NTA-modified QDs, with potential control of the attachment site and QD: ligand molar ratios. (Reprinted from Gao, X., Yang, L., Petros, J. A., Marshall, F. F., Simons, J. W., Nie, S. *In vivo* Molecular and Cellular Imaging with Quantum Dots. *Current Opinion in Biotechnology* 2005, 16: 63–72. With permission from Elsevier.)

the fluorescence exhibited by the quantum dots. Analysis by mass spectroscopy reveals the identity of the ligand.

9.4.4 Photodynamic therapy

Another interesting use of quantum dots in biological systems is as a therapeutic agent. A quantum dot can be coated with a dye and either directly injected into a target region such as a tumor or conjugated with an antibody such that it can recognize and bind to its target (Zhang et al., 2007). If light at a wavelength that excites the quantum dot is shone on the organism, the quantum dot is excited and transfers that excitation to the dye via methods such as those discussed in Section 9.3. The dye can then transfer energy and/or electrons to the medium, forming reactive oxygen species (ROS, also known as free radicals) that destroy the target cell. Such an approach is promising for treating cancerous tumors while avoiding damage to healthy tissue (Lucky et al., 2015). A schematic of photodynamic therapy is shown in Figure 9.19.

9.5 NANOWIRES

A nanowire is a nanostructure that has a diameter on the scale of a nanometer and an unrestricted length. An example of a nanowire is shown in Figure 9.20. Nanowires have been synthesized to be as long as one millimeter (10^{-3} m), but are more commonly made to be about one

Figure 9.19 Schematic of photodynamic therapy. Light excites a quantum dot, which is attached to a target cancer cell with an antibody. The quantum dot transfers energy to a dye, which in turn reacts with its environment to produce reactive oxygen species that destroy the cancer cell.

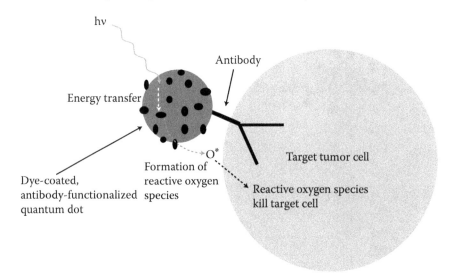

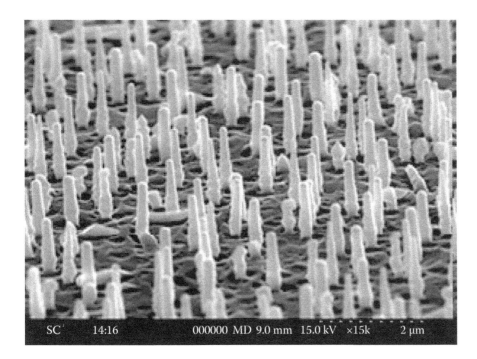

Figure 9.20 A vertical silicon nanowire array on a silicon substrate. (Image provided by Professor Hari Reehal, London South Bank University.)

micrometer (10^{-6} m). Regardless, the typical nanowire has a width-to-length ratio of 1:1000 or more, allowing the structure to be viewed as one-dimensional (1D). Combined with the comparatively small size of nanowires, the 1D structure is responsible for the unique properties that grossly distinguish nanowires from the average (three-dimensional) wire.

The potential applications for nanowires are immense. As technology continues to advance, the trend in industry is to generate smaller but more powerful electronic devices than previously sold on the market. As a result, computer components such as resistors, capacitors, and circuits continue to shrink. To link these components, increasingly small wires are desired. There is also great interest in nanowires from the battery industry, since the high surface area of nanowire arrays relative to bulk metals could lead to more efficient anodes and cathodes. Industries also see the potential for nanowires in such applications as chemical composites, field emitters, and biomolecular nanosensor leads. Moreover, nanowires' conductance can be controlled by synthesizing the wire from any one of a variety of substrates, allowing for nanowires to have more specialized roles. As a result, nanowires are categorized as (in order of increasing conductance ability) insulators (SiO_2, TiO_2), semiconductors (Si, GaN, InP), or metallics (Au, Cu, Pt, Ni, etc.). With such major applications, research relating to the general properties and synthetic strategies of nanowires is in high demand.

9.5.1 Quantum effects on conductivity of nanowires

Since most applications of nanowires are related to electronics, the effects of quantum confinement on nanowires' conductance is of particular interest. As we have previously discussed, energy is quantized—molecules are limited to discrete energy states. However, in bulk materials, these states form nearly continuous bands of allowed energies (Section 9.2.2). In a conductor (partially filled conduction band), quantum mechanical effects can generally be neglected when describing charge transport through the system. In contrast with a semiconductor, in which conductivity is limited by the number of electrons that can be excited across the band gap, conductors have virtually no energetic penalty for charge transport. Bulk conductors can be described by classical models such as the **Drude model**, which treats a conductor as noninteracting electrons diffusing through a field of cations (atomic nuclei and their remaining strongly bound electrons), such that the conductivity

$$\sigma = \frac{Ne^2 \tau_c}{m_e} \tag{9.21}$$

is a function of the density of electrons (N) and the system and is the average time between electron-cation collisions (τ_c), both of which can be derived from the structure of the material. The conductance of a wire can be determined by its conductivity and dimensions,

$$G = \frac{\sigma A}{L} \tag{9.22}$$

where L is the length of the wire and A is its cross-sectional area. The ease at which current passes through a wire is more often represented as resistance than as conductance; resistance is the inverse of conductance ($R = 1/G$).

Since nanowires exhibit discrete quantum confinement, they have measurable, discrete energy levels, and their resistance can be calculated quantum mechanically. The conductivity of a single state in a nanowire, often called a "channel" (Ciraci et al., 2001), can be estimated using the Landauer formula for a 1D conductor connected at both ends,

$$G = \frac{2e^2}{h} T \tag{9.23}$$

where h is Planck's constant and T is the electronic transmission coefficient for the material. The transmission coefficient is derived by analogy to transmission and reflection of light, but refers to penetration of a wavefunction into a finite barrier. Essentially, it is equivalent to the Marcus T_{DA}. Note that as T approaches 1, the conductance of a channel in the nanowire approaches a quantized limit of $2e^2/h$, or 0.0387 kΩ^{-1}.

Multiple states can participate toward a nanowire's conductivity; the number of available channels depends on the density of states between the Fermi level and the energy corresponding to the voltage applied to the system $E_F < E < eV$. The conductance for the nanowire is the conductance of a single channel/state times the number of participating states; for an ideal 1D nanowire (e.g., particle-in-a-1D-box), it reduces to

$$G = N \frac{2e^2}{h} \tag{9.24}$$

Therefore, increasing the voltage applied to the system of nanowires will cause the conductance to increase in discrete increments of $2e^2/h$ as the applied voltage increases and the system exhibits quantized conductivity. Equations 9.23 and 9.24 have the implication that for a nanowire exhibiting quantized conductivity, conductance depends on the number of available channels but no longer depends on the length of the nanowire, unlike the classical relationship in Equation 9.22.

9.5.2 Electron transport in nanowires

As discussed in the previous section, conductivity of nanowires is heavily influenced by quantum effect. Yet, conductivity varies significantly depending on the composition and width of the nanowire, which influence the number of available channels for conduction. The extremely large length-to-width ratio of a nanometer would suggest that they have a very high resistance (low conductance) compared to that of a bulk phase wire with the same chemical composition (see Equation 9.22). For most nanowires, this is indeed the case even though conductance is quantized.

Much of the reduction in conductance in nanowires is due to a reduction in the number of channels available for conductivity; however, the nanowire's narrow width contributes in a number of ways to increasing resistance. For one, decreased width increases the relative effect of the wire's surface molecules. The atoms or molecules at the surface of the wire are much less tightly bound than atoms or molecules found in the interior of the wire. As well, surface molecules are in contact with far fewer of the wire's molecules than those found in the interior of the wire. The overall consequence is that molecules found at the surface of a wire have greater difficulty propagating electrons to other wire molecules. Therefore, high surface-area-to-volume ratios tend to increase the resistance. The basic concept of width effects on resistance is related to what is known as the mean-free path.

Mean-free path relates to the distance an electron travels between subsequent collisions with other moving particles. Unlike for chemical reactions, where collisions are necessary for a reaction to proceed, collisions are not favorable for electron transport through nanowires since the electron can be deflected in any direction (Figure 9.21), reducing T (Equation 9.23) and impacting the wire's forward flow of electrons. Collisions occur most frequently when the mean-free path of the electron is smaller than the width of the nanowire; these collisions reduce the conductivity of the nanowire. In the opposing case, a nanowire is said to be a "ballistic transporter" when the mean-free path is much longer than its width, such that electrons experience only quantum confinement as opposed to collisions, and electrons can flow unimpeded ($T \sim 1$). Ballistic transporters (typically metal nanowires) have very high conductance values compared to those whose wire lengths fail to exceed the mean-free path (Takayanagi, 2001).

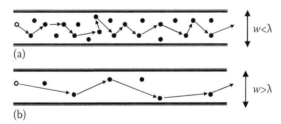

Figure 9.21   A schematic demonstrating the effects of mean-free path. The open circle indicates the electron and the solid black circles represent larger atoms. w is the width of the wire, and λ is the mean-free path of the conduction electron. (a) Diffusive transport occurs when the width of the wire is smaller than the mean-free path of the electron, and (b) ballistic transport occurs when the width of the wire is greater than the mean-free path.

9.5.3 Nanowire synthesis

Two general concepts exist for nanowire synthesis, each with their own unique synthetic strategies. The "top-down" approach is one such concept. In this approach, nanowires are synthesized starting from a bulk substrate from which material is removed until only a field of nanowires remain. Nanolithography and electrophoresis are examples of the top-down techniques employed to make nanowires. However, it is difficult using current instruments to synthesize ideal one-dimensional nanowires via top-down methods. The instruments and techniques are currently limited by the length and width of the nanowires they can synthesize as well as the general accuracy they exhibit when working on a substrate. For these reasons, the top-down approach is presently less commonly employed than its alternative.

The "bottom-up" approach synthesizes nanowires by continuously extending a thread of bound molecules. Various techniques have developed using the bottom-up approach including suspension-based techniques, where a nanowire is elongated in a vacuum using chemical etching or ion-bombardment techniques, and vapor–liquid–solid (VLS) techniques, where a nanowire is synthesized via a catalyst converting a vapor into a solid below the catalyst, continuously elongating the chain of solid molecules (Crossland et al., 2007). Solution-based bottom-up techniques hold particular promise since they allow scientists to synthesize nanowires in much larger numbers than alternative methods. One particularly intriguing solution-based technique uses copolymers to form templates inside which nanowires can be synthesized (Figure 9.22). Certain copolymers are ideal for template formation because, at the correct temperature and electric field strength, they form evenly spaced hexagonal pores. Once the substrate is added, these pores serve as the locations for nanowire synthesis. Mild reagents can then be used to degrade the template and isolate the nanowires.

9.5.4 Summary

Nanowires have great potential. As the field grows, we will begin to see new synthetic techniques capable of producing highly elongated nanowires in large quantities. These techniques will be exceedingly precise, allowing for wire shape and function to be maintained during the entirety of the nanowire synthesis. As the study of nanowires continues to progress, new imaging devices will be forced to develop in parallel to

Figure 9.22 Gold nanowires were synthesized using a solution-based bottom-up technique utilizing a copolymer template. Images of (a) Ag nanowires of uniform diameter, (b) a thin slice of the nanowire shown in (a), (c) end of an individual nanowire, and (d) Ag nanowires functionalized with a thiol. Images were taken using a scanning electron microscope. (Reprinted with permission from Chen, J., Wiley, B. J., and Xia, Y. One-Dimensional Nanostructures of Metals: Large-Scale Synthesis and Some Potential Applications. *Langmuir* 8, 2007: 4120–4129. © 2007 American Chemical Society.)

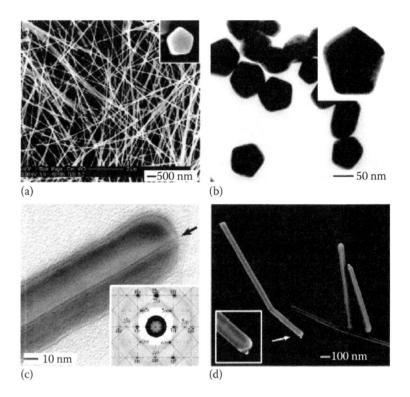

nanowire synthetic techniques. These new devices will allow nanowires to be seen in higher magnification and clarity, allowing for more accurate analysis of experimental results. Studies will continue to be done on synthesized nanowires to help define the nanoparticle's unique properties so it can be better applied to products in the industry, including batteries, computers, and chemical composites.

9.6 CARBON NANOTUBES

Carbon nanotubes are another important class of a 1D nanomaterial. They are constructs of carbon that have diameters on the order of a few nanometers and lengths that can be significantly longer. Carbon nanotubes are extremely strong; the tensile strength of a single-walled carbon nanotube is many times stronger than steel. They also have unique electrical and optical properties and are efficient conductors of heat. These properties have made the study of carbon nanotubes surge in recent years (Dai, 2002).

9.6.1 Carbon nanotube structure

Carbon nanotubes are a member of the fullerene family of carbon structures; they may contain hexagonal, pentagonal, and sometimes heptagonal rings of pure carbon. It may be helpful to imagine a flat "sheet" of such rings that can be rolled up to make a single-walled nanotube. Regardless of the number of carbons within a ring, each shares a bond with three adjacent carbons. This makes every carbon within the tube sp^2 hybridized. Carbon nanotubes are often capped with a carbon hemisphere that is also composed of sp^2 hybridized carbons in rings. Carbon nanotubes can come in single-walled or multiwalled form. The multiwalled nanotube structure can be thought of as concentric single-walled nanotubes encompassing one another.

Because differing carbon nanotubes may consist of similar sp^2 hybridized carbons, one might assume they all will have similar properties. This is an incorrect assumption. Differences in how the carbon atoms are structurally organized play a significant role in determining the properties of the carbon nanotube as a whole. For example, one way of organizing the carbons can make the nanotube metallic and conducting, whereas another organization scheme will make it a semiconductor. Because of this, the atomic structure of a nanotube is often described in terms of its chirality, which can be broken down into a chirality vector, C_h, and a chiral angle, θ (Figure 9.23).

The length of the chirality vector is related to the diameter of the nanotube it describes and can be easily determined if the interatomic spacing of the carbon atoms is known. The interatomic spacing of carbon atoms is often a known parameter because it is the same for all nanotubes with a

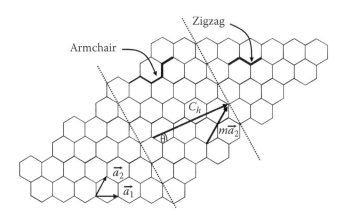

Figure 9.23 Schematic diagram showing how a hexagonal sheet of graphite can be rolled to form a carbon nanotube. The properties of the nanotube will depend on the chiral vector and the chiral angle. The unit vectors are shown in the lower left-hand side of the diagram.

similar lattice or unrolled "sheet" of rings. An equation for describing the chiral vector is given in Equation 9.25 (Thostenson et al., 2001):

$$C_h = na_1 + ma_2 \qquad (9.25)$$

In this equation n and m are the number of steps in the direction of the unit vectors, a_1 and a_2, in their respective directions across the sheet of unrolled carbon rings. Figure 9.23 is a graphical depiction of this along a graphene sheet composed of only hexagonally ordered carbon atoms. The chiral angle, θ, is a measurement of the resulting twist in the "rolled-up" nanotube and is defined as the angle between the chiral vector and the unit vector a_1. This angle can be determined by relating the number of steps n and m as shown in Equation 9.26 (Saito et al., 1998):

$$\cos\theta = \frac{(2n + m)}{[2(n^2 + m^2 + nm)]} \qquad (9.26)$$

Nanotubes with chiral vector steps $n = m$ are known as "armchair" nanotubes and have a chiral angle of 30°. Nanotubes with one of the step integers equaling zero are referred to as "zigzag" nanotubes and have a chiral angle of 0°. Both of these are achiral nanotubes because their mirror images are identical to the original structure. In all other cases, where $n \neq m$, the nanotubes are chiral, with a chiral angle between 0° and 30°.

9.6.2 Some properties of nanotubes

Interestingly, the relationship of steps n and m can be used to determine if the nanotube will be conducting or semiconducting. If $n - m$ is a positive multiple of three, or zero, the nanotube has no band gap and therefore is a conductor. If $n - m$ is not a positive multiple of three, or zero, then the nanotube will have a small band gap and therefore is a semiconductor.

The ability of carbon nanotubes to conduct heat is dependent on temperature and is comparable to that of diamond or a graphite sheet. The peak thermal conductivity for a single-walled nanotube is reached at around 100 K and is around 37,000 $Wm^{-1}K^{-1}$. From there the thermal conductivity drops as the temperature increases to around 3000 $Wm^{-1}K^{-1}$ at a temperature of 400 K.

The precise values for mechanical and elastic properties of carbon nanotubes are currently disputed. They are thought to be the strongest and stiffest materials in terms of tensile strength and elastic modulus. It has been shown that they are many times stronger than steel and that they can return to their original state after being bent. Nanotubes have also been found to be stable at very high temperatures, nearly 3000 K in a vacuum and 700 K in air.

9.6.3 Methods for growing nanotubes

There are three primary methods for nanotube production: arc discharge, laser ablation, and chemical vapor deposition (Baddour and Breins, 2005). Arc discharge uses a high-voltage arc to vaporize a carbon source. This frees carbon, which then accumulates into nanotubes in the growth chamber. Following accumulation the nanotubes are collected and purified. Laser ablation is similar to the arc discharge method, but it makes use of a high-energy laser instead of an electric current to vaporize the carbon source. Chemical vapor deposition relies on the decomposition of a carbon-containing gas in the presence of a metal catalyst, such as iron, cobalt, or nickel, which helps facilitate carbon accumulation and nanotube growth. Each method has many variations with their own strengths and weaknesses that should be considered when choosing which to use.

9.6.3.1 Arc discharge

In the arc discharge method, two carbon rods are set up as an anode and cathode and are placed in a closed growth chamber. The chamber is filled with an inert buffer gas such as helium, argon, nitrogen, or hydrogen and kept at relatively low pressures. Interestingly, the inert gas in the chamber has been found to affect the diameter of grown nanotubes while the pressure in the growth chamber does not. Keidar and Waas (2004) found that when mixtures of helium and argon are used to fill the chamber, the molar fraction of argon determines the nanotube diameter. Once the chamber is filled with gas, a direct current is applied until a stable electric arc is formed between the carbon rods. Once the anode reaches a critical temperature, it vaporizes into a gaseous carbon source. The gas in the chamber causes the carbon to deposit. The structures found in these deposits are nanotubes, which need to be purified after collection. This method has been used for the creation of both single-walled and multiwalled nanotubes. In order to produce substantial amounts of single-walled carbon nanotubes a metal catalyst needs to be added to the electrodes.

Unfortunately, once collected, the nanotubes must be manipulated onto substrates for further use, which can be difficult for large quantities of such small structures. Arc discharge also requires a large amount of power and is operated at a high temperature, making it an unfavorable method for the production of large quantities of nanotubes.

9.6.3.2 Laser ablation

Laser ablation uses a high-energy laser beam to vaporize a carbon source. The carbon source is sealed off in a quartz tube and heated to temperatures ranging from a few hundred degrees to just over a thousand degrees Celsius. The tube is filled with an inert gas that flows toward a cooled collector at its end. As the laser vaporizes the carbon source, the flow of inert gas carries it down the tube, where it deposits. The nanotubes are then collected from the deposits and purified.

In both arc discharge and laser ablation, the vaporized carbon is released as low molecular weight carbon, which coalesces to form larger molecular weight structures. The carbon begins to condense very quickly after vaporization. Using metal catalyst particles doped into the carbon source prevents carbon from closing into a cage structure and decreases the amount of carbon deposited in nonnanotube form by aggregating with the carbon quickly. This results in more single-walled nanotubes being formed when a metal catalyst is used. Nanotube length can be controlled by changing the time the carbon spends in the high temperature area of the growth chamber, which can be controlled by using heavier or lighter gases and altering the flow rate. Research is undergoing into changing growth conditions, which may make laser ablation a viable option for the large-scale production of nanotubes.

9.6.3.3 Chemical vapor deposition

Chemical vapor deposition relies on the decomposition of a carbon-containing gas, such as methane, ethane, benzene, acetylene, or ethylene, which then aggregates on a catalyst or substrate to form nanotubes. Studies have shown that gases with saturated carbon bonds produce structures with more walls than those grown from unsaturated gases. Therefore, methane and ethane are preferred for single-walled carbon nanotube growth, and ethylene, benzene, and acetylene are preferred for multiwalled carbon nanotube growth. Because high temperatures create more high-quality crystals by reducing defects and increasing crystallization, the source gas must also be one that does not thermally decompose into amorphous carbon and reduce the purity of nanotubes. The amorphous carbon can accumulate on the aggregation sites and inactivate them from forming nanotubes. The aggregation sites are typically metal catalysts on a substrate, but can be a dust or gas if raw material not on a substrate is desired. The size of the catalyst on the substrate determines the diameter of the grown nanotubes.

The nanotubes produced by chemical vapor deposition are similar to those produced by other methods; however, because the carbon is already in the gas phase it does not have such a limited carbon source size as laser ablation or arc discharge methods. There is also more control over the aggregation sites where the carbon nanotubes are grown, meaning that they can be grown as a raw material or on a substrate. Unfortunately, the lower temperatures of chemical vapor deposition result in nanotubes that are not as high quality as those produced in the laser ablation and arc discharge methods, although techniques have been designed to help overcome this drawback and make this method viable for the large-scale production of high-quality nanotubes.

9.6.4 Catalyst-induced growth mechanism

Nanotube growth does not start immediately after carbon gas comes into contact with the metal catalyst. The carbon vapor first dissolves into the metal catalyst particles. Once the particles are saturated, the carbon atoms assemble into n sp^2 hybridized structures at a less reactive surface. The carbon then moves out of the metal, adding more carbon atoms to the nanotube growing on the less reactive surfaces at the edges of the

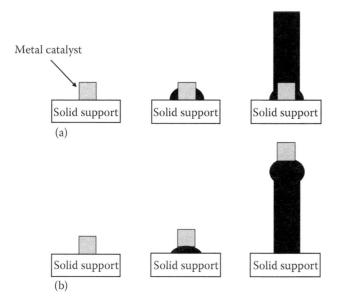

Figure 9.24 The two suggested models of carbon nanotube growth through chemical vapor deposition onto a metal catalyst, root growth (a) and tip growth (b). In both models, the carbon grows around the edges of the metal catalyst. In both models, the carbon grows around the edges of the metal catalyst.

particle. This is why the diameter of the metal particle determines the diameter of the nanotube being grown.

There are two suggested growth mechanisms for nanotube growth from a metal catalyst. The first method is root growth in which the nanotube grows upward from the metal catalyst particle, which stays attached to the substrate. The second suggested growth mechanism is tip growth in which the nanotube grows from the catalyst downward into the substrate. This pushes the catalyst up and detaches it from the substrate (Sinnot et al., 1999). The two mechanisms are illustrated in Figure 9.24.

9.7 GRAPHENE

Carbon nanotubes are not the only important form of nanostructured carbon. Networks of sp^2 carbon can also be assembled in large, planar sheets that can be macroscopic dimension along two axes but are only one atom deep, essentially a single layer of graphite. Such atomically thin layers of carbon are called **graphene**. As electrons are confined along only one axis, graphene is a 2D nanomaterial.

Graphene can be synthesized by vapor deposition techniques that are similar to those used for carbon nanotubes, but on catalysts with a planar surface geometry. It can also be produced by exfoliation, in which a single layer is lifted off a piece of bulk graphite after adhesion to a substrate, which can be as simple as a piece of cellophane tape. Finally, graphene can also be synthesized in solution by reducing graphene oxide (produced from partial oxidization of graphite) or merging other types of polycyclic aromatic hydrocarbons (Allen et al., 2010). Synthesis of graphene from graphene oxide is shown in Figure 9.25.

Like carbon nanotubes, graphene has many potential uses. It has very high tensile strength and controllable electrical and thermal conductivity. Conductive varieties of graphene can be used to form thin, transparent electrodes for semiconductor devices in place of typical inorganic transparent electrode materials such as indium tin oxide (ITO). Graphene surfaces can also be functionalized with other compounds via partial oxidation to graphene oxide. Graphene can also be formed into strong but highly porous structures, making it useful as a dielectric for high-energy capacitors. It has also been examined as a gate material for nanoscale field-effect transistors; an example of such a field-effect transistor produced by Kaner and coworkers (Tung et al., 2008) is shown in Figure 9.26.

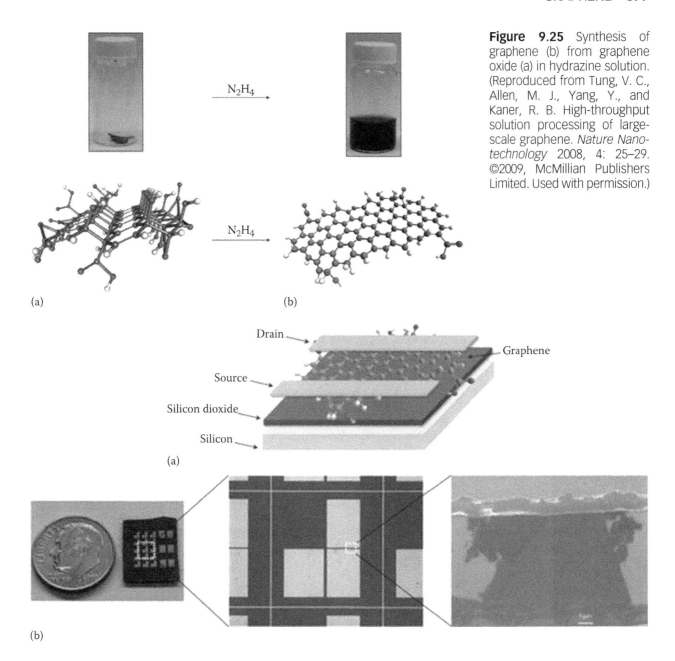

Figure 9.25 Synthesis of graphene (b) from graphene oxide (a) in hydrazine solution. (Reproduced from Tung, V. C., Allen, M. J., Yang, Y., and Kaner, R. B. High-throughput solution processing of large-scale graphene. *Nature Nanotechnology* 2008, 4: 25–29. ©2009, McMillian Publishers Limited. Used with permission.)

Figure 9.26 A graphene-based FET produced by the Kaner group. Panel (a) shows a schematic of the device and panel (b) shows a photograph of the prototype device along with both optical and electron micrographs showing the graphene gate on the FET. (Reproduced from Tung, V. C., Allen, M. J., Yang, Y., and Kaner, R. B. High-throughput solution processing of large-scale graphene. *Nature Nanotechnology* 2008, 4: 25–29. ©2009, McMillian Publishers Limited. Used with permission.)

Another interesting property of graphene is its ability to exhibit quantized conductance when subjected to a strong magnetic field. This effect is known as the quantum Hall effect and had been previously observed in other two-dimensional materials at low temperatures but can occur in graphene at room temperature (Novoselov et al., 2007). Materials exhibiting the quantum Hall effect exhibit discrete states with different conductivities in a manner similar to nanowires; the conductance of a 2D system exhibiting the quantum Hall effect is the von Klitzing equation (Klitzing et al., 1980):

$$G = \frac{ne^2}{h} \tag{9.27}$$

where n is an integer ($n = 1, 2, 3, ...$). The equation is often written in terms of resistance (R), which, as previously stated, is the inverse of conductance.

The unusual electronic properties of graphene, combined with its high mechanical strength, make it an interesting 2D material for a variety of applications. However, at present, commercial development of graphene is substantially less mature than development of nanoparticles or nanotubes.

9.8 NANOFRAMEWORKS

Three-dimensional (3D) nanomaterials have not been heavily discussed in this text so far, but are also an active area of research. Nanoframeworks, which consist of repeating arrays of nanoscale pores separated by walls of bonded atoms, are one important type of 3D nanomaterial. Such frameworks can consist of organic and/or inorganic components.

Frameworks containing both metals and organic components are called metal-organic frameworks or MOFs. For example, zeolitic imidazole frameworks (or ZIF) consist of metal atoms such as zinc complexed by the nitrogen atoms on imidazole groups on organic molecules. The structure of ZIF-8, a type of ZIF, is shown in the left panel of Figure 9.27. MOFs have been considered for a variety of applications; for example, ZIF-8 is capable of adsorbing a substantial quantity of carbon dioxide, making it of interest for removing CO_2 from exhaust (Tian et al., 2014) and a Zr-based MOF (UIO-66-NH_2) has shown promise for separating hazardous components out of nuclear waste (Banerjee et al., 2016).

In addition to metal-organic frameworks, all-inorganic nanoframeworks can also be developed for a variety of applications and can even be found

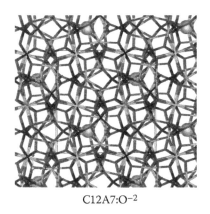 

C12A7:O^{-2} ZIF-8

Figure 9.27 Two examples of nanoframeworks: C12A7 (left) and ZIF-8 (right). Oxygen atoms are shown confined in some of the cages in the C12A7.

in nature. The mineral mayenite, found in certain types of cement, is formed from a 12:7 ratio of CaO and Al_2O_3 and is commonly called C12A7. It consists of a framework of aluminum and calcium ions linked by oxygen atoms into nanoscale cages. However, the stoichiometry is such that each cage has a net charge and requires the presence of counterions to neutralize this charge. In natural mayenite, one out of every six cages contains an oxygen (O^{-2}) anion. However, these anions can be replaced with other small ions (Matsuishi et al., 2005) or even by electron anions, which are electrons that have been separated from other ions and confined by their surroundings (analogous to solvated electrons in ammonia). The excess electrons add additional delocalized energy levels adjacent to the conduction band and allow the material to conduct electricity (Matsuishi et al., 2003). The ability to confine and exchange substituent ions makes C12A7 promising for catalysis and in electronics, and functional modifications to C12A7 have been extensively explored by Hosono and coworkers at the Tokyo Institute of Technology.

End of chapter questions

1. Marcus theory provides a framework for analyzing the effect of reaction thermodynamics and environment on an electron transfer process. While calculating absolute electron transfer rate constants is beyond the scope of this text, we can consider the relationship between ΔG and λ. Calculate the *relative* rates compared to $\Delta G = 0$, holding all other parameters constant, for the following situations, (a) $\Delta G = \lambda$, (b) $\Delta G = 1/2\lambda$, (c) $\Delta G = 2\lambda$, (d) $\Delta G = 1/10\lambda$, and (e) $\Delta G = 10\lambda$. Plot the rates. What trend do you see? Assume $T_{DA} = 1$ and $T = 300$ K.

2. Consider the density of states shown for YLD-124 in Figure 9.5. What kind of material (conductor, insulator, semiconductor, n-type semiconductor, or p-type semiconductor) is YLD-124? Explain your reasoning.

3. Solar cells are a source of clean electricity that produce no CO emissions beyond those produced by the materials used in their manufacturing. They also have no moving parts and harness an abundant, renewable resource (sunlight). These properties have made them highly appealing for future energy generation.

   a. Global energy consumption in 2010 was roughly 17 TW. How large of an area of 20% efficient photovoltaic cells would be needed to meet this demand, assuming average sunlight of 1000 W/m and neglecting any storage or transmission losses? Compare this to the total land area of the earth and the land area of your country/state/region.

   b. How might nanomaterials be useful for producing photovoltaic cells able to cover very large areas?

   c. Recall that the power produced by a photovoltaic cell is proportional to the product of V and I. The efficiency of a photovoltaic cell is the power produced by the cell divided by power delivered by the sunlight striking the cell. Could you increase the efficiency of a solar cell by making V very large? Consider how V is related to the band gap of the system and the meaning of the band gap in terms of which photons can be absorbed. Looking up an emission spectrum for sunlight may be useful, but is not required. What would happen if you made V very small instead?

4. The rate constant k for Förster transfer decays as the sixth power of the distance between the donor and acceptor. Using what you learned about electrostatic interactions in Chapter 5, explain where this r dependence is observed.

Cited references

Allen, M. J., Tung, V. C. and Kaner, R. B. Honeycomb Carbon: A Review of Graphene. *Chemical Reviews* 2010, 110: 132–145.

Anikeeva, P. O., Halpert, J. E., Bawendi, M. G., and Bulovic, V. Quantum Dot Light-Emitting Devices with Electroluminescence Tunable over the Entire Visible Spectrum. *Nano Letters* 2009, 9: 2352–2356.

Baddour, C. E. and Breins, C. Carbon Nanotube Synthesis: A Review. *Int. J. Chem. React. Eng.* 2005, 3: 1–20.

Banerjee, D., Xu, W., Nie, Z., Johnson, L. E., Coghlan, C., Sushko, M., Kim, D., Schweiger, M., Kruger, A., Doonan, C., and Thallapally, P. K. Zirconium-Based Metal–Organic Framework for Removal of Perrhenate from Water. *Inorganic Chemistry* 2016, 55: 8241–8243.

Ciraci, S., Buldum, A., and Batra, I. P. Quantum effects in electrical and thermal transport through nanowires. *Journal of Physics: Condensed Matter* 2001, 13: R537–R568.

Crossland, E. J., Ludwigs, S., Hillmyer, M. A., and Steiner, U., Freestanding Nanowire Arrays from Soft-Etch Block Copolymer Templates. *Soft Matter*, 2007, 3: 94–98.

Dai, H., Carbon Nanotubes: Opportunities and Challenges, *Surf. Sci.*, 2002, 500: 218–241.

Danyal, K., Shaw, S., Page, T. R., Duval, S., Horitani, M., Marts, A. R., Lukoyanov, D., Dean, D. R., Raugei, S., Hoffman, B. M., Seefeldt, L. C., and Antony, E. Negative cooperativity in the nitrogenase Fe protein electron delivery cycle. *Proceedings of the National Academy of Sciences* 2016, E5783–E5791.

Gao, X., Yang, L., Petros, J. A., Marshall, F. F., Simons, J. W., and Nie, S. *In Vivo* Molecular and Cellular Imaging with Quantum Dots. *Curr. Opin. Biotechnol*. 2005, 16: 63–72.

Gan, Q., Bartoli, F. J., and Kafafi, Z. H. Plasmonic-Enhanced Organic Photovoltaics: Breaking the 10% Efficiency Barrier. *Advanced Materials* 2013, 25: 2385–2396.

García-López, V., Chen, F., Nilewski, L. G., Duret, G., Aliyan, A., Kolomeisky, A. B., Robinson, J. T., Wang, G., Pal, R., and Tour, J. M. Molecular machines open cell membranes. *Nature* 2017, 548: 567–672.

Hagfeldt, A., Boschloo, G., Sun, L., Kloo, L., and Pettersson, H. Dye-Sensitized Solar Cells. *Chemical Reviews* 2010, 110: 6595–6663.

Huang, P.-S., Boyken, S., and Baker, D. The Coming of Age of De Novo Protein Design. *Nature* 2016, 537: 320–327.

Keidar, M. and Waas, A. M. On the Conditions of Carbon Nanotube Growth in the Arc Discharge. *Nanotechnology* 2004, 15: 1571.

Klitzing, K. V., Dorda, G., and Pepper, M. New Method for High-Accuracy Determination of the Fine-Structure Constant Based on Quantized Hall Resistance. *Phys. Rev. Online Arch.* 1980, 45: 494–497.

Kudernac, T., Ruangsupapichat, N., Parschau, M., Maciá, B., Katsonis, N., Harutyunyan, S. R., Ernst, K.-H. and Feringa, B. L. Electrically driven directional motion of a four-wheeled molecule on a metal surface. *Nature* 2011, 479: 208–211.

Lucky, S. S., Soo, K. C., and Zhang, Y. Nanoparticles in Photodynamic Therapy *Chemical Reviews* 2015, 115: 1990–2042.

Marcus, R. A. On the Theory of Oxidation-Reduction Reactions Involving Electron Transfer I. *Journal of Chemical Physics* 1956, 24: 966–978.

Marcus, R. A. and Sutin, N. Electron Transfer in Chemistry and Biology. *Biochimica et Biophysica Acta* 1985, 811: 265–322.

Matsuishi, S., Hayashi, K., Hirano, M., and Hosono, H., Hydride Ion as Photoelectron Donor in Microporous Crystal. *Journal of the American Chemical Society* 2005, 127: 12454–12455.

Matsuishi, S., Toda, Y., Miyakawa, M., Hayashi, K., Kamiya, T., Hirano, M., Tanaka, I., and Hosono, H. High-Density Electron Anions in a Nanoporous Single Crystal: $[Ca_{24}Al_{28}O_{64}]^{4+}$ $(4e^-)$. *Science* 2003, 301: 626–629.

Mazzio, K. A. and Luscombe, C. K. The future of organic photovoltaics. *Chemical Society Reviews* 2015: 44, 78–90.

National Renewable Energy Laboratory. *Best Research-Cell Efficiencies*. 2017. https://www.nrel.gov/pv/assets /images/efficiency-chart.png. Accessed 07/31/2017.

Nirmal, M., Brus, L. Luminescence Photophysics in Semiconductor Nanocrystals. *Accounts of Chemical Research* 1999, 32: 407–414.

Novoselov, K. S., Jiang, Z., Zhang, Y., Morozov, S. V., Stormer, H. L., Zeitler, U., Maan, J. C., Boebinger, G. S., Kim, P., and Geim, A. K. Room-Temperature Quantum Hall Effect in Graphene. *Science* 2007, 315: 1379.

Saito, R., Dresselhaus, G., and Dresselhaus, M. S. *Physical Properties of Carbon Nanotubes*. Imperial College Press: London, 1998.

Shirai, Y. Osgood, A. J., Zhao, Y., Kelly, K. F., and Tour, J. M. Directional Control in Thermally Driven Single-Molecule Nanocars. *Nano Letters* 2005, 5: 2330–2334.

Sinnot, S. B., Andrews, R., Qian, D., Rao, A. M., Mao, Z., Dickey, E. C., and Derbyshire, F. Model of Carbon Nanotube Growth through Chemical Vapor Deposition. *Chem. Phys. Lett.* 1999, 315: 25–30.

Takayanagi, K. Suspended Gold Nanowires: Ballistic Transport of Electrons. *JSAP Int.* 2001, 3: 3–8.

Thejokalyani, N. and Dhoble, S. J. Novel approaches for energy efficient solid state lighting by RGB organic light emitting diodes – A review. *Renewable and Sustainable Energy Reviews* 2013, 32: 448–467.

Tian, F., Cerro, A. M., Mosier, A. M., Wayment-Steele, H. K., Shine, R. S., Park, A., Webster, E. R., Johnson, L. E., Johal, M. S., and Benz, L. Surface and Stability Characterization of a Nanoporous ZIF-8 Thin Film. *Journal of Physical Chemistry C* 2014, 118: 14449–14456.

Tung, V. C., Allen, M. J., Yang, Y., and Kaner, R. B. High-throughput solution processing of large-scale graphene. *Nature Nanotechnology* 2008, 4: 25–29.

Zhang, P., Steelant, W., Kumar, M., and Scholfield, M. Versatile Photosensitizers for Photodynamic Therapy at Infrared Excitation. *Journal of the American Chemical Society* 2007, 129: 4526–4527

References and recommended reading

Gompper, G. and Schick, M., Eds. 2006, *Soft Matter*. John Wiley & Sons, Weinheim, Germany. The first volume in this book explores polymer melts and mixtures. The second volume focuses on complex colloidal suspensions. Both volumes are heavy on theoretical studies and are recommended for graduate students seriously interested in computational chemistry.

Kuhn, H. and Försterling, H.-D. *Principles of Physical Chemistry: Understanding Molecules, Molecular Assemblies, Supramolecular Machines*, 2000, John Wiley & Sons, Chichester, West Sussex, UK. Chapters 22 (Organized Molecules Assemblies) and 23 (Supramolecular Machines) are crucial reads for anyone interested in the physical chemistry of self-assembly and supramolecular processes. These chapters are very well written and accessible to undergraduates.

Hanson, G. W., *Fundamentals of Nanoelectronics*. 2008, Prentice-Hall, Upper Saddle River, NJ. This book gives a nice quantum mechanical treatment of nanoelectronics. Chapter 9, Nanowires and Nanotubes, is particularly useful. This book assumes a strong background in physics and is recommended for students interested in free and confined electrons in nanomaterials.

Hoffman, P. *Solid State Physics*. 2008, Wiley-VCH. Weinheim, Germany. This concise undergraduate textbook provides an accessible introduction to the properties of solid materials, including band theory and dielectric functions. While some concepts, such as working in reciprocal space for periodic materials, may be less intuitive for chemists used to working with molecular structures, they are essential for those working or interacting with materials scientists or solid-state physicists.

Lehn, J. M., *Supramolecular Chemistry*. 1995, VCH Weinheim. An essential reference to supramolecular chemistry written by the Nobel Laureate who coined the term.

Messenger, R. A. and Ventre, J. *Photovoltaic Systems Engineering*. 2004, CRC Press, Boca Raton, FL. This is an engineering text that includes not only extensive information on the structure and physics of solar cells but also on their implementation for power generation.

Rao, C. N. R., Müller, A., and Cheetham, A. K., Eds. *The Chemistry of Nanomaterials: Synthesis, Properties, and Applications* (Volumes 1 and 2), 2005, John Wiley & Sons, Weinheim, Germany. This book contains some of the best reviews on quantum dots, nanotubes, and nanowires (synthesis and properties). The book also contains a good chapter on oxide nanoparticles.

Rao, C. N. R., Müller, A., and Cheetham, A. K., Eds. *Nanomaterials Chemistry: Recent Developments and New Directions*, 2007, John Wiley & Sons, Weinheim, Germany. This book provides good coverage of mostly inorganic nanomaterials. The book describes the use of nanomaterials for some interesting applications such as supercapacitors, molecular machines, and transistors.

Simon, J. and André, J. J. *Molecular Semiconductors*. 1985, Springer-Verlag, Berlin, Germany. While dating from the early days of organic semiconductors, this text has good explanations of charge transfer and electronic structure for organic systems.

Turro, N. J., Ramamurthy, V., and Scaiano, J. C. *Principles of Molecular Photochemistry: An Introduction*. 2009, University Science Books, Sausalito, CA. A recently updated classic graduate text on photochemical processes in molecules, including charge transfer and excitation transfer. While written primarily for graduate students and researchers, it is written without extensive dense mathematics and is accessible for interested undergraduates.

CHAPTER 10

Fabrication, Properties, and Applications of Thin Films

CHAPTER OVERVIEW

This final chapter discusses the fabrication and applications of thin films, a category of 2D nanomaterials with diverse compositions and applications. Thin films can range from simple layers of amphiphiles or model biological membranes to solid layers of silicon, oxides, and metals in integrated circuits, or even combinations of organic soft matter and "hard" inorganic structures. As film and device fabrication and surface functionalization are highly active fields of research, this chapter provides a survey of key technologies and recent primary literature intended to assist students in transitioning from "textbook" material to active research in the field.

10.1 LANGMUIR–BLODGETT FILMS

First demonstrated by Katharine Blodgett in 1934 (Blodgett, 1935), the **Langmuir–Blodgett** (LB) technique has been by far the oldest and most extensively studied organic nanofilm formation method. Essentially, a monolayer at the air–water interface (the Langmuir film) is mechanically transferred to a solid substrate generating the LB film. Although interest waned until the 1970s, the potential application of LB films in the field of optics and material science has revived LB research. This was fueled by the results that showed an LB film comprised of a dye to be the first multilayer film to produce interesting NLO effects. Today, research toward the synthesis of tailor-made molecules with intrinsic nonlinear susceptibilities and their incorporation into LB films is flourishing.

10.1.1 Langmuir films

The first step of LB multilayer assembly is to form a **Langmuir film** by dissolving surface-active organic molecules, such as surfactants or amphiphiles, in a nonpolar, volatile solvent and then spreading the solution onto a polar liquid surface, usually water. Once the volatile solvent evaporates, the remaining amphiphilic molecules are oriented at the air–water interface so that the hydrophilic headgroups are buried in the bulk aqueous phase while the hydrophobic tail groups are directed upward into the air. However, for this orientation to occur, two conditions must be met: (1) the hydrophobic tail must be long enough to prevent its dissolution in the aqueous media, and (2) the hydrophilicity of the headgroup must be strong enough to prevent the formation of thicker multilayer films at the interface or evaporation of the surfactant molecules. Moving barriers on either side sweep the water surface and force the amphiphiles to pack, forming an ordered, compressed monolayer of the Langmuir film (Figure 10.1).

The packing density of a Langmuir film is crucial in determining the final structure of an LB film. As we have already seen, surface pressure is a measure of the change in the surface tension of a pure liquid due to the presence of a surfactant. Basically, it is defined as the difference between

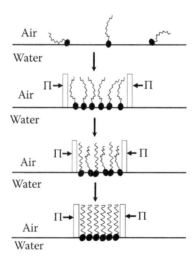

Figure 10.1 A Langmuir film. Amphiphilic molecules in chloroform solvent are deposited on water surface. The Langmuir film is formed by confining the amphiphiles at the air–water interface with movable barriers.

the surface tension of the pure water and the surface tension of the aqueous solution containing amphiphilic solutes. This phenomenon is especially important in Langmuir films. If the number of molecules comprising the Langmuir film is known, then surface pressure as a function of the area occupied by each molecule may be studied. The resulting plot is known as a pressure–area (Π–A) isotherm, and its shape is unique to the molecule used to form the film. Figure 10.2 illustrates a typical isotherm for a long-chain carboxylic acid [$CH_3(CH_2)_nCOOH$] at the air–water interface. At constant temperature the Π–A isotherm is given by the changes in surface tension as the monolayer is compressed.

The Π–A isotherm may consist of several regions, although molecules usually exhibit only a few of them. When the molecules are first deposited onto the film and no external pressure has been applied, the monolayer behaves most like a 2D gas (G), which can be approximated by the 2D ideal gas equation of state,

$$\Pi A = k_B T \qquad (10.1)$$

In this equation, Π is the surface pressure, A is the molecular area, k_B is the Boltzmann constant, and T is the temperature. In the **G phase**, all interactions with the water are attractive, so even the hydrophobic tail is in intimate contact with the water. Transition to the **liquid-expanded**

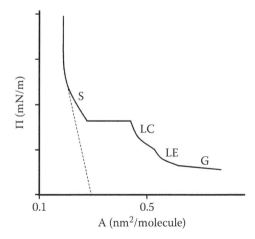

Figure 10.2 A simplified Π–A isotherm for a simple long-chain fatty acid. Extrapolating the slope of the S phase to zero pressure enables one to obtain the molecular area per molecule at zero pressure (see Example 10.1).

phase (LE) occurs as monolayer compression begins and the tail starts to lift away from the surface. The transition states form a plateau because the existence of two phases gives the monolayer only one degree of freedom and constant temperature gives constant pressure, Π (Knobler, 1990).

The **liquid-condensed phase** (LC) forms with further compression. The hydrocarbon chains are highly aligned and a high number of *trans* conformations appear, indicating the achievement of long-range order due to van der Waals interactions. Finally, application of great external pressure leads to the **ordered solid** (S) phase, and the area per molecule is comparable to the closely packed chains of the amphiphile's 3D crystal. Further compression leads to the collapse pressure, Π_c, where the molecules are ejected randomly throughout the film.

Molecular dimensions can be obtained through the Π–A isotherm by extrapolating the slope of the S phase to zero pressure to get the molecular area per molecule at zero pressure, A_0. A_0 can be compared to theoretical values of the cross-sectional area of hydrocarbon chains, and a close correlation indicates that a closely packed monolayer has been formed with the hydrophobic tails oriented normal to the surface. Furthermore, the Π–A isotherm gives information on the stability of the monolayer at high pressures.

Example 10.1 Determining the Limiting Surface Area per Molecule

Use Figure 10.2 to determine the area per molecule for the amphiphile described by the isotherm.

Solution By extrapolating the slope of the S phase to zero pressure (the dashed line in Figure 10.2), one obtains an area of 0.2 nm^2 for each molecule.

10.1.2 Langmuir–Blodgett films

The LB film is subsequently formed by the immersion of a substrate into the water to break the Langmuir film and transfer the monolayer onto the substrate. Additional immersions result in the fabrication of the multilayer film, whose thickness depends largely on the molecular chain length and number of dippings. The LB deposition technique is illustrated in Figure 10.3.

Deposition of tightly packed molecular monolayers with unilateral molecular directionality normal to the substrate can be induced through

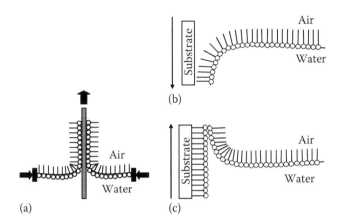

Figure 10.3   (a) The LB method is used to transfer a film at the air–water interface to a solid substrate. (b) The downstroke results in the monolayer being transferred to the hydrophobic substrate. (c) Withdrawing the substrate generates a bilayer as a Y-type film.

annealing. Annealing involves repeated compression and then expansion of the Langmuir film. Studies have found that the calculated area per molecule obtained through the Π–A isotherm actually decreases under such a procedure, leading to a closely packed and well-aligned monolayer.

Film deposition onto the hydrophilic or hydrophobic substrate occurs only when the direction of the motion coincides with the curvature of the meniscus at the solid–liquid interface. If a hydrophobic substrate is used, the meniscus will curve downward upon immersion into the subphase, resulting in deposition on the downstroke and adhesion of the tails to the hydrophobic surface. If this is the first immersion, subsequent monolayer deposition behavior determines the LB film type. If withdrawal of the substrate results in subsequent layer formation in a head-to-head fashion, then a Y-type LB film is formed. On the other hand, if monolayer transfer occurs only upon insertion of the hydrophobic substrate, then an X-type LB film is formed. A third type of LB film, Z-type, is formed on a hydrophilic substrate where the amphiphilic headgroup adheres to the substrate and transfer occurs only upon withdrawal of the substrate.

The Y-type film is the most common type of LB film formed. Since monolayer transfer occurs on both the down- and upstrokes, multilayer films may also deposit on hydrophilic substrates with the exception that no monolayer adsorbs in the first insertion. Due to the head-to-head and tail-to-tail conformation, the film formed, if composed of only one kind of amphiphilic molecule, is completely centrosymmetric. Such a film possesses no dipole moment, rendering the assembly useless for NLO applications. However, if fatty acid spacers, such as stearic or arachidic acid, are incorporated between the other amphiphilic species, a net dipole moment may result. This procedure requires two separate LB compartments for each type of molecule. The substrate is inserted through a layer of the amphiphile of interest, moved to the other compartment under water, and withdrawn out of the fatty acid spacer Langmuir film. Research has shown a relatively large SHG signal for such a polar-ordered arrangement. Conversely, X- and Z-type films are inherently arranged in an acentric manner with a net dipole moment due to the head-to-tail molecular conformation. In fact, studies by some researchers have found that Z-type LB films generate a strong SHG signal.

Despite the virtues of the LB technique, it also has its shortcomings. One limitation is the unstable nature of the adsorbed films. In practice, although the transfer process occurs very slowly, the layers are not always transferred onto the substrate as would be desired. After deposition, the molecules have been shown to rearrange to more stable conformations, which could lead to loss of the acentric order that gives the films their NLO activity. Another disadvantage is the occurrence of chromophore randomization. It has recently been shown that intermixing between LB layers is present. This produces structural defects and chromophore tilting that becomes amplified when the number of LB layers is increased. Thus, it is not surprising that experimental results show that the SHG signal resulting from the head-to-head, polar-ordered (Y-type) LB films decreased with thickness (Johal et al., 1999). Moreover, the fatty acid spacers used to make acentric alignment more energetically favorable could lead to even more disorder in thicker LB films. Stearic acid has been found to form two-dimensional crystalline structures that result in a heterogeneous monolayer filled with defects. Finally, staggered molecular structures may form due to repulsive interaction between strong dipoles. The effect of intermolecular interactions on the order of the LB film are still sparsely understood, and current research addresses the effects of dipolar interactions on film formation. With the limitations of the LB technique and other aforementioned methods, the introduction of the electrostatic self-assembly method was readily welcomed.

10.2 POLYELECTROLYTES

Polyelectrolytes are polymers, or chains of molecules, that contain ion-izable functional groups that exist in a charged state in solution, allowing the polymer to function as an electrolyte. Acidic functional groups such as carboxylic or sulfonic acid groups can give up a proton, becoming nega-tively charged, and basic functional groups such as amines can accept a proton, becoming positively charged. In order to maintain charge of neutrality, these charges must always be countered by charges of opposite sign, whether on polymers or on small free ions in solution. Figure 10.4 shows some common polycations and polyanions. The amount of charge on a polyelectrolyte determines whether it is classified as strong or weak.

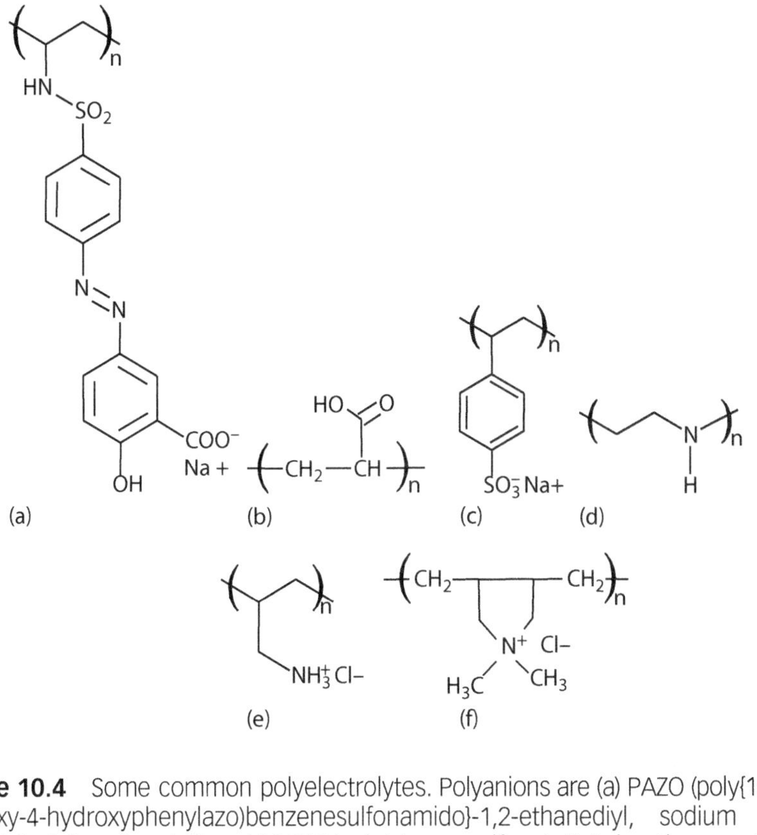

Figure 10.4   Some common polyelectrolytes. Polyanions are (a) PAZO (poly{1-[4-(3-carboxy-4-hydroxyphenylazo)benzenesulfonamido}-1,2-ethanediyl, sodium salt]), (b) PAA [poly(acrylic acid)], and (c) PSS [poly(styrenesulfonate)]. Polycations are (d) PEI [poly(ethylenimine)], (e) PAH [poly(allylamine hydrochloride)], and (f) PDDA [poly (diallyldimethyl ammonium chloride)].

Strong polyelectrolytes are fully ionizable, whereas weak polyelectrolytes, with fractional charge, are only partially ionizable. Typically, polyelectrolyte conformation is "flat" because the charges on a linear polymer repel each other due to Coulombic repulsion. With the addition of salt ions, however, this conformation can be altered to a more coiled or collapsed state due to screening of the charges on the polymer by the other ions in the solution. This ability to change polyelectrolyte conformation can change their function and is very useful in materials chemistry.

Synthetic organic chemists have been able to control the characteristics of polyelectrolyte assemblies in many different manners, most simply by altering monomeric units, electrolyte group, and polymer length. As a result, polyelectrolytes make a very tunable building block for complex materials. They have been used in a number of applications ranging from drug delivery agents, to components in conducting films, to facilitating colloidal suspensions, and to creating mimics of biological molecules. Recently, polyelectrolyte microcapsules have found use as microreactors for catalyst activity, precipitation reactions, crystallization reactions, and polymerization reactions.

10.2.1 Electrostatic self-assembly

One useful feature of polyelectrolytes is that many of them can interact with each other in a process known as **electrostatic self-assembly** (ESA). As a simple example, Figure 10.5 illustrates the complexation of two oppositely charged polyelectrolytes in the bulk phase. Polyelectrolytes can also adhere to oppositely charged surfaces, forming nanofilms. Because ESA relies on attractive interactions between materials of alternating charge, a number of different kinds of molecules can be used, such as small inorganic molecules, proteins, dendrimers, and polyelectrolytes. Our exploration of ESA films will focus on polyelectrolytes. The flexibility of the polyelectrolytes used for ESA, as well as the ease of the self-assembly process, gives it advantages over other methods of creating thin films, such as the LB technique. In contrast to ESA, the LB technique requires far more instrumentation, time, and cost to create films that are limited in long-term stability.

Films formed through ESA techniques can be composed of many adjacent layers and are generally well ordered on the supramolecular scale depending on the reaction conditions in which they are made. Both the amount of polyelectrolyte as well as the order in which they are deposited can be easily controlled. In the formation of the first layer, a polyion

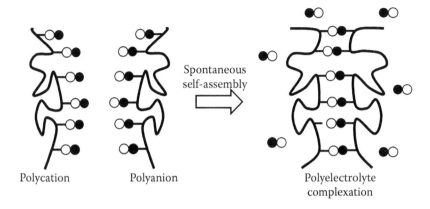

Polycation   Polyanion   Polyelectrolyte
                                   complexation

Figure 10.5 The complexation reaction between oppositely charged polyelectrolytes. The ion-pair formation (or ion exchange reactions) occurs due to a large entropic gain when counterions are expelled into solution. Filled circles represent anionic sites and open circles represent cationic sites.

adsorbs to an oppositely charged substrate surface because of favorable electrostatic attraction between the charged groups on the polyion and those on the surface. The counterions that previously neutralized the charges on both components are displaced into solution. The process is irreversible, barring a large change in pH, due to the low probability of simultaneously breaking all of the ionic interactions between the polyelectrolyte and the surface. The amount of polyelectrolyte deposited can be controlled by varying the strength of the interaction (e.g., by changing the polyion, pH, or ionic strength of the system).

A commonly used system is a positively charged polyelectrolyte on an oxidized silica or silanol surface. The silica substrate surface is negatively charged due to deprotonated silanol groups. When exposed to a cationic solution, the negative charge on the substrate quickly attracts the cations to the surface. ESA methods typically expose a solid substrate surface to the desired adsorbent by "dipping" the substrate into an aqueous solution of the adsorbent for about 10 minutes. In order to build multilayered assemblies, this "dipping" procedure is repeated with alternating polycation and polyanion aqueous solutions, with a rinse step with either water or slightly ionic solution in between that increases the ability of the next layer to be well ordered and removes the counterions that had been displaced in the previous assembly step. The rinse step removes weakly adsorbed molecules that are not a part of the film and primes the surface

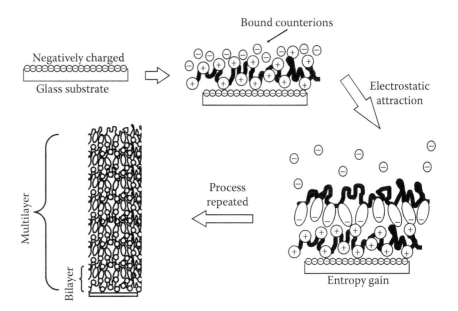

Figure 10.6   Polyelectrolyte self-assembly. The polycation adsorbs from aqueous solution to the negatively charged substrate. This is followed by exposure of this film to an aqueous solution of the polyanion. Adsorption of the polycation to the polyanionic film results in a bilayer film. The process can be repeated many times to produce a multilayer nanofilm.

for the oppositely charged adsorbent. This approach is generally known as the layer-by-layer (LbL) technique.

Repetition of this process can be used to make many layered films. To better understand the overall formation of a multilayer assembly we must break down the mechanisms of ESA further into adsorption of the polyelectrolyte layer to the substrate film, interpenetration of polyelectrolyte materials within the film, and the complexation of molecules that occurs within the film. Figures 10.6 and 10.7 illustrate how ESA can be used to construct polyelectrolyte multilayer assemblies.

10.2.2 Charge reversal and interpenetration

Within a multilayer film, there are two main regions: the surface, at which **charge overcompensation** or reversal occurs, and the bulk area, in which interpenetration occurs. The surface of the film is repeatedly exposed to polyelectrolyte solutions of alternating polyion charge. The bulk is not

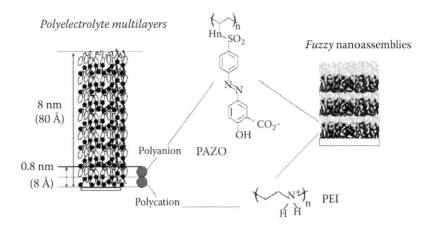

Polyelectrolyte multilayers

Fuzzy nanoassemblies

8 nm
(80 Å)

0.8 nm
(8 Å)

Polyanion   PAZO

Polycation

PEI

Figure 10.7 Dimensions of a typical multilayer constructed using the polycation PEI and the polyanion PAZO. The thickness of a bilayer as measured by ellipsometry is around 1 nm. A 10-bilayer film has a thickness between 100 and 150 nm. Interlayer interpenetration is illustrated as the "fuzzy" nanoassembly on the right. The degree of interpenetration is drastically underexaggerated.

exposed to the aqueous environment. This greatly impacts the differences in chemistry that occurs in these two main regions.

At the surface, charge reversal, or charge overcompensation, helps drive the formation of layers within the film. Typically, polyelectrolytes are assumed to take a structure with 1:1 charge stoichiometry in order to minimize their electrostatic energy. However, in order to drive the formation of multilayers, an overall surface charge must be exposed prior to the next deposition step in an oppositely charged polyelectrolyte solution. Charge reversal of the surface enables the electrostatic interactions that drive the self-assembly process.

The charge-overcompensated region of the terminal layer is matched by oppositely charged salt counterions in solution (See Chapter 5 for details). Upon exposure to the next polyionic solution, the counterions are displaced by the polyion. This displacement is driven primarily by entropic gain from displacement of the counterions. Entropic gain combined with electrostatic interactions results in a net negative free energy of adsorption that allows the polyelectrolyte to remain adsorbed to the surface. Further adsorption of the polyelectrolyte will stop when repulsive interactions from material already adsorbed prevail. As a result, approximately reproducible depositions occur in the formation of each subsequent layer. Once a polyelectrolyte has fully adsorbed to the surface, charge overcompensation occurs again at the terminal layer surface, yielding a reversal in surface charge compared to the film prior to deposition. As happened for the previous layer, the charged surface is balanced by positive counterions from the polyelectrolyte solution. An assembly with one polycationic and one polyanionic layer comprises a single **bilayer**.

The presence of counterions at the charged surface can be better explained through the use of image forces. Image forces result from the differences in the dielectric constants of water and the film interface. Stronger dielectric environments polarize to a greater extent in response to an applied electric field, providing more effective screening of that electric field than a weaker dielectric environment. Water has a very high dielectric constant (~80), while organic polymers typically have relatively low dielectric constants (2–10). The difference in the dielectric environment causes a repulsion of water charges from the surface, allowing a polyelectrolyte counterion to bind at the surface instead. These counterions will subsequently be displaced by a polyelectrolyte with the same charge opposite to the surface—a process driven by gains in entropy because more salt ions are displaced from the surface than the number of polyelectrolyte molecules that adsorb.

To maintain the overall neutral charge of the film, charge compensation—either intrinsic or extrinsic—must be present. Extrinsic compensation refers to polymer charges that are balanced by extrinsic materials, such as the salt counterions. It has been found that there is actually very little salt concentration observed within the film. Instead, there is a 1:1 stoichiometric ratio of polycation to polyanion charge. As a result, the overall charge balance within the multilayers must be attributed primarily to intrinsic compensation. However, some extrinsic compensation has been found to occur at the substrate surface, where charge overcompensation is neutralized by salt counterions.

Within the bulk of the film, we might expect clearly defined layers of alternating charge. This is not what occurs. Upon adsorption the polyelectrolyte does not simply "lie" on the surface. It has been commonly observed that after adsorption, the outer polyelectrolyte then diffuses slightly into the inner polyelectrolyte layer. This effect results in blurred or "fuzzy" distinctions from layer to layer (Figure 10.7). The effects of interpenetration underscore the lack of organized structure observed within the bulk film.

The degree to which interpenetration occurs, while not completely understood, is linked to ionic concentration and charge density of the polyelectrolyte. In general, ionic concentration is correlated with interfacial overlap. The salt ions cause conformational changes of the polyelectrolytes by shielding the charges on the polyelectrolytes, enabling a polymer to "extend." As a result, although it becomes more difficult for the polyelectrolyte to diffuse deeper into the assembly, the presence of salt ions opens up more polymer segments to competitive ion exchange with counterions. Polyelectrolyte charge—whether a polyelectrolyte is strong or weak—also contributes to the extent of overlap observed. With a weaker polyelectrolyte or a polyelectrolyte with lower charge density, a flat/extended conformation of the polymer cannot be achieved. As a result, the polyelectrolyte cannot penetrate the outer film layer to as great a degree. Overall, ionic concentration plays a significant role (in addition to the charge density of the polyelectrolyte) in contributing to the interpenetration of the polyelectrolyte within the film (Radeva and Petkanchin, 1997).

Interpenetration results in the bulk of the film appearing to be more homogeneous, while the surface of the terminal layer has an overall charge. Interfacial overlap also results in a lack of crystalline structuring. This limits the applications of ESA films in producing materials that require ordering for functionality purposes, such as chromophore ordering for nonlinear optical responses. However, the organizational structure of ESA films has been found to be useful for the construction of conductive polymeric materials. The overlap from interpenetration allows electrons to flow easily through the film, which would not be the case if the layers were stratified into organized layers.

10.2.3 Multilayer formation

Charge reversal, interpenetration, and complexation all contribute to the structure and formation of a multilayer. Broader factors to consider for successful multilayer formation are film stability and film thickness. The first, film stability, requires that a completed multilayer assembly be stable in structure when removed from solution so that the film does not change or degrade following assembly. The second, film thickness, can be tailored based on ESA conditions and number of multilayers.

Film stability requires that the polyelectrolyte assembly be irreversible. As a result, spontaneous desorption of material from the solid substrate film cannot occur in a well-formed assembly. Desorption due to small ion competitors or desorption due to exposure to polyelectrolyte solution all cannot occur in order to maintain the irreversibility of the film. It has been determined that desorption of material occurs with such slow kinetics that the ESA film can be considered irreversible.

Control of multilayer thickness is determined by a variety of factors, such as polymer chain length, polymer charge, and strength of ionic solution. The most obvious way to control film thickness is to control the number of multilayers formed on the substrate surface. A greater number of multilayers will result in a thicker overall film, whereas fewer multilayers will decrease the film thickness. In addition to the number of multilayers, layer thickness is highly dependent on ionic concentration with an almost linear relationship. Nearly linear multilayer growth has been observed, up to 100 bilayers (Reveda and Petkanchin, 1997). Low salt concentrations will allow the polymer to assume a flat, extended conformation. Higher salt concentrations result in greater shielding of the charges on the polyelectrolyte, allowing the charges to be in closer proximity and resulting in a more coiled structure. Low ionic concentration screens charges on the polyelectrolyte and the surface charges, allowing for a flat conformation. The looped and coiled conformation that results from increased ionic concentration creates thicker layers and thus an overall thicker film, although it will not necessarily change the amount of polymer adsorbed to the surface. Charge density on the polymer can also affect film thickness. A lower charge density correlates with thicker layer formation. Similar to the results at high ionic concentration, a polymer with lower charge density will assume a more coiled conformation due to fewer electrostatic interactions between the substrate surface and the weakly charged polyelectrolyte. Perhaps contrary to intuition, it is important to note that polymer molecular weight and polymer branching do not cause significant changes in multilayer thickness.

10.3 MODEL PHOSPHOLIPID BILAYER FORMATION AND CHARACTERIZATION

Biological membranes are complex structures that contain a large amount of proteins and are difficult to study in vivo. By using model phospholipid bilayers, we can study how specific membrane components interact by incorporating only these specific components into the model membrane (Castellana et al., 2006). Furthermore, phospholipid bilayers closely resemble cell membranes in many ways. They retain two-dimensional fluidity and are capable of hosting membrane proteins. Therefore, model membranes have historically been used to study membrane properties in vitro. They have also been used for the investigation of biological processes involving membranes that occur at the cellular level, such as ligand-receptor interactions, viral attack, and cellular signaling events.

A host of techniques are used to study these membrane systems. In this book, discussion will be limited to widely used fluorescence techniques such as FRAP, FRET, and FLIC, as well as a few surface-sensitive techniques such as AFM and QCM. There are a number of other important methods, not covered in this section, for studying model bilayers such as electron microscopy, which allows for the direct visualization of membranes in vivo, impedence spectroscopy, which allows for electrical measurements of membranes that are important for understanding neurons, as well as neutron and x-ray scattering techniques that are used to probe the structure and periodicity of membranes. Several types of model lipid bilayers can be fabricated and will be discussed in the next few sections.

10.3.1 Black lipid membranes

Black lipid membranes are lipid bilayers formed in solution over a small aperture, usually less than 1 mm in diameter. The hole is formed in a hydrophobic material such as Teflon and is usually part of a wall separating two compartments that can be filled with aqueous solution (see Figure 10.8a). The two most popular methods for black lipid formation involve the painting of the lipid solution over the aperture or the formation of a folded bilayer. Painting is carried out with a small artist's paintbrush. The formation of folded lipid bilayers requires a container with two compartments separated by a small aperture, and the solution levels in each compartment must be controlled independently (Figure 10.8b). The desired solution is filled into each compartment and a monolayer of phospholipid material is spread on top of one of the compartments. The solution level containing the monolayer can be raised and lowered over the aperture to deposit the bilayer.

Black lipid membranes have been used to study various biophysical processes. Black lipid membranes are suspended in solution, where there are no unwanted interferences of the membrane with an underlying support. The absence of a support allows for transmembrane proteins to be incorporated within the phospholipid bilayer such that they remain fully mobile and active. The ability to insert single protein pores into black lipid membranes allows for the creation of potential nanodevices. This has been accomplished by Gu et al. (1999) through the use of genetically modified α-hemolysin. α-Hemolysin mutants that can noncovalently capture cyclodextrin molecules within their pores were created. A current change is measured due to the restriction of the pore by cyclodextrin.

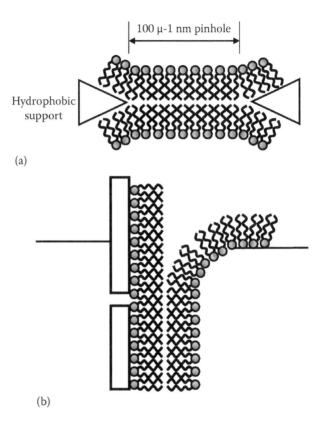

Figure 10.8 (a) Illustration of a black lipid membrane. (b) The formation of a folded lipid bilayer. Reprinted from Castellana, E. T. and Cremer, P. S. Solid Supported Lipid Bilayers: From Biophysical Studies to Sensor Design. *Surface Science Reports* 2006, 61(10):429–444. Copyright 2006, with permission from Elsevier.

The binding and unbinding of small organic molecules at the single molecule level can be measured using this process.

However, the lack of support limits the lifetime of the bilayer due to poor stability. Furthermore, the methods of detection and characterization of black lipid membranes are also limited. Because there is no support, surface-sensitive techniques cannot be used for characterization.

10.3.2 Solid supported lipid bilayers

Supported bilayers, as illustrated in Figure 10.9, are more robust and stable than black lipid membranes and can also be analyzed by surface-specific analytical techniques such as AFM, QCM, DPI, and SPR (Richter et al., 2006).

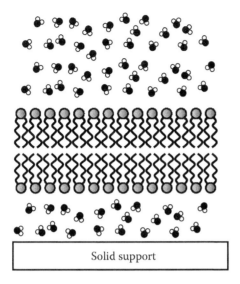

Figure 10.9 Schematic diagram of a supported phospholipid bilayer. (Reprinted from Castellana, E. T. and Cremer, P. S. Solid Supported Lipid Bilayers: From Biophysical Studies to Sensor Design. *Surface Science Reports* 2006, 61 (10):429–444. Copyright 2006, with permission from Elsevier.)

In these systems fluidity is maintained by a 10–20 Å layer of trapped water between the substrate and the bilayer (Groves and Boxer, 2002).

In order to support a high-quality membrane the surface should be hydrophilic, smooth, and clean (Tamm and McConnell, 1985). A few commonly used substrates for bilayer formation are silica, mica, gold, and titanium oxide. There are three general methods for the formation of supported phospholipid bilayers on planar supports. The first involves the deposition of a lower leaflet of lipids from the air–water interface by the Langmuir–Blodgett technique (Figure 10.10a). This is followed by the transfer of an upper leaflet by the Langmuir–Schaefer procedure, which involves horizontally dipping the substrate to create the second layer. A second method of supported bilayer formation is the adsorption and fusion of vesicles from an aqueous suspension to the substrate surface (Figure 10.10b). The third method, a combination of the first two methods, can be employed by first depositing a monolayer via the Langmuir–Blodgett technique followed by vesicle fusion to form the upper layer (Figure 10.10c). Each of the three methods contains particular

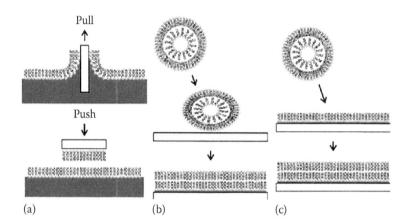

Figure 10.10   Common methods for forming supported bilayers. (a) Langmuir–Blodgett technique to deposit monolayer followed by pushing the substrate horizontally through another lipid monolayer. (b) Vesicles in solution adsorb and fuse to the surface to form a bilayer. (c) A combination of (a) and (b). (Reprinted from Castellana, E. T. and Cremer, P. S. Solid Supported Lipid Bilayers: From Biophysical Studies to Sensor Design. *Surface Science Reports* 2006, 61(10): 429–444. Copyright 2006, with permission from Elsevier.)

advantages and disadvantages. The first and third methods are useful for the formation of asymmetric bilayers.

The adsorption and fusion of **small unilamellar vesicles** (SUVs) is one of the easiest and most versatile means for forming supported bilayers. Small unilamellar vesicles can be prepared many different ways. Lipids are often stored in chloroform to prevent degradation. Small unilamellar vessels can be fused by withdrawing a known quantity of lipids from chloroform suspension and placing them in a small glass container with a known surface area. The chloroform can be evaporated with an inert gas such as nitrogen. This makes a thin coating of lipids in the container; residual chloroform can be removed by drying under vacuum. The dried lipids can be rehydrated to multilamellar vesicles by exposing the film to whatever solvent is desired to be trapped within the vesicle, then sonicating them. Extrusion of these multilamellar vesicles through porous polycarbonate membranes at high pressure will result in SUVs with a size distribution dependent on the size of the polycarbonate membrane's pores.

Parts of the mechanism of vesicle adsorption and fusion to form a bilayer have been elucidated by QCM studies (Figure 10.11). The process begins with the adsorption of SUVs from the bulk solution onto the substrate.

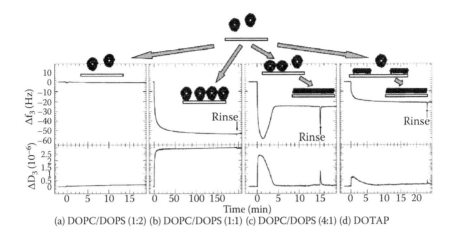

(a) DOPC/DOPS (1:2) (b) DOPC/DOPS (1:1) (c) DOPC/DOPS (4:1) (d) DOTAP

Figure 10.11 Lipid deposition pathways measured by QCM-D on silica. (a) Vesicles do not adsorb. (b) Vesicles adsorb and remain intact, forming a supported vesicular layer (SVL). (c) Vesicles adsorb and remain initially intact. At high vesicular coverage an SLB is formed. (d) Vesicles adsorb and rupture instantaneously to form an SLB. (Reprinted with permission from Richter, R. P., Berat, R., and Brisson, A. R. Formation of Solid-Supported Lipid Bilayers: An Integrated View. *Langmuir* 2006, 22(8): 3497–3505. © American Chemical Society.)

In the early stages after adsorption, SUVs may fuse with one another to form larger unilamellar vesicles. The vesicles then rupture, forming a supported bilayer in a process that depends on the destabilizing inter-actions affecting the vesicle, such as the osmotic pressure due to a salt concentration within the vesicle that differs from the salt concentration outside the vesicle, or the attraction between the individual lipids and substrate that causes vesicles to deform from their more stable spherical shape into a more oblong shape. Figure 10.11 shows four ways that lipid vesicles interact with the solid support as followed by QCM-D. The QCM-D technique is a valuable tool for screening the overall properties of the deposited lipids. The dissipation parameter allows for distinguishing between intact, adsorbed vesicles (high dissipation) and bilayer patches (low dissipation). As shown in Figure 10.11, vesicles either do not adsorb (Figure 10.11a); adsorb and remain intact, giving rise to a supported vesicular layer (Figure 10.11b); or form a supported bilayer (Figure 10.11c and d). As observed by QCM, supported bilayer formation can occur via two scenarios with distinct kinetics. In one case, the vesicles rupture quickly upon interaction with the solid support (Figure 10.11d). In another case, a large amount of intact vesicles are adsorbed at an inter-mediate stage of the process (Figure 10.11c).

The main disadvantage of supported bilayers is that the supported bilayer is not truly decoupled from the underlying substrate. The layer of hydration at the bilayer-support interface allows for membrane fluidity, but this layer is also too thin to prevent transmembrane proteins from interacting unfavorably with the underlying substrate. Such interactions can denature the protein or cause the membrane to lose its fluidity.

10.3.3 Polymer cushioned phospholipid bilayers

Figure 10.12 shows the same system from Figure 10.11 in the presence of a lipopolymer support that prevents denaturation. The addition of a polymer layer effectively decouples the membrane from the surface and still allows for investigation by an array of surface science techniques. A well-designed polymer cushion should behave much like a cytoskeleton present in eukaryotic cells. In physisorbed systems, weak interactions between the phospholipid bilayer and the polymer support can result in an unstable system. This can be overcome by covalently attaching the polymer to the substrate. Next, anchor lipids or alkyl side chains capable of inserting into the phospholipid bilayer are employed in order to further increase stability. These effectively tether the membrane to the underlying support and allow for the incorporation of membrane proteins without denaturing them. Unfortunately, this ability to study membrane proteins comes at the cost of membrane fluidity. In general, it is desirable for the polymer cushion to be soft, hydrophilic, and not too strongly charged.

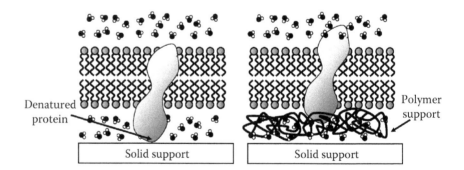

Figure 10.12 Peripheral domains of transmembrane proteins can become immobilized and denatured on a solid support. A polymer cushion can help shield the protein from the substrate. (Reprinted from Castellana, E. T. and Cremer, P. S. Solid Supported Lipid Bilayers: From Biophysical Studies to Sensor Design. *Surface Science Reports* 2006, 61(10): 429–444. Copyright 2006, with permission from Elsevier.)

Two classes of polymers, polyelectrolytes and lipopolymers, are emerging as popular choices for cushion material. Polyelectrolyte cushions can be directly adsorbed from solution to a variety of substrates by means of layer-by-layer deposition. Lipopolymers consist of a soft hydrophilic polymer layer presenting lipidlike molecules at their surface that can insert into a phospholipid membrane and tether it to the polymer spacing.

10.3.4 Fluorescence recovery after photobleaching

FRAP is an optical technique capable of quantifying the two-dimensional lateral diffusion of fluorescently labeled lipids in a bilayer (Figure 10.13a). This technique is one of the most common techniques performed in order to characterize lipid bilayers. The technique begins by saving a background image of the fluorescing bilayer. Next, a light source is focused onto a small patch of the viewable area either by switching to a higher magnification objective or shrinking the field of view via a pinhole. The fluorophores in this region are quickly photobleached, which is the photochemical destruction of a fluorophore, and an image reveals a noticeable dark spot surrounded by undamaged fluorophores in the bilayer (Figure 10.13b). As Brownian motion proceeds, the still-fluorescent lipids will diffuse throughout the sample (Figure 10.13c) and replace the destroyed lipids in the bleached region. This diffusion proceeds in an ordered fashion that can be modeled by the diffusion equation (Smith et al., 2008),

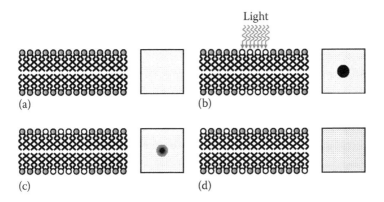

(a) (b) (c) (d)

Figure 10.13 Principle of FRAP. (a) The bilayer is uniformly labeled with a fluorescent tag. (b) The label is selectively photobleached. (c) The intensity within the photobleached region is monitored as a function of time. (d) Eventually uniform intensity is restored.

which shows how the fluorescence intensity $f(t)$ recovers (increases) over time, t:

$$f(t) = e^{-2\tau_D/t} \left[I_0 \left(\frac{2\tau_D}{t} \right) + I_1 \left(\frac{2\tau_D}{t} \right) \right] \tag{10.2}$$

where I_0 and I_1 are known as modified Bessel functions, and $\tau_D = r^2/4D$ where r is the radius of the bleached area at $t = 0$ and D is the fission coefficient. The diffusion constant D can be simply calculated from a fit of this equation. Diffusion constant values for lipids tend to be in the range of 0.5–5 $\mu m^2/s$. Figure 10.14 shows what a typical FRAP profile looks like (a plot of fluorescence intensity versus time). Fitting this curve to Equation 5.15 allows the determination of D.

10.3.5 Fluorescence resonant energy transfer

FRET, previously discussed in Chapter 9, is a useful tool for quantifying molecular dynamics in nanoassemblies of biophysical and biochemical importance. For monitoring the locations of two molecules relative to one another, one of them is labeled with a donor and the other with an acceptor fluorophore. When they are not near each other, only the donor emission is detected upon donor excitation. However, when the donor and acceptor are in proximity (1–10 nm) due to the interaction of the two molecules, the acceptor emission is predominantly observed because of the intermolecular transfer of energy from the donor to the acceptor.

FRET has been used by Lei and Macdonald (2003) in order to monitor the fusion of two lipid vesicles. Video images of two vesicles not fusing and two vesicles fusing are shown in Figure 10.15. As the vesicles come into contact, the contact area becomes a different color (depending on the

Figure 10.14 Illustration of a typical FRAP recovery curve with corresponding images. The dashed line represents a fit to the data using a 2D diffusion model equation.

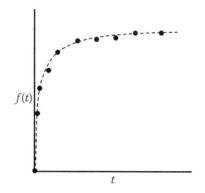

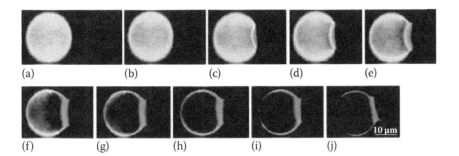

Figure 10.15 FRAP images showing the fusion of two vesicles. The red and green color is shown in gray scale. (a–j) Shows snapshots of the vesicle fusion process from no contact (a) to 660 ms after contact (j) in roughly 30-ms intervals. (Reprinted from Lei, G. and MacDonald, R. *Biophysical Journal* 2003, 85: 1585–1599, Biophysical Society, Elsevier. Copyright 2003, with permission from Elsevier. See Lei and Mac-Donald for the color image.)

fluorophores used—red and green in this case), corresponding to simultaneous emission of both red and green from the same area (shown in gray scale in the figure). Thereafter there are dramatic color and intensity changes (Figure 10.15j), which are interpreted as hemifusion; the color of the contact area changes, the contact area becomes slightly fatter, and then red color can be seen beginning to diffuse from the contact area over the surface of the positive vesicle, reducing the intensity of the green color in the process. This donor–acceptor relationship was examined in more detail in Section 9.3.

10.3.6 Fluorescence interference contrast microscopy

In FLIC, the proximity of fluorescent probes to a reflective plane leads to the modulation of the fluorescence intensity, an analysis of which can provide nanometer-scale topographic information. FLIC can have resolution of a few nanometers to hundreds of nanometers. The principle underlying FLIC is that in the presence of a reflective surface, light can travel to a fluorophore along a direct path or a reflected path; the difference in optical path length between these two routes leads to interference (Figure 10.16). The resulting intensity is a function of this path-length difference, which is a function of the height of the fluorophore above the reflective plane. Interference occurs independently for the excitation and the emission light, each of which must be considered separately. The typical setup makes use of a silicon wafer as the reflector. A transparent oxide (SiO_2) layer acts as a spacer. The system of interest (i.e., a lipid bilayer with fluorescent probes) is placed on top of the oxide layer.

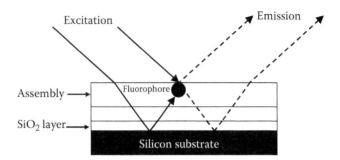

Figure 10.16   Basic principles of a FLIC experiment. The resulting intensity is a function of the path length difference between the direct and reflected light paths, which is a function of the height of the fluorophore above the reflective plane.

FLIC can also be used to look at the topography of a secondary membrane. A secondary lipid bilayer can be assembled atop a solid supported bilayer via the rupture of giant lipid vesicles with diameters that approach tens of microns. The secondary membrane is separated from the primary membrane by a confined layer of water and is therefore free to exhibit nanometer-scale height fluctuations.

10.3.7 Solvent assisted lipid bilayer formation

Recently the solvent-assisted lipid bilayer (SALB) formation method was discovered as a simpler alternative to the vesicle fusion method described in Section 10.3.2 (Tabaei, S. R. et. al., 2014). Furthermore, SALB formation does not involve vesicles and is thus not susceptible to the limiting step of vesicle formation, which is exclusive to certain lipid compositions. Therefore, SALB formation is applicable to a much wider range of lipid compositions than vesicle fusion and can thus mimic more bilayer types (e.g., mammalian, bacterial, and cholesterol-rich bilayers).

The SALB method involves the formation of a bilayer at the solid–liquid interface. This occurs with a change of bulk solutions passed over a silica surface, which induces multiple lyotropic phase transitions. The lipids are initially dissolved in water-miscible alcohols (e.g., isopropanol or ethanol). The solid support surface is exposed to a water-based buffer solution (10 mM Tris, 150 mM NaCl, pH 7.5) followed by isopropanol. Then, the lipid mixture at a final concentration of 0.5 mg/mL is introduced. Finally, the water-based buffer is reintroduced. As the water fraction increases, the lyotropic phase transitions occur and induce the lipids to self-assemble into a planar bilayer.

10.4 SELF-ASSEMBLED MONOLAYERS

A **self-assembled monolayer** (SAM) is formed when a particular group on a molecule has a strong affinity for a specific surface. Thiols and silanes are examples of such molecules, and these are discussed in the next two sections. Spontaneous chemisorptions of these adsorbate molecules may occur from either the vapor or liquid phase and often leads to the formation of an organized monolayer. The adsorbate, being a self-assembling molecular building block, often has a relatively simple structure with the reacting group attached to an alkyl chain and terminated by some other inert group, although similar processes can be used for more complex molecules. The chemisorption step is usually rapid, lowering the surface energy of the substrate. This is followed by slow two-dimensional organization of the alkyl chains. This organization step is dominated by inter-chain hydrophobic interactions.

Surface organization itself typically occurs in stages, where each stage represents a distinct phase. First, a low-density phase is formed comprised of adsorbate molecules randomly dispersed on the surface of the substrate. This is followed by a transition to a second intermediate phase with randomly dispersed molecules lying flat on the substrate. The final transition to a high-density phase involves the molecules orienting themselves normal to the substrate surface. Exactly how these phase transitions proceed depends on temperature. If lateral interactions are ignored, chemisorption follows the Langmuir adsorption isotherm. The overall kinetics of the process are first-order and are approximately described by Equation 10.3:

$$d\theta/dt = k(1 - \theta) \tag{10.3}$$

where θ is proportional to the amount of area occupied and k is the rate constant.

10.4.1 Thiols on gold

Thiols are compounds having the structure RSH (where R $\neq$ H). These molecules are also referred to as mercaptans. Some commercially available thiols are shown in Figure 10.17. The thiol group is extremely reactive to gold and other noble metals, resulting in the formation of a strong metal–sulfur covalent bond having a bond strength on the order of 100 kJ mol^{-1}. The resulting monolayer on the metal surface is thermally stable and is resistant to various solvents and electrolytes.

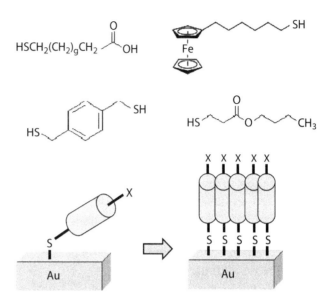

Figure 10.17   Top: Some examples of thiols used to functionalize a gold surface. Bottom: Thiol bound to a gold substrate via a strong Au–S bond. The gray tube is typically a hydrocarbon chain of specified length. *X* represents some surface functional group.

Gold is usually the preferred metal for thiol deposition, mainly because it is inert, biocompatible, and can withstand harsh chemical cleaning treatments. Gold is also easy to pattern using lithographic methods. Low molecular weight thiols can be deposited on gold surfaces by vapor deposition. Most other thiols can be deposited using a simpler method. First, the thiols are dissolved in an organic solvent such as chloroform or ethanol. The gold is then immersed in this dilute solution (~1 mM) for 12 to 72 hours at room temperature. The metal is then removed, washed with appropriate solvents, and dried in a stream of nitrogen. It should be stressed that the gold surface needs to be free of impurities before being immersed in the thiol solution. The surface is usually cleaned in appropriate solvents and then exposed to intense UV light in the presence of oxygen for about 30 minutes. The production of oxygen atoms and ozone during the UV exposure causes organic impurities on the gold surface to be oxidized.

The final structure of SAMs constructed using thiols depends on the curvature of the substrate. Planar substrates have no curvature. However, SAMs on nanoparticles such as colloidal particles and nanocrystals tend to stabilize the reactive surface of the particle and can provide specific organic functional groups at the particle–solvent interface. This surface functionalization is particularly useful for applications, such as immunoassays, that are dependent on the chemical composition of the surface.

10.4.2 Silanes on glass

Silanes are generally used to construct SAMs on nonmetallic oxide surfaces, such as those found on silicon and glass surfaces (SiO_2). The most commonly used silane to produce a SAM is octadecyltrichlorosilane (also known as OTS). The structure of OTS is shown in Figure 10.18. It is an organometallic compound in which a long octadecyl chain is connected to the reactive trichorosilane group (R-$SiCl_3$). OTS reacts violently with water and is sensitive to air. Like thiols, the substrates (glass) are immersed in a dilute solution of OTS, where the solvent is organic.

OTS reacts with the Si–OH groups found on the surface of clean glass or silicon. The reaction is illustrated in Figure 10.18. In order for the reaction to proceed efficiently, the glass substrates must be cleaned thoroughly. This is usually achieved by first immersing the glass slides into a hot mixture of concentrated sulfuric acid and 30% hydrogen peroxide for about 30 minutes. This mixture, known as piranha etch solution, is a powerful and extremely hazardous oxidizing agent that rapidly removes organic impurities from the glass surface. Silicon is usually cleaned by exposing the surface to high-intensity UV light for about 30 minutes.

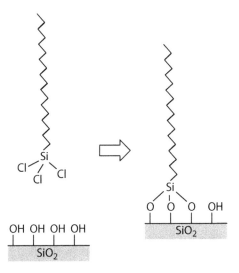

Figure 10.18 Octadecyltrichlorosilane (or OTS) chemisorbed on a clean glass substrate resulting in a close-packed SAM.

As mentioned previously, the production of oxygen atoms and ozone causes organic impurities on the silicon to be oxidized.

OTS and other silanes such as dodecyltrichlorosilane are used in the semiconductor industry to form nanofilm SAMs on silicon dioxide substrates during fabrication of integrated circuits. More specifically, these molecules act as thin insulating gates in metal–insulator semiconductors. OTS is also used in conjunction with conducting polymers in organic-substrate LCD displays.

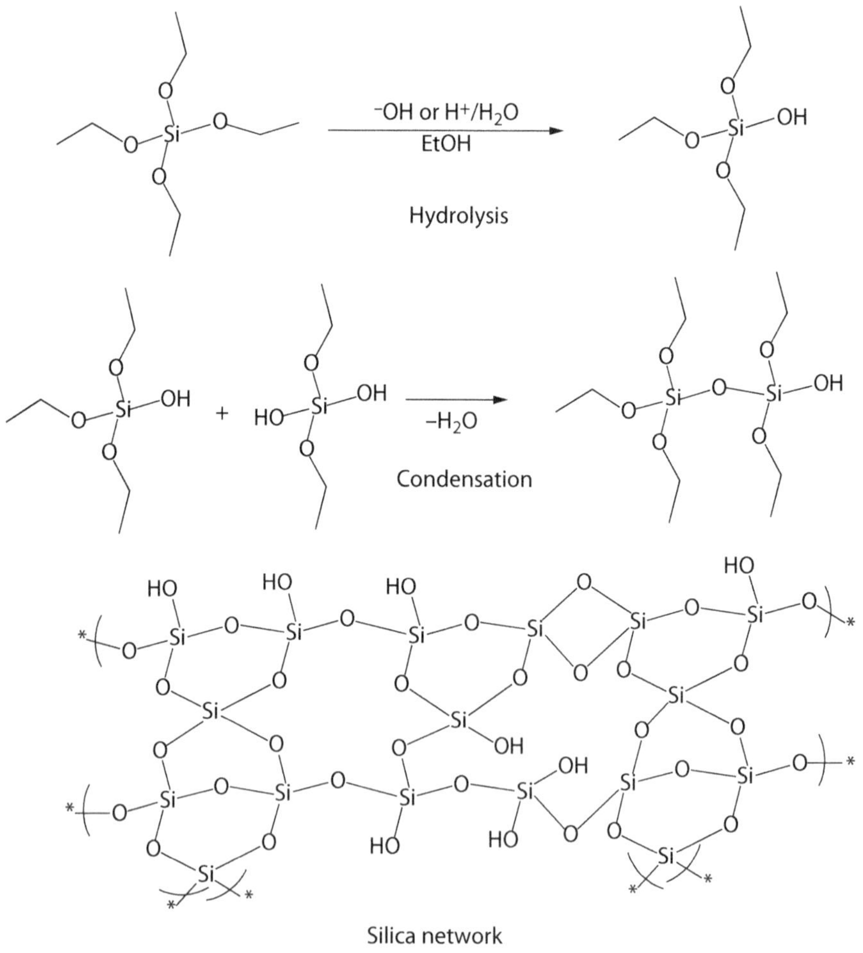

Figure 10.19 The sol-gel process. Silanes are highly reactive to OH groups and $(R\text{-}O)_3Si\text{–}OH$ molecules and can self-react in a condensation process leading to the silica network containing nanopores.

An interesting extension of the silane chemistry discussed above is the sol-gel processing method used to prepare highly porous silica networks. As an example, let's consider ethoxysilane (Figure 10.19), which can easily be hydrolyzed to the $(R-O)_3Si-OH$ form. Since silanes are highly reactive to OH groups, $(R-O)_3Si-OH$ molecules can self-react in a condensation process leading to the silica network containing nanopores.

10.5 POLED AMORPHOUS FILMS

The methods that we have discussed thus far have involved formation of films that are self-assembled by various types of interactions, whether electrostatics, the hydrophobic effect, or covalent bonds. These types of film are intrinsically ordered to some degree based on the interactions used to build them. However, in some situations, thicker (but still nanoscale) films need to be quickly assembled or it is desirable to fabricate a film from materials that do not intrinsically order well. This can be done through a process called **poling**, where an external potential such as an applied voltage is used to order molecules, which are then locked in place by reducing the temperature and/or chemical crosslinking.

10.5.1 Spin-coating

To produce a poled film, the material of interest first needs to be coated on to the substrate. This is often done using a process called spin-coating, where a droplet of a solution containing the material in a volatile solvent is placed on the substrate, which is then rotated at high speeds. The (apparent) centrifugal force of the substrate rotating pulls the solution outward, resisted by the surface tension of the liquid, creating a uniform film with a thickness that depends on the solvent and rotational speed of the spin-coater. A schematic of spin-coating is shown in Figure 10.20. Spin-coating is extensively used in the semiconductor industry for applying photoresists for lithography (see Section 10.6.1).

10.5.2 Poling

Unlike films produced using self-assembly methods, spin-coated films are not intrinsically ordered. If acentric order is desired, for example, for a nonlinear optical material, the film must be poled to orient the molecules in a desired direction. If the film is composed of polar molecules, an electric field can be used for **poling**. The film is heated to near its glass transition

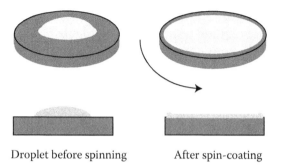

Droplet before spinning   After spin-coating

Figure 10.20 Schematic of the spin-coating process. A droplet of a solution is placed on a substrate (left). Spinning the substrate at high speed pulls material toward the edge of a substrate and produces a uniform film as the solvent evaporates (right).

temperature (T_g, the point at which it transitions from a solid to a rubbery, viscous liquid) and placed under a strong electric field. The dipoles of the molecules reorient to align with the field, resisted by thermal energy, which causes them to randomly rotate. For a weakly-interacting system (for example, a dye at low concentrations in a polymer film), the order with respect to the electric field can be approximated by the **Langevin model**.

$$\langle \cos \theta \rangle \approx \frac{\mu E_0}{3 k_B T}$$

$$\langle \cos^3 \theta \rangle \approx \frac{\mu E_0}{5 k_B T}$$

(10.4)

where θ is the angle between the dipole and the electric field, μ is the dipole moment of the molecule being poled, and average E_0 is the local electric field acting on each dipole, approximated as

$$E_0 = \left(\frac{3\varepsilon}{2\varepsilon + n^2} \right) E_{\text{applied}}$$

(10.5)

where ε is the dielectric constant of the material, n is the refractive index of the material, and E_{applied} is the applied field strength, which is simply the applied voltage divided by the thickness of the film (V/d). Note the similarities of Equations 10.4 and 10.5 to the Onsager model discussed in Chapter 5; both models rely on the same assumption that the response of a molecule to an electric field (dielectric response) can be modeled by freely rotating dipoles that only interact with their surrounding environment in an averaged manner.

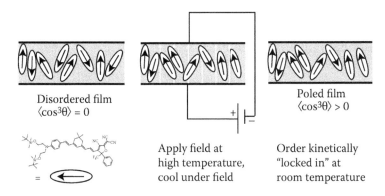

Disordered film
$\langle \cos^3\theta \rangle = 0$

Poled film
$\langle \cos^3\theta \rangle > 0$

$=$ ⟨←⟩

Apply field at
high temperature,
cool under field

Order kinetically
"locked in" at
room temperature

Figure 10.21 Electric field poling. A film of a material containing polar molecules is deposited on a substrate; the polar molecules are initially randomly oriented (left). When the material is heated and placed under a strong electric field, the dipole moments of the polar molecules align themselves with the field (center). Cooling the material kinetically locks in the alignment of the dipoles even after the field is removed (right).

Once the film has had time to order under the field (usually seconds to minutes), the temperature is reduced while the field is maintained. The film resolidifies and the molecules become kinetically trapped in an ordered state. A schematic of the poling process is shown in Figure 10.21. The poled film is not thermodynamically stable in the absence of the applied field, but the activation energy E_a of the molecules rerandomizing is too large for the molecules to rerandomize at room temperature. The film can then be used in applications such as non-linear optical devices.

10.6 PATTERNING

As the field of nanotechnology expands, the demand for surface manufacturing techniques that are cheaper, more flexible, and allow greater feature resolution will continue to increase. The word "nanolithography" can refer to a wide variety of nanoscale surface manufacturing techniques for creating patterned surfaces, such as those found in semiconductor circuits. There are many different ways that such patterns can be fabricated. Here we present a survey of these methods, their advantages, their disadvantages, and possible applications.

10.6.1 Optical and electron beam lithography

One important method is **optical lithography** (or photolithography), which is a method capable of producing nanoscale patterns using very short wavelength lasers. The wavelength of light determines the resolution of the nanopattern. For example, using a single exposure and single lithography mask in vacuum, UV lithography at 193 nm can reach a resolution of about 100 nm, while x-ray lithography can reach a resolution of ~15 nm by using light of wavelength ~1 nm. However, resolution can be enhanced via a variety of techniques, and lithography on features below the diffraction limit is an active field of research.

Photolithography can be performed with either a photomask placed directly on a substrate (contact lithography) or using light focused through a photomask onto a substrate using a specialized projector called a stepper. Higher precision and smaller features can be obtained with a stepper than with contact lithography; however, steppers are very expensive, such that contract lithography is more common for laboratory-scale fabrication. For either type of lithography, the region to be patterned is typically a flat substrate such as a silicon wafer coated with an imaging photoresist layer, as shown in Figure 10.22. A photoresist is essentially a light-sensitive material which, depending on the type of photoresist, will either harden or degrade when exposed to light. The general concept of photolithography consists of projecting electromagnetic radiation on a surface through openings in a mask template that protects the surface so that only areas not covered by the mask are exposed. Then a chemical that functions as a developing solution is used to remove any degraded or nonhardened photoresist, leaving either a "negative" or "positive" imprint of the mask. Further chemical treatment can be used to etch the film or substrate that is not covered by photoresist, and then to also remove the photoresist once the etching is complete. The process can be repeated

Figure 10.22 Patterning a substrate using UV light. The light passes through a photomask and degrades exposed regions of the photoresist.

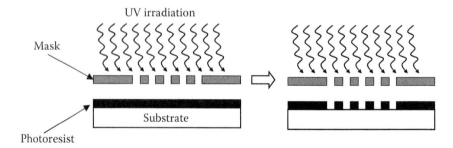

many times, in conjunction with deposition of multiple materials, in order to build complex structures such as integrated circuits. The technique excels at creating regular, patterned surfaces, and the process produces surfaces fairly quickly, but the fabrication of the mask can be difficult.

As mentioned above, the resolution of the pattern is determined by the wavelength of light. Usually UV light is used to create patterns down to about 50 nm. Equation 10.6 shows how the minimum feature size (R) is related to the wavelength (λ) of light used.

$$R = k \frac{\lambda}{n \sin \theta} = \frac{k\lambda}{NA} \qquad (10.6)$$

The constant k in Equation 10.6 is usually around 0.25 (Sanders, 2010), and NA is the numerical aperture of the lens seen from the substrate. This equation tells us that the resolution can be improved by using a smaller wavelength and a larger numerical aperture. Maximum feature resolution, d, as a function of wavelength is $\lambda/2$ for simple lithography. The lower limit for d is around 100 nm for typical laser systems used for semiconductor fabrication (most commonly 193 nm). Increasing resolution by decreasing laser wavelength deeper into the UV is challenging due both sample concerns and optical complexity. However, feature sizes down to 14 nm (42 nm distance between features) have been achieved at commercial scale (Borkar, 2014) by using ultrashort wavelength extreme UV (EUV) light (e.g., 13.5 nm), by performing lithography under water or higher refractive index fluids to increase the numerical aperture (Sanders, 2010), and/or using multiple photomasks, multiple exposures, and computational procedures called optical proximity correction to exploit the principles of interference to produce smaller features (ITRS, 2013). However, these techniques require expensive equipment that is typically only accessible by large semiconductor manufacturing companies; plant costs can run into the billions of dollars for state-of-the-art processes.

Various types of radiation can be used for lithography depending on one's purpose. Electron beams can also be used for lithography instead of photons; their short wavelength allows for the fabrication of very small features without diffraction becoming an issue (analogous to electron versus optical microscopy, as discussed in Chapter 8). The highest-performance electron-beam lithography techniques reported can create features down to a scale of 1 nanometer (Manfrinato, 2017). Additionally, electron beams have been used to create structures on a silicon surface without intermediate developing stages; the beams themselves etch the surface. Electron-beam lithography suffers the drawback of low through-put, making scaling to industrial use difficult, and it requires costly electron microscopy instrumentation to perform.

10.6.2 Soft lithography

Another common patterning technique is soft lithography. This method refers to a family of lithographic techniques for patterning using molds and elastomeric stamps. The stamps are usually pieces of polydimethyl-siloxane (PDMS) that have been patterned usually against a master to form a relief pattern.

As an example, a pattern of octadecane thiol (ODT) molecules can be transferred to a gold substrate (Figure 10.23). First, the stamp is placed in a solution containing ODT. ODT molecules adsorb onto the PDMS surface forming the "ink" component of the stamp. The solvent is removed and the PDMS stamp is put in contact with a gold substrate, where the ODT molecules spontaneously chemisorb onto the gold surface. Then, the pattern from the stamp is transferred for the gold via the ODT ink. Further details on the chemisorption of thiols onto gold can be found in Section 10.4.1.

10.6.3 Nanosphere lithography

Nanosphere or colloidal lithography uses a packed array of nanospheres as the mask for the surface to be patterned. Chemicals can be deposited in the

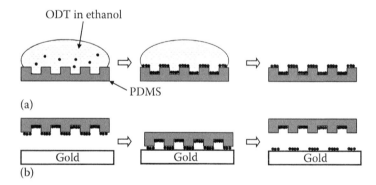

Figure 10.23   (a) A solution of ODT is placed over the PDMS stamp. The solvent is removed and ODT molecules assemble on the stamp. (b) The PDMS stamp with the ODT "ink" is placed on a gold substrate. When the stamp is removed, the ODT molecules in contact with the gold surface chemisorb to the substrate, transferring the pattern from the stamp to the gold via the ODT ink.

exposed areas between the spheres or the surface in these exposed areas can be etched by using ion beams. The pattern produced can be manipulated in several ways. For example, chemical deposition can be done at different angles, or heating the packed array of nanospheres at different temperatures can cause it to deform in desired ways. It is also possible to create multiple layers of sphere arrays where the upper layers can act as a mask for the lower layers, leaving a complex three-dimensional structure. It is worth noting that the resolution is dependent on the size of the nanospheres used in the mask. By using sufficiently small spheres, one can achieve resolutions below 100 nm.

This technique has the benefit of being simple and inexpensive compared to other nanolithographic techniques. Nanosphere lithography can also produce structures that are difficult or impossible to create with other nanolithographic techniques. Unfortunately, it is difficult to produce a flawless surface with the method; errors occur as the result of difficult-to-avoid disorder in the nanosphere mask, which may lead to small differences between masks that are formed in similar conditions. Furthermore, the space between features and their size cannot be changed independently, as both are directly determined by the size of the nanospheres used. The number of structures that can be created by this method is also limited because structures created by nanosphere lithography must be based on the unoccupied interstices of a packed sphere array. For example, it is unlikely that a field of square nanosphere pillars could be produced by a mask of patterned spheres.

10.6.4 Patterning using AFM

In addition to being used as an imaging tool, an AFM can be used to manipulate surfaces on a very small scale. In standard use, the AFM images a surface in three dimensions by measuring the deflection of a small, very sharp tip as it moves along a surface. For patterning, however, the tip can be used to interact with a surface and alter it in order to produce very small structures.

Some nanolithographic techniques that employ AFM may be termed "destructive"—they etch or otherwise damage a surface to create the intended surface structures. The tip can be used to break apart surfaces on a very small scale or to move very small objects around on a surface. The AFM tip can also be modified to irreversibly alter the surface in various ways. For example, a catalyst can be added to the tip to initiate reactions. Another common method for surface alteration is the initiation of very localized redox reactions by running a current through a conductive AFM tip. UV light can be used to control this localized redox process because it affects electron-hole pairs in a way that changes their resistance to current flow. This means that a low-voltage current running through the tip will be able to oxidize a surface only when UV light is shined on the surface at the same time. This method can produce smooth surface features with a height of about 40 nm and a lateral length of about 1 μm. Surface-destructive AFM is thus fairly flexible, but its use can be limited by expense and low production speed.

Other nanolithographic techniques that use AFM may be termed "constructive" because they build up the intended structure on the surface. As with destructive AFM nanolithography, there are several different possible implementations. A common constructive method is the use of AFM as a "dip pen." The AFM tip is covered in some molecule to be used as "ink," which is then deposited on the surface at the meniscus between the tip and the surface. A resolution below 50 nm is currently possible with dip-pen methods.

Metals can be transferred to the tip by applying pressure to the tip on a metal surface, heating it, and allowing it to cool. The substance can be transferred from the tip to a surface by reheating the tip as it is dragged over the surface. Using this technique with indium has been successful in the creation of conductive nanoscale wires, with widths as low as 50 nm.

There are many different molecules that can be used as ink for this method, giving a great deal of flexibility in its uses for surface patterning. For example, a surface can be patterned with chemicals that will protect it against a later etching treatment, allowing for an interesting combination of constructive and destructive lithographic techniques to make more complex patterns. Another possible use of this method is the patterned deposition of particles that can act as "seeds" for new pattern creation, either by catalyzing reactions or by binding to other molecules with interesting shapes that form the basis for a new pattern.

An advantage of dip-pen nanolithography is that the technique can also be scaled up by using an array of several AFM tips instead of just one, allowing something analogous to desktop printer functionality for nanofabrication. Using an array of tips would allow for the creation of patterns at greater speeds and over larger areas. This has shown great potential for the use of AFM in the creation of nanoscale electronic components such as transistors.

Dip-pen techniques also have drawbacks, the most unfortunate of which is that the created patterns are not perfectly reproducible. A number of different factors affect molecule deposition: temperature, humidity, the type of molecule used as ink, the characteristics of the AFM tip, the write speed, and the pressure used can all affect the pattern ultimately produced on the surface. For example, altering the humidity can change the size and structure of the meniscus where ink is deposited, resulting in a larger or smaller area of the surface onto which the same amount of ink should have been deposited.

"Fountain-pen" methods, in which the tip is placed within a nanoscale tube that delivers a constant flow of "ink" molecules to the tip, are a related surface fabrication method with similar advantages and drawbacks. The resolution here is worse than dip-pen methods, as the fountain-pen methods can produce only patterns with feature widths above 200 nm.

AFM tip hammering nanolithography is yet another variation on AFM nanolithography. In AFM tip hammering nanolithography, the AFM tip is used to imprint the surface. Wang et al. (2009) give an example of the technique being used on a polystyrene-*block*-poly(ethylene/butylenes)-*block*-polystyrene (SEBS) copolymer. The AFM tip deforms the polystyrene sphere component of the copolymer, resulting in an imprinted area. AFM can also be used for embossing as well as printing; one simply imprints the area around the area to be embossed to get a raised structure.

AFM imprinting has several advantages. For example, the surface imprint is reversible. Depending on the copolymer used as the surface, heating it at high temperatures that allow for reorganization can erase the imprint. This does not mean that the imprint is unstable at all temperatures. At room temperature, below the threshold energy required for reorganization of many molecules, the patterns are fairly stable. For the SEBS copolymer, it was found that the imprint is erased in about 5 minutes at 70°C and that around 50% of the original surface contrast remains after 70 days at 25°C. More stable copolymers would require a higher temperature to reverse the patterning and clear the imprint, but they would likely also be more stable for longer at room temperature. As well, the AFM tip hammering nanolithography has a very good feature resolution and is currently able to produce patterns with features as small as 13 nm for imprinting and 18 nm for embossing. The combination of feature resolution, surface stability, and reversibility makes the method attractive for high-density data storage. The method also has drawbacks, the primary one being surface specificity. The surface chosen must be imprintable with a reasonable level of force but also stable enough that the imprinting will survive for a desirable amount of time.

10.6.5 Summary

Nanolithographic methods can be used individually or combined with other fabrication methods to amplify the range of possible surface structures and their uses. For example, photolithography can be used to create a pattern on a surface on which nanowires can be grown. The combination of these two techniques would generate a surface with multiple levels of roughness, a surface property with practical applications in superhydrophobicity. Nanolithography is thus a quickly developing field encompassing many techniques for the fabrication of very useful nanoscale patterned surfaces.

10.7 DNA AND LIPID MICROARRAYS

Over the past three decades, advances made in our understanding of genetics have led to the creation of DNA microarrays. DNA microarrays are surfaces that have been coated with specific oligonucleotide sequences and have been particularly useful in gene expression studies. The number of oligonucleotides on a DNA microarray can vary greatly depending on the microarray's purpose; diagnostic microarrays may have tens of oligonucleotides, while microarrays for research or screening purposes may have hundreds of thousands of oligonucleotides on a single array. This section will address the basic protocol of a DNA microarray experiment, the various methods of array fabrication, optimization of an array, and their applications. The student is encouraged to review basic biochemistry of DNA before reading this section.

10.7.1 Using a DNA microarray

Each DNA microarray is unique to a certain purpose, such as diagnosing cardiovascular disease or monitoring the entire genome of a strain of *Escherichia coli* (*E. coli*); as such, a microarray must be fabricated or purchased for each experiment. Depending on the size of the microarray and the availability of equipment, it is often more feasible to purchase prefabricated microarrays. Fabrication techniques will be described in more detail later in this section.

Each spot on a DNA microarray contains thousands of oligonucleotide molecules called probes that complement their corresponding mRNA. A sample is prepared for DNA microarray analysis by extracting the RNA from the cells. The RNA can be purified by treating it with poly(A) polymerase followed by oligo(dT) chromatography. The purified RNA is then labeled for detection; in the past, ^{33}P has been used, although radioactive labeling has mostly been replaced by the use of fluorescent markers such as the cyanine dyes Cy3 and Cy5. The labeled RNA "targets" are then spotted on the DNA microarray probes to which the RNA binds in a process called "hybridization." Before measuring fluorescence, the DNA microarray is washed to remove any excess or unhybridized targets. Fluorescence can be quantified by confocal laser scanning microscopy or other fluorescence microscopy techniques. Using internal standards, fluorescence can be used to calculate the number of targets that hybridize to the probes, which can then be used to determine the change in gene expression in a normal state compared to an experimental state. Running a DNA microarray experiment is relatively quick and easy; the biggest factor detracting from more widespread use is the lengthy fabrication process.

10.7.2 Array fabrication

There are three primary methods for DNA microarray fabrication: in situ synthesis, contact printing, and noncontact printing (Dufva et al., 2005). In situ synthesis creates the oligonucleotides directly on the microarray and is capable of producing the highest quality microarrays, although these high-quality arrays require expensive robotics systems and the use of clean room techniques. All methods of in situ synthesis use nucleosides that have already been modified with a protecting group to ensure that the correct sequences of oligonucleotides are present on completion. In the first method (Figure 10.24a), nucleosides are modified with the photolabile reagent 2-nitrophenyl propoxycarbonyl (NPPOC). These protecting groups are removed using light, which allows the growing oligonucleotides on the array react with available nucleosides in solution. Synthesis is controlled by the use of various masks that are placed on top of the microarray, allowing only specific regions to react. After subjecting a mask to light, the appropriate nucleoside is added, the microarray is washed, and then the process repeats. The downside to this method is that it requires a different mask for each step of the synthesis. Another method (Figure 10.24b) involves similar chemistry, yet instead of masks, a series of mirrors is used to direct light to the spots to be deprotected. Yields of 77% for 25 base pair oligonucleotides have been achieved using NPPOC as the protecting group.

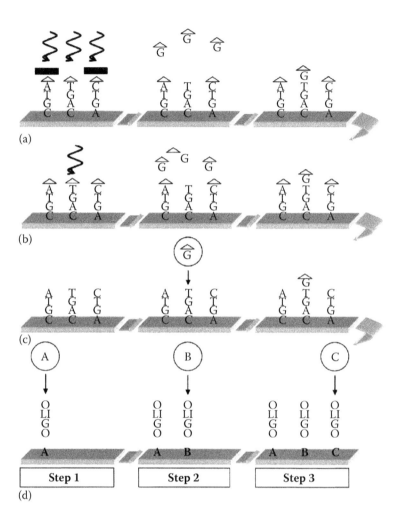

Figure 10.24 Methods of fabrication: (a) in situ with masks, (b) in situ with mirrors, (c) in situ printing, and (d) printing presynthesized oligonucleotides. (Reprinted from Dufva, M. Fabrication of High Quality Microarrays. *Biomolecular Engineering* 2005, 22: 173–184. Copyright 2006, with permission from Elsevier.)

Because they are cheap and still feasible, printing methods are generally the preferred methods of DNA microarray fabrication. Contact printing uses a robotic system of high-definition pins to dispense quantities of probe solutions at given coordinates. Noncontact printing differs in that instead of pins, inkjets or similar technologies are used to dispense picoliter quantities of solution. Noncontact printing is preferred over contact printing as noncontact printing creates smaller spots, and thus more spots can be fit on a microarray. Noncontact printing has been used in one method of in situ synthesis (Figure 10.24c). In this method, all bases are deprotected followed by site-specific dispensing of phosphoramidite-protected nucleosides; cycles of deprotection and the addition of nucleosides are repeated to complete oligonucleotide synthesis. Perhaps the most practical methods for fabrication of DNA microarrays are the printing methods in which all oligonucleotides are synthesized beforehand. Via contact or noncontact printing, known quantities of each individual oligonucleotide are dispensed at their desired position (Figure 10.24d).

10.7.3 Optimization of DNA microarrays

A great deal of optimization is required before a DNA microarray can be used effectively. The quality of a DNA microarray is greatly influenced by the method in which it was fabricated. For example, spot density, the number of spots that can be placed in a given area of a microarray, varies greatly among *in situ* and noncontact printing and contact printing methods. Using *in situ* methods, spot sizes with diameters less than 10 μm have been created, whereas printing methods can create spot sizes between 20 and 30 μm. Also affecting spot density is the array geometry. It is important for spots to be arranged in a geometry that maximizes spot density while preventing any overlap; additionally, precise array geometry is significant for future data analysis. Currently, our ability to create high spot density arrays has surpassed our ability to accurately interpret fluorescence. Current fluorimeters are capable of resolving fluorescence from spots as small as 30 μm. The development of higher resolution fluorimeters will be required for further miniaturization of DNA microarrays.

Spot morphology is another parameter that must be monitored. A homogeneous spot is preferred for data analysis. Poor spot morphology is less of an issue with light-directed *in situ* syntheses. Spot morphology of printing methods can be improved by controlling humidity and temperature as well as by adjusting the spotting buffer.

Perhaps the most important factors to be controlled for optimization are probe density and hybridized density. These two factors are closely related, as probe density is the number of probes in a given spot and hybridized density is the fraction of these probes that hybridize when saturated with target molecules. In order to get the greatest fluorescent signal, hybridized density must be maximized; however, a number of issues arise. If probe density is too high, probe molecules may interact with other probe molecules on the surface and become unable to hybridize. Additionally, high probe density leads to high steric hindrance, which prevents targets from hybridizing. A number of troubleshooting methods have been developed to optimize these parameters. In many cases, the microarray surface and the probe molecules can be modified to maximize functional probe density.

Another method to optimize hybridized density uses linker molecules to move the probes farther from the surface to minimize or prevent any unwanted interactions. Innovation of the linker has resulted in a dendrimeric linker, which has been shown to greatly increase probe density and hybridized density.

10.7.4 Applications of DNA microarrays

DNA microarrays are excellent tools for studying gene expression and have valuable uses in both research and industry. One of the first uses of DNA microarrays was in the study of the growth of *E. coli*. Samples of stationary-phase *E. coli* and log-phase *E. coli* were analyzed by DNA microarray. The data found from microarray analysis led to the discovery of multiple growth-regulating genes and an overall improved understanding of this significant microorganism (Ye et al., 2001).

Industrially, DNA microarrays are of great significance in optimizing biocatalysis. In particular, fermentation processes are often optimized using DNA microarrays. As biological fuel cells gain more attention, it is likely that DNA microarrays will be of great importance in optimizing these systems.

DNA microarrays have great potential to impact health care. Intelligent drug design has benefited from the use of DNA microarrays as it allows drug developers to anticipate adverse effects, or alternatively, to discover new drug targets. Other kinds of microarrays are also effective in diagnosing diseases. DNA microarrays have been designed to screen for inherited genetic diseases; also, changes in the expression of certain genes can be attributed to cancer and other chronic illnesses, allowing for earlier detection. Perhaps the most exciting possibility with DNA microarrays is the development of personalized medicine. Using a DNA microarray to better understand an individual's genotype will allow doctors to more effectively treat each patient, ultimately resulting in a higher treatment efficiency and substantial improvements in patient outcomes.

10.7.5 Arrays of supported bilayers and microfluidic platforms

Spatially addressed microarrays have found use as extremely rapid and powerful means of data collection. Unlike DNA-, protein-, or peptide-based arrays, phospholipid bilayer systems must remain hydrated at all times in order to retain the desired supramolecular structure (Figure 10.25). This requirement creates a significant challenge for creating arrays of supported bilayers.

The first method for patterning surfaces with solid supported phospholipid bilayers was created by Groves et al. (1997). A typical formation procedure involved the patterning of photoresist on fused quartz wafers by means of standard photolithographic techniques. SUVs can then be

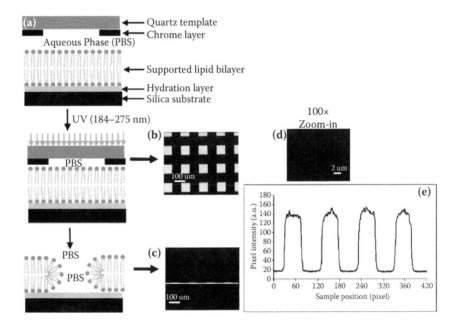

Figure 10.25 Direct patterning of void arrays within bilayer membranes using UV photolithography. (a) A schematic diagram of the key process steps. (b) Bright-field image, where bright squares are quartz and dark regions reveal the chrome background. (c) Epifluorescence images revealing resultant fluorescence patterns. (d) A high-magnification fluorescence image of the sharp boundary between the UV-exposed and UV-protected regions. (e) Fluorescence intensity profile across an arbitrarily chosen line spanning four alternating UV-exposed and unilluminated bilayer regions. (Reprinted with permission from Yee, C. K., Amweg, M. L., and Parikh, A. N. Direct Photochemical Patterning and Refunctionalization of Supported Phospholipid Bilayers. *Journal of the American Chemical Society* 2004, 126 (43):13962–13972. © American Chemical Society.)

fused onto the substrate between the barriers, creating a lithographically patterned array of identical planar-supported membranes.

Arrays of supported membranes can also be fabricated by selectively destroying regions of a continuous supported bilayer. This is achieved by high-intensity UV illumination through a photomask under aqueous conditions (Yee et al., 2004). The UV radiation generates both ozone and singlet oxygen in highly localized regions. These species react with and degrade the lipids to form water-soluble components. These patterns of holes or of corralled bilayers display long-term stability, retaining their geometric shapes, sizes, and distribution as well as their relative position on the substrate surface.

As shown in Figure 10.26, sharp corners of geometrical features on the mask always result in smoothly curved or rounded edges. This observation is attributed to steric crowding and line-tension effects of the phospholipids. This approach is applicable for producing void patterns of arbitrary shapes, sizes, and densities at predetermined regions within the bilayer.

A neat feature of this patterning method is that the voids can be backfilled by subsequent exposures to the same or a different vesicle solution.

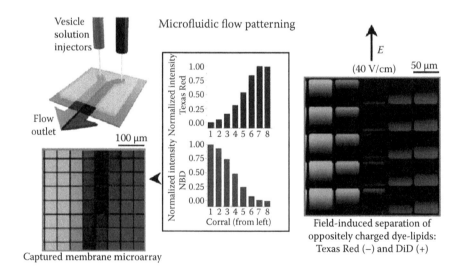

Figure 10.26 Addressing by laminar flow in a microfluidic channel. Diffusive mixing in a microchannel under laminar flow conditions provides a concentration gradient of different dye-labeled vesicles. (Reprinted with permission from Groves, J. T. and Boxer, S. G. Micropattern Formation in Supported Lipid Membranes. *Acc. Chem. Res.* 2002, 35: 149–157. © American Chemical Society.)

This makes it possible to manipulate membrane compositions and dynamically probe lipid–lipid diffusive processes. POPC bilayer samples patterned using the photochemical method shown in Figure 10.24, when exposed to small unilamellar vesicles, causes the nonfluorescent voids to be filled with lipids from the secondary vesicles. Depending on the composition of the secondary vesicles, the pattern can be erased or retained. If the lipids in the secondary vesicles are the same as the initial patterned bilayer, the lipids will diffuse and homogenize quickly, thereby erasing the pattern. If the lipids in the secondary vesicles have significantly different translational mobility, they will retain the backfilled pattern longer. This suggests the possibility of creating a fluid bilayer background with patterned microdomains at specific locations.

The use of laminar flow inside microfluidic channels is also an effective means of producing composition arrays of supported phospholipids bilayers in which two distinct chemical components can be varied simultaneously along a one-dimensional gradient. This allows for the addressing of patterned substrates by the flow of concentration gradients of SUVs formed by the diffusion mixing of two different SUV solutions. Figure 10.27 demonstrates the process of forming a one- or two-component composition array by laminar flow in microfluidic channels. A drawback to this method is the limited number of distinct components that can be simultaneously addressed as well as the lack of control over the positioning of the bilayers.

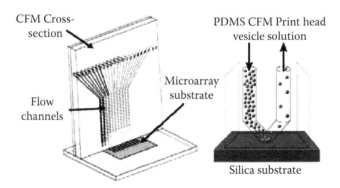

Figure 10.27   Schematic of the CFM apparatus and close-up of the CFM print head in contact with a silica substrate used for bilayer formation. (Reprinted with permission from Smith, K. A., Gale, B. K., and Conboy, J. C. Micropatterned Fluid Lipid Bilayer Arrays Created Using a Continuous Flow Microspotter. *Analytical Chemistry* 2008, 80(21): 7980–7987. © American Chemical Society.)

More recently, Smith et al. (2008) used a 3D continuous flow microspotter (CFM) system for the preparation of multianalyte lipid bilayer arrays. This method is capable of producing higher-density multicomponent arrays compared to traditional 2D microfluidics. The PDMS microspotter consists of a series of inlet and outlet wells connected by pairs of microfluidic channels embedded within the polymer. When the PDMS print head contacts the substrate, one continuous channel is formed between the inlet and outlet pairs, resulting in the continuous flow of solution over the substrate. Each channel is individually addressable, allowing for the production of 2D bilayer arrays. A prepatterned substrate is not necessary because the bilayers are effectively corralled into discrete micrometer-sized domains by the residual PDMS deposited on the silica substrate from the PDMS print head, which prevents the lipids from spreading. The packing of the lipids within a $400 \cdot 400$-μm^2 area with spacing of 400 µm between areas resulted in well-behaving bilayers.

10.7.6 Summary

DNA and lipid microarrays are fairly new technologies whose potential utility has yet to be reached. Widespread use is currently limited by the length of time involved in oligonucleotide synthesis and array fabrication techniques, which make the cost of a microarray experiment rise significantly. As technology advances, microarrays are likely to be an essential tool in any biochemist's or molecular biologist's arsenal and play a key role in clinical diagnostics.

10.8 NANOSCALE HYBRID PHOTONICS

State-of-the-art computers and other data processing systems are based on nanoscale integrated circuits that are able to condense incredible densities of components on a silicon wafer. When electronic computers were first developed in the 1940s, they were composed of bulky relays or vacuum tubes, and room-sized systems were required to perform what would now be incredibly simple calculations. However, by 1971, sufficient engineering advances had been made to use photolithography to make transistors at a scale of 10 microns and to combine thousands of transistors in a single package that were capable of equaling the power of a 1940s-era computer in less than a square inch (Intel, 2010). As discussed earlier, advances in nanoscience have now made integrated circuits with components that are a thousand times smaller per side commonplace, allowing vastly more powerful integrated circuits containing billions of components.

A similar transition is presently underway in photonics, or performing information processing using light. While optical fibers have been used to efficiently transmit information for decades, the components used for converting signals between electronic systems and optical systems have been bulky and expensive. While substantial strides have been made toward smaller and more efficient photonic circuitry, photonics have not yet reached the scale where complex electronic and photonic circuits can be combined on the same chip, taking advantage of the small size of electronic systems and the high operating speeds/bandwidths of optical circuitry (Heni, 2017).

10.8.1 The electro-optic effect

One of the principal barriers to chip-scale integration has been developing materials with adequate nonlinear optical activity that can be integrated into conventional chip-making processes. Converting between electronic signals requires the ability to generate short pulses of light (representing digital ones and zeros) at incredibly high speeds (gigahertz to terahertz), which requires materials that can control their optical properties (e.g., the real or imaginary components of their refractive index) at short timescales (Dalton, 2015). Of these devices, ones based on changing the real component of the refractive index are of interest due to lower signal loss and distortion than devices based on absorbance of light.

Changing a material's refractive index in response to an electric field is a nonlinear optical process akin to sum-frequency generation, but where one of the two interacting electric fields is an electrical signal and the other is a beam of light. Such effects are called **electro-optic** (EO) effects. While several electro-optic effects exist, we will focus on the Pockels effect, which is the strongest EO effect in materials that are noncentrosymmetric; for example, have net alignment of their dipoles in a single direction (see Chapter 8). The strength of the Pockels effect is proportional to the applied field (E) and a quantity known as the electro-optic coefficient (r_{33}):

$$\Delta n = \frac{n^3 r_{33} E}{2} \qquad (10.7)$$

The effect and implications of changing the refractive index (e.g., change in phase or propagation angle of light) are similar to those discussed in

Chapter 8 for methods such as SPR or DPI; however, for the EO effect, the change in index comes from the applied field instead of from the addition or removal of an adsorbate.

The electro-optic coefficient is in turn proportional to the second-order susceptibility (χ_2), which, for a molecular material, is proportional to the concentration (N), hyperpolarizability (β), and degree of acentric order $\langle\cos^3\theta\rangle$ of the material:

$$r_{33} = \frac{-2\chi_2(\omega,0)}{n^4(\omega)} \propto N\beta\langle\cos^3\theta\rangle \tag{10.8}$$

Thus, a well-ordered material with a large hyperpolarizability yields a large electro-optic coefficient, which allows obtaining the same refractive index shift using a weaker electric field than in a lower-performing material.

10.8.2 Organic–inorganic hybrid electro-optics

Present electro-optic systems are typically based on crystalline inorganic materials such as $LiNbO_3$ $(r_{33} = 31$ pm/V), which is adequate for macroscopic devices. However, ferroelectric crystals such as $LiNbO_3$ are not readily integrated into nanoscale integrated circuits due to their structure and the requirement of using high voltages to obtain a sufficient refractive shift in a small device (Dalton, 2015). One of the more common types of electro-optic modulators is the Mach–Zehnder interferometer (Figure 10.28), which uses similar interferometry principles to the sensor in a DPI instrument.

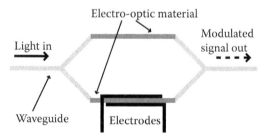

Figure 10.28 Schematic of a Mach–Zehnder electro-optic modulator. An optical waveguide is split into two arms containing a material with a high electro-optic coefficient (r_{33}). Applying a voltage to the electrodes shifts the refractive index in that arm, changing the speed at which the light travels through that arm and causing destructive interference of the light if the voltage is equal to the switching voltage V_π.

In a Mach–Zehnder interferometer, a beam of light traveling through a waveguide is split into two parallel waveguides of equal length and then recombined (similar to in a DPI instrument). Applying an electric field to one arm (or electric fields of opposite polarity to both arms) changes the refractive index in that arm and causes the light to travel at a different speed than in the other arm, changing the phase of the light and causing their electric fields to interfere when they recombine. If the applied voltage is sufficient to induce a phase shift of 180° (π radians), then the waves perfectly destructively interfere and no light is emitted. This critical voltage is called V_π, and digital signals can be encoded by rapidly switching between zero voltage and V_π.

One of the most critical parameters in integrating photonic systems with electronics is the parameter $V_\pi L$, or the product of the switching voltage and the device length. This parameter is in turn proportional to the electro-optic coefficient r_{33}, which makes a high electro-optic coefficient critical for nanoscale devices, such that the device can be run at a voltage consistent with the electronic components. High hyper-polarizability organic dyes have been demonstrated to give bulk electro-optic coefficents in excess of 500 pm/V, with substantial room for improvement (Dalton, 2015). Furthermore, organic dyes such as YLD-124 (shown in Chapter 8) can be integrated into devices using thin-film formation methods such as those that we have discussed, and generate their refractive index shift by electronic polarization (a very fast process) instead of vibrational motion (much slower), as in inorganic materials.

Integrating organic dyes in a nanoscale silicon (silicon–organic hybrid) or metal-coated (plasmonic–organic hybrid) waveguide has allowed shrinking device waveguide widths to 50–200 nm, with electro-optic coefficients up to about 350 pm/V, operating bandwidths of >100 GHz, operating voltages of <1 V, and total device footprints with total device dimensions on the order of microns (Heni, 2017). An example of plasmonic–organic hybrid modulator is shown in Figure 10.29.

While nanoscale photonic circuitry is currently in its infancy compared to electronics, continued advances will allow processing data hundreds or thousand times as fast as in pure electronic circuits, although realization of this objective will require further advances in materials design, understanding of interfacial science, and development of nanofabrication methods, building on the amazing advances that have already occurred in the electronic domain.

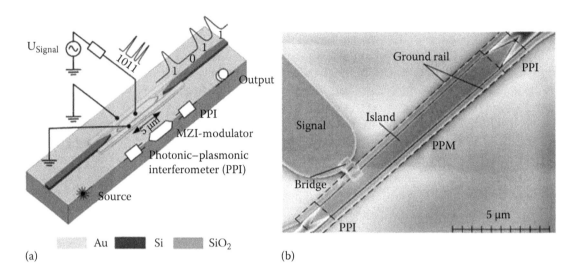

Figure 10.29 An example plasmonic–organic hybrid modulator, in which surface plasmons in gold are used to couple the electronic and optical signals in the device, working at a smaller length scale than the diffraction limit for the light passing through the waveguide. Along with silicon–organic hybrid devices, plasmonic–organic hybrid devices are a promising technology for integrating photonic and electronic circuitry at the nanoscale. (Reprinted with permission from Heni, W., Kutuvantaivida, Y., Haffner, C., Zwickel, H., Kieninger, C., Wolf, S., Lauermann, M., Fedoryshyn, Y., Tillack, A. F., Johnson, L. E., Elder, D. L., Robinson, B. H., Freude, W., Koos, C., Luthold, J., and Dalton, L. R., *ACS Photonics* 2017, 4: 1576–1590., Copyright 2017, American Chemical Society.)

End of chapter questions

1. (a) Using antibodies A and D, which bind specifically to an antigen X at different sites, and a UV–vis spectrophotometer, design a series of experiments to quantify the concentration of X in solution. Explain all steps/calculations.

(b) What kind of repulsive and attractive interactions would you expect between two zwitterionic lipid bilayers? Between a zwitterionic lipid bilayer and solid support? How are they different?

(c) Consider a situation where quantum dots are injected into a cell for imaging—will the size of quantum dots interfere with cell functions or pose any problems in this experiment?

2. Consider the three molecules shown in Question 11, Chapter 5. Langmuir films that can be constructed from these molecules show hysteresis in their Π–A isotherm. Hysteresis occurs when the compression isotherm differs from the expansion isotherm. Hysteresis in the isotherm for these molecules is largest for the one with the strongest end group dipole moment. Explain this observation.

3. Using the following materials and your knowledge of self-assembled monolayers, design a way to immobilize an antibody (containing primary amines) on a gold surface. The surface is functionalized through a process of incubation in the given solutions, and QCM is then used for the immobilization of the antibody itself. Give details as to why and how each material binds to the underlying layer. Qualitatively predict what the QCM mass versus time profiles will look like for each deposition step. Materials: (a) An aqueous solution of 1-ethyl-3-[3-dimethylaminopropyl]carbodiimide hydrochloride (or EDC), (b) mercaptoundecanoic acid in anhydrous ethanol, (c) an aqueous solution of (NHS), and (d) a gold-coated substrate.

4. Consider the fabrication of a polyelectrolyte multilayer composed of PEI and PAZO on a glass substrate. After each deposition step, the mass of the film can be measured by QCM and thickness by ellipsometry. Predict how these values change for a 10-bilayer film as the following parameters are varied: (a) Decreasing the pH of the polyelectrolyte solution from 7 to 4, (b) increasing the salt concentration from 0 to 100 mM, and (c) increasing the deposition temperature from 25°C to 30°C. For (a) you will need to look up various pK values for groups on each polyelectrolyte.

Cited references

Blodgett, K. B. Films Built by Depositing Successive Unimolecular Layers on a Solid Surface. *J. Am. Chem. Soc*. 1935, 57: 1007–1022.

Borkar, R., Bohr, M., and Jourdan, S. Advancing Moore's Law on 2014. http://www.intel.com/content/dam/www/public/us/en/documents/presentation/advancing-moores-law-in-2014-presentation.pdf.

Castellana, E. T. and Cremer, P. S. Solid Supported Lipid Bilayers: From Biophysical Studies to Sensor Design. *Surf. Sci. Rep.* 2006, 61(10):429–444.

Dufva, M. Fabrication of High Quality Microarrays. *Biomol. Eng.* 2005, 22: 173–184.

Groves, J. T. and Boxer, S. G. Micropattern Formation in Supported Lipid Membranes. *Acc. Chem. Res*. 2002, 35: 149–157.

Groves, J. T., Ulman, N., and Boxer, S. G. Micropatterning Fluid Lipid Bilayers on Solid Supports. *Science* 1997, 275 (5300):651–653.

Gu, L. Q. et al. Stochastic Sensing of Organic Analytes by a Pore-Forming Protein Containing a Molecular Adapter. *Nature.* 1999, 398(6729):686–690.

Heni, W., Kutuvantaivida, Y., Haffner, C., Zwickel, H., Kieninger, C., Wolf, S., Lauermann, M., Fedoryshyn, Y., Tillack, A. F., Johnson, L. E., Elder, D. L., Robinson, B. H.,

Freude, W., Koos, C., Luthold, J., and Dalton, L. R., *ACS Photonics* 2017, 4: 1576–1590.

Intel Corporation. *The Story of the Intel 4004: Intel's First Microprocessor*. 2010, https://www.intel.com/content/www/us/en/history/museum-story-of-intel-4004.html.

International Technology Roadmap for Semiconductors (ITRS) 2013, http://www.itrs.net/Links/2013ITRS/Home 2013.htm.

Johal, M. S., Parikh, A. N., Lee, Y., Casson, J. L., Foster, L., Swanson, B. I., McBranch, D. W., Li, D. Q., and Robinson, J. M. Study of the Conformational Structure and Cluster Formation in a Langmuir–Blodgett Film Using Second Harmonic Microscopy and FTIR Spectroscopy. *Langmuir* 1999, 15: 1275.

Knobler, C. M. Seeing Phenomena in Flatland: Studies of Monolayers by Fluorescence Microscopy. *Science* 1990, 249: 870.

Lei, G. and MacDonald, R. Lipid Bilayer Vesicle Fusion: Intermediates Captured by High-Speed Microfluorescence Spectroscopy. *Biophysical Journal* 2003, 85: 1585–1599.

Manfrinato, V. R., Stein, A., Zhang, L., Nam, C.-Y., Yager, K. G., Stach, E. A., and Black, C. T. Aberration-Corrected Electron Beam Lithography at the One Nanometer Length Scale. *Nano Lett.* 2017, doi:10.1021/acs.nanolett.7b00514.

Möbius, D. Z. Abstandsgesetz des Energieübergangs aus Messung der sensibilisierten Fluoreszenz. *Naturforsch*. 1969, 24a: 251.

Radeva, T. and Petkanchin, I. Electrical Properties and Conformation of Polyethylenimine at the Hematite-Aqueous Interface. *J. Colloid Interface Sci*. 1997, 196: 87.

Richter, R. P., Berat, R. and Brisson, A. R., Formation of Solid-Supported Lipid Bilayers: An Integrated View. *Langmuir* 2006, 22(8):3497–3505.

Sanders, D. P. Advances in Patterning Materials for 193 nm Immersion Lithography. *Chem. Rev*. 2010, 110: 321–360.

Smith, K. A., Gale, B. K., and Conboy, J. C. Micropatterned Fluid Lipid Bilayer Arrays Created using a Continuous Flow Microspotter. *Anal. Chem*. 2008, 80(21):7980–7987.

Tabaei, S. R. et al. Formation of Cholesterol-Rich Supported Membranes Using Solvent-Assisted Lipid Self-Assembly. *Langmuir* 2014, 30: 13345–13352.

Tamm, L. K. and McConnell, H. M. Supported Phospholipid-Bilayers. *Biophys. J*. 1985, 47(1):105–113.

Wang, Y. et al. AFM Tip Hammering Nanolithography. *Small* 2009, 5(4):477–483.

Ye, R., Wang, T., Bedzyk, L., and Croker, K. Applications of DNA Microarrays in Microbial Systems. *J. Microbiol. Meth*. 2001, 47: 257–272.

Yee, C. K., Amweg, M. L., and Parikh, A. N. Direct Photo-chemical Patterning and Refunctionalization of Supported Phospholipid Bilayers. *J. Am. Chem. Soc*. 2004, 126(43): 13962–13972.

References and recommended reading

Dalton, L. R., Günter, P., Jabinsek, M., Kwon, O-Pil, Sullivan, and Philip, A. *Organic Electro-Optics and Photonics: Molecules, Polymers, and Crystals*. 2015, Cambridge University Press, Cambridge, UK. This book provides an excellent review of organic nonlinear optical materials and their applications in data processing, including both poled organic films and self-assembled and crystalline organic materials.

Decher, G. and Schlenoff J. B. (Eds.). *Multilayer Thin Films: Sequential Assembly of Nanocomposite Materials*. 2003, John Wiley & Sons, Weinheim, Germany. This is the best source of information on electrostatic self-assembly of polyelectrolytes. This book is a crucial read for those interested in the layer-by-layer assembly of polyelectrolytes. The book provides excellent coverage of both fundamental principles and potential applications of polyelectrolyte multilayers.

Gompper, G. and Schick, M. (Eds.). *Soft Matter*. 2006, John Wiley & Sons, Weinheim, Germany. The first volume in this book explores polymer melts and mixtures. The second volume focuses on complex colloidal suspensions. Both volumes are heavy on theoretical studies and are recommended for graduate students seriously interested in computational chemistry.

Hamley, I. W. *Introduction to Soft Matter*, Revised Edition. 2007, John Wiley & Sons, Chichester, West Sussex, UK. Chapter 6 provides an excellent read on lipid membranes, proteins, and other macromolecular assemblies. Since material on liquid crystals is not covered in detail, it is recommended that the student read Chapter 5 in this book. This book also has some excellent problems.

Kuhn, H., and Försterling, H.-D. *Principles of Physical Chemistry: Understanding Molecules, Molecular Assemblies, Supramolecular Machines*. 2000, John Wiley & Sons, Chichester, West Sussex, UK. Chapters 22 (Organized Molecules Assemblies) and 23 (Supramolecular Machines) are crucial reads for anyone interested in the physical chemistry of self-assembly and supramolecular processes. These chapters are very well written and accessible to undergraduates.

Lehn, J. M. *Supramolecular Chemistry*. 1995, VCH Weinheim. An essential reference to supramolecular chemistry written by the Nobel Laureate who coined the term.

Prasad, P. N. *Introduction to Biophotonics*. 2003, John Wiley & Sons, Hoboken, NJ. This book, while not focusing on nanomaterials, provides some very interesting examples of biomaterials that are used in nanofilms. These include materials for biosensors and microarray technology for genomics and proteomics. There is also an interesting chapter on bionanophotonics.

Prasad, P. N. *Nanophotonics*. 2004, John Wiley & Sons, Hoboken, NJ. This is a highly recommended read. There is excellent coverage on nanolithography and nano-photonics for biotechnology and nanomedicine.

Glossary

Absorption (spectroscopy): Absorption processes are processes in which energy from a photon is transferred to a molecule or material, promoting it to an excited state.

Activity (chemical): The availability of a species to participate in a reaction. Activity is proportional to concentration.

Adiabatic process: A process in which there is no heat transfer between the system and surroundings.

Adsorbate: A term used to describe an atom, ion, or molecule that is either deposited onto a surface or is the species to be adsorbed.

Adsorbent: Surface upon which an adsorbate is deposited.

Adsorption isotherm: A plot of how the amount of adsorbate on a given adsorbent changes with pressure (for gas-phase deposition) or concentration (for liquid-phase deposition).

Aggregation: Buildup of material at an interface. The term also describes the formation of clusters or an increased local concentration of molecules (e.g., micelle) in bulk phases.

Aggregation number: Number of molecules that comprise a single micelle.

Amphiphile: Molecule (typically a surfactant) containing both hydrophilic and hydrophobic domains.

Anisotropic: Having a directional dependence. An assembly where all molecules point the same direction is an example of an anisotropic system, as is a material in which the refractive index depends on the orientation of the electric field of light propagating through the system.

Anti-Stokes shift: An increase in energy between absorbed and emitted (or scattered, in the context of Raman spectroscopy) photons due to the material simultaneously dropping to a lower-energy vibrational state and the energy from the vibrational transition being added to the photon.

Atomic force microscopy (AFM): A sensitive imaging technique that exploits a small piezoelectric tip monitored by a laser to detect minute changes in surface topology.

Attenuation: The gradual loss in intensity through a medium; for example, absorption processes attenuate light as it passes through a medium.

Ballistic transport: A type of electron transport that occurs in nanostructures whose mean-free path is less than its width. Ballistic transporters (typically metal nanowires) have very high conductance values compared to those whose wire widths fail to exceed the mean-free path.

Band theory: A theoretical model in solid state physics in which as the number of quantum mechanical states with similar energies become sufficiently large, those states can be treated as a continuous region of states called a band instead of individual states.

BET isotherm: An adsorption isotherm that describes multilayer film formation. This isotherm assumes that physisorption on the substrate is infinite, that no interlayer interactions exist, and that each layer can be described by the Langmuir model.

Black lipid membrane: Lipid membrane assembled on an aperture in a hydrophobic surface. The membrane is referred to as "black" because it is dark when exposed to light as photons reflected off the back half of the bilayer destructively interfere with those bouncing off the front half.

Bottom-up synthesis: In this approach, nanostructures are built by continuously extending a thread of bound molecules. Subsequent layers are built atop previous ones exploiting intermolecular forces.

Boundary conditions: Constraints that define the value of a function at the boundary of a region. In quantum mechanics, boundary conditions typically define the region in which a particle is confined (e.g., **spatial quantization**).

Bulk phase: Solvent region above or below an interface. The term is used in context to adsorption. For example, molecules move from the bulk phase and aggregate on the surface.

Cassie–Baxter wetting: A wetting state in which water rests upon nano- or micro-sized "pins" with air spaces in between.

Catalyst: A species that when added to a reaction mixture speeds up the rate of the reaction by reducing the activation energy of the reaction.

Charge overcompensation: Refers to the tendency of a charged layer to attract excess counter ions during self-assembly. In this way, layers can be subsequently built atop one another with the overcompensated charge acting as a medium for continued growth of molecules of alternating charge.

Charge reversal: Integral to the iterative process of electrostatic self-assembly. A charged molecule can aggregate atop a preexisting surface of opposite charge. The resulting surface is the opposite charge of the original layer, an instance of charge reversal. See also *charge overcompensation*.

Charge transfer: Transfer of electrons or holes (unfilled orbitals) between different molecules or materials.

Chemical equilibrium: The state in which the rates of forward and reverse reaction for a process are equal and the Gibbs energy change of the system is zero, such that the concentrations of all species in the reaction remain constant.

Chemical kinetics: The study of the rates of reactions and the implication of these rates.

Chemical potential: The partial molar Gibbs energy of a species.

Chemisorption: Process of using chemical bonds to tether molecules to a surface.

Chromophore: The region of a molecule that absorbs light and undergoes an electronic transition.

Classical physics: A system of physical laws (Newton's laws, Maxwell's equations, etc.) developed before the twentieth century, in which the values for physical properties of a system are continuous functions and simultaneously measurable. While superseded by newer theories, classical physics is still useful for characterizing macroscopic systems and as an approximation for some nanoscale systems.

Clausius–Mossotti equation: A mathematical relation between the polarizability of individual molecules and the refractive index of a material.

Closed system: A system that exchanges energy but not matter with its surroundings.

Collision frequency: The rate at which molecules in a reaction collide with each other and are available to react, but do not necessarily react.

Column chromatography: Method of purifying and separating one chemical compound from another.

Complex refractive index: A frequency-dependent complex function that describes the interaction of light and matter, where the real component (**refractive index**) represents refraction/changes in the speed of light through the material and the imaginary component (**extinction index**) represents absorption.

Conduction band: In band theory, the lowest energy band with empty states, analogous to the LUMO in a molecular system.

Conjugation: Alternating single and double bonds in linear carbon chains. All carbon atoms in the chain are sp^2 hybridized and the electrons in the π-bonds are delocalized along the entire carbon chain. See also π-bonding.

Constructive interference: A phenomenon in which light waves combine to produce a more intense wave or brighter light area than the individual waves. This type of interference occurs when the waves are in-phase.

Contact angle: Angle formed between a drop of liquid and a solid interface.

Contact angle hysteresis: The difference between a droplet's receding and advancing contact angle.

Converse piezoelectric effect: The ability of a crystal to oscillate due to applied alternating current.

Cooperative adsorption: An adsorption processes whereby the presence of one adsorbate enhances the adsorption of another. Also refers to the property set of positively and negatively charged molecules, which do not deposit on a surface individually, to stabilize electrostatic interactions and allow for the deposition of both molecules on a surface simultaneously.

Cooperative binding: Binding of a ligand (small molecule) to a molecule with multiple binding sites increases the rate of other ligands binding to the macromolecule. In contrast, **anticooperative** binding is when a ligand binding decreases the rate of other ligands binding to the macromolecule.

Coulombic energy: The energy of interaction between charges.

Critical micelle concentration (CMC): Concentration of surfactant molecules in solution above, in which micelles begin to spontaneously form.

Conductor: A material in which the highest-energy band with occupied quantum states also contains unoccupied states, such that electrons can move freely. Conductors conduct electricity well and strongly absorb or reflect light at all wavelengths.

Conservation of energy: Internal energy of system and surroundings must remain constant, although energy can be transferred between the system and surroundings via either heat or work.

Debye length: The characteristic decay length of the electrostatic potential for the Debye–Hückel model of electrostatic screening. Also used as a rough approximation for the size of the electric double layer.

Degenerate: Multiple quantum states that have the same energy are degenerate. The **degeneracy** of a state is the number of states that have the same value.

Degrees of freedom: Degrees of freedom are ways in which a molecule or particle can move (e.g., translate, rotate, or vibrate).

Density of states: A function used to represent the number of quantum mechanical states (filled or unfilled) with particular energies. The density of states for a molecular system consists of a series of discrete states, but the density of states of a system with a large number of electrons (including many nanosystems) exhibits bands containing many states (see Band Theory).

Desorption: Process of a molecule bound to a surface becoming detached from that surface and returning to the bulk phase.

Destructive interference: A phenomenon in which light waves combine to produce a less intense wave or dimmer light area than the individual waves. This type of interference occurs when the waves are completely out-of-phase.

Dielectric constant: A dimensionless quantity, also known as the relative permittivity, that represents the ability of a bulk system to polarize in response to an electric field.

Dielectric screening: The reduction in strength of electrostatic interactions due to the presence of a dielectric.

Diffusion constant: The constant of proportionality between the flux (amount of material moving through region) and concentration gradient in the region.

Diffusion-controlled reaction: A reaction in which the rate-limiting step is the transport of reactants and products through a system until they can collide with each other and initiate the remaining steps of the reaction.

Dimensionality: The number of dimensions in which a system is not constrained; can apply to translational or rotational degrees of freedom as well as to the extent of a system/number of dimensions in which the system repeats indefinitely.

Dipole: The simplest multipole, formed by a positive and a negative charge. The strength of a dipole is called a **dipole moment**.

Dispersion forces: Force between two instantaneously induced dipoles.

Dissipation (QCM): For an oscillating system such as a QCM crystal, dissipation is the ratio of the energy lost to the environment per cycle over the total energy stored by the oscillator.

DNA microarray: DNA microarrays are surfaces that have been coated with specific oligonucleotide sequences and have been particularly useful in gene expression studies. DNA microarrays vary in the number of oligonucleotides as is determined by the nature of the study; diagnostic DNA microarrays generally use tens of oligonucleotides, whereas those for research and screening can have hundreds of thousands of oligonucleotides on a single microarray.

Doping: Adding impurities to a material to change its properties, often referring to altering the properties of semiconductors by adding atoms with different numbers of valence electrons than those in the base material or oxidizing or reducing some of the molecules in the system.

Drude model: A model for conductivity that assumes that mobile electrons in a material do not interact with each other and instead only interact with the cations (atomic nuclei + strongly bound electrons) in the material.

Eigenfunction: A function that is regenerated when acted upon by an operator (see *eigenvalues*).

Electrical double layer: The diffuse layer of counter ions in a solution, which are associated with a charged surface.

Elementary steps: Individual transformations between reactants, intermediates, and products during the course of a reaction.

Ellipsometry: An optical method for measuring the thickness and refractive index (real and/or imaginary, depending on instrument type) of materials by measuring change in the phase and the reflection coefficients of elliptically polarized light.

Emission (Spectroscopy): Emission processes are processes in which a molecule or material emits a photon of light with an energy that corresponds to the difference between two quantum states, such that the molecule drops from a higher-energy state to a lower one. Fluorescence and phosophorescence are emission processes.

Enthalpy: Heat transferred in an isobaric process.

Entropy: A state function representing the possible number of configurations available to a system or to the surroundings. Systems with more states available are less ordered and have energy spread out over more states.

Evanescent field: Standing waves formed at the boundary between two media with different wave motion properties. They decay exponentially with distance from the boundary in which the waves are formed.

Excitation transfer: Transfer of energy from one molecule to another (e.g., one drops to a lower energy quantum state and the other is excited to a higher energy quantum state).

Exciton: A "quasiparticle" consisting of one or more excited electrons and unfilled orbitals (holes) that are electrostatically bound to each other. Excitons can diffuse through a system and can absorb light at different wavelengths than individual electrons depending on how strongly they are bound and whether they are oscillating in-phase or out-of-phase.

Extensive property: A thermodynamic variable (e.g., mass or volume) that is proportional to the size of the system.

Extinction coefficient: Also called molar absorptivity, it is a parameter that helps define how strongly a substance absorbs light. It is often given at a given wavelength per mass unit or per molar concentration. The extinction coefficient is proportional to the imaginary component of the material's refractive index.

Ferroelectric: A material that has a permanent electric dipole moment (e.g., a positively charged surface and a negatively charged surface) analogous to a permanent magnet. Ferroelectric materials are non-centrosymmetric.

Field-effect transistor: A semiconductor device that allows current to pass between two electrodes when a voltage is applied to a third electrode, causing the energy levels of states in the region between the electrodes to shift.

Fluorescence: A relaxation process in which the molecule relaxes by reemitting light, generally of lower energy than the light it absorbed. More specifically, it refers to the light emitted when an electronic transition occurs between electronic states of the same spin multiplicity (e.g., singlet-singlet transitions).

Fluorescence interference contrast microscopy (FLIC): Monitors the calculable modulation of fluorescence intensity due to the interaction between a reflecting surface and fluorescent objects in order to attain nanometer-accurate height measurements. In this method, a film on a reflective surface is capped with a fluorescently tagged entity and the assembly's size is quantified based on the specific interaction between the surface and fluorophore.

Fluorescence resonance energy transfer (FRET): Mechanism of energy transfer between chromophores whereby an excited donor transfers an electron over a short (<10 nm) distance to an acceptor molecule through dipole–dipole coupling.

Fluorophore: Fluorescent unit of a molecule, which absorbs and emits energy at a specific wavelength. Fluorophores are a type of chromophore.

Förster transfer: A type of excitation transfer involving interactions between transition dipoles in two different states.

Fractal: Geometric shape that can be split into sections that are at least partially identical to a smaller replica of the entire structure.

Free electron model: In this model, the electron does not exist as a discrete particle moving along the line. Rather, it resembles a standing wave whose exact form depends on the value of n.

Functional nanoassembly: A nanoscale structure consisting of multiple molecules/materials or large molecules with multiple components that is capable of performing a task when interacting with another molecule and/or provided with energy (e.g., light or electrical energy).

Fundamental vibrational frequency: The lowest vibrational frequency for a system (e.g., the $v = 0$ to $v = 1$ transition for a harmonic oscillator). It is the natural frequency of vibration of a system undergoing harmonic motion.

Gibbs energy: A state function representing the energy of a system at constant pressure. Systems at constant pressure are at equilibrium when $\Delta G = 0$ and processes at constant pressure are spontaneous when $\Delta G < 0$.

Graphene: Atomically thin layers of graphite, consisting of sheets of sp^2 carbon atoms arranged in hexagonal rings.

Gravimetry: Measurement techniques that characterize a material's response to gravity (weight), which in turn allows determination of their mass.

H-type aggregate: One-dimensional molecular assembly in which the dipole moments are aligned parallel to each other but perpendicular to the line joining their centers. This is sometimes referred to as the "face-to-face arrangement."

Half-life: The time taken for the concentration of a reactant to reach half of its initial value during a reaction, equal to $\ln(2)/k$, where k is the rate constant for the reaction.

Hamiltionian: An operator that describes the kinetic and potential energy of a system in terms of coordinates in space.

Hard sphere model: A way to determine atomic radius by assuming atoms in a solid are hard spheres and pack closely together.

Harmonic oscillator: A model quantum mechanical system in which particles are confined to a region defined by a quadratic function. Typically used to represent molecular vibrations.

Heisenberg uncertainty principle: A key principle in quantum mechanics stating that for certain pairs of variables, such as position and momentum or time and energy, it is impossible to determine the values of both values to infinite precision. Beyond a certain limit, an increase in the precision in which one variable is known decreases the precision to which the other variable may be measured.

Hill equation: An equation that relates the equilibrium constant of a reaction involving many small molecules (ligands) binding to a large molecule (macromolecule) with many potential binding sites to the concentration of the ligand and the concentration of binding sites, under the assumption that all of the binding sites are identical but that binding at one site *may* affect binding at others (cooperative binding).

HOMO: Highest occupied molecular orbital.

Hydrodynamic radius: The radius of a sphere that diffuses at the same rate as the molecule. Because most molecules are not spherical, this radius is often smaller than the effective rotational radius.

Hydrogen bond: An especially strong dipole–dipole interaction between hydrogen atoms bonded to an electronegative atom (e.g., O, N, halogens) and another electronegative atom.

Hydrophobic effect: The entropically driven effect in which hydrophobic molecules/components of molecules cluster together when in water or another polar solvent such that the number of favorable interactions between the polar solvent molecules is maximized.

Induced dipole: A dipole formed by the interactions of the electrons in a molecule and an electric field, which may be from other molecules in the system or an external source (e.g., applied voltage).

Induced dipole interactions: Force existing between a permanent dipole and a neighboring induced dipole.

Insulator: A material in which the energy difference between the highest occupied quantum state in the valence band (or HOMO) and the lowest unoccupied state in the conduction band (LUMO) is larger than 3–4 eV. Insulators do not significantly conduct electricity or absorb visible light.

Intensive property: A thermodynamic variable (e.g., temperature or density) that does not depend on the amount of material present.

Interface: The two-dimensional region of space at which two different phases contact each other.

Interferometry: The study of the ways in which light waves interact or interfere with each other. Interferometric techniques use changes in the interference of light waves with each other to determine properties of a material.

Intermediate: A species that is both produced and consumed during the course of a reaction.

Intermolecular force (F): Forces acting between sets of molecules, such as hydrogen bonding or dipole–dipole interactions.

Intermolecular potential: A function that defines the potential energy of a chemical system as a function of distance between components.

Many intermolecular potentials are defined as the distance between pairs of components; these are called **pair potentials**.

Internal energy: The energy contained within the system.

Interpenetration: In polyelectrolyte multilayer films, this is the tendency of polycations and polyanions to commingle to form highly homogeneous assemblies rather than distinct, stratified layers. This process shields the excess charge within the distinct layers, allowing for tighter packing.

Ion–ion forces: Attractive or repulsive interaction between ionic species.

Isobaric process: A process in which the pressure within the system is unchanged.

Isolated system: A system that exchanges neither energy nor matter with its surroundings.

Isothermal process: A process in which there is no change in temperature of the system or the surroundings.

J-type aggregate: One-dimensional molecular assembly in which the dipole moments of the individual monomers are aligned parallel to the line joining their centers. This is sometimes referred to as the "end-to-end arrangement."

Laminar flow: The flow of fluid in parallel layers.

Langevin model: A model for the response of dipoles to an electric field that assumes no interaction between the dipoles, similar to the Onsager model.

Langmuir adsorption isotherm: Equation relating the concentration or pressure of an adsorbate to the degree of deposition on an adsorbent.

Langmuir–Blodgett deposition: The transfer of a monolayer to a substrate using a trough, which compresses molecules (often surfactants) on the surface of a liquid buffer. The hydrophobic tails of the molecules will assemble on the surface of the substrate as it is passed through the compressed layer of surfactant. Multilayer films can be created by switching direction of the substrate entering the compressed layer.

Layer-by-layer deposition: Process of building nanoscale assemblies by exploiting intermolecular or chemical attraction forces.

A common type of layer-by-layer assembly (LbL) is electrostatic self-assembly, where, for example, a positively charged polyelectrolyte assembles spontaneously on an oxidized silica or silanol surface.

Lennard-Jones potential: A common intermolecular pair potential representing the interaction between atoms or molecules as a dispersion term proportional to the sixth power of distance and a overlap repulsion proportional to the 12th power of distance. The curve describes both attractive and repulsive interactions between two neutral particles and the distance at which these interactions cancel each other out.

Light-emitting diode: A semiconductor device that produces light when an electric current is passed through it in a particular direction. Also known as LEDs.

Lipopolymer: Consists of a soft hydrophilic polymer layer with lipidlike molecules at their surface. These structures can insert into a phospholipid membrane and tether to the polymer spacing.

Lithography: See *nanolithography*.

London dispersion forces: Intermolecular forces that involve quantum mechanical correlation between electrons in materials instead of electrostatic properties that can be represented with classical mechanics.

LUMO: Lowest unoccupied molecular orbital.

Marcus theory: A theoretical model for electron transfer processes in which electrons tunnel through a barrier between overlapping donor and acceptor states and the kinetics can be described in terms of the difference in Gibbs energy between states, the overlap of the states, and the response of the solvent/environment to the movement of the electron.

Mean-free path: Relates to the distance an electron travels between subsequent collisions with other moving particles.

Mechanical work (w): Occurs when a system exerts a force on its surroundings (negative) or the surroundings exert a force on the system (positive). A product of unbalanced forces and is one of two ways energy can be transferred (the other is heat).

Micelle: Spherical structures formed from the aggregation of surfactant molecules above the critical micelle concentration (CMC).

Microemulsion: A clear solution of liquid water, oil, and surfactant without phase separation of hydrophobic and hydrophilic entities.

Molecular orbital (MO): Mathematical function that describes the wavelike behavior of electrons in a molecule.

Molecular polarizability: The degree of electron density distortion due to an electric field.

Molecularity: The number of molecules involved in an elementary step. A **unimolecular** step involves a single molecule, a **bimolecular** step involves two molecules colliding, and so on.

Multipoles: Distributions of two or more charges that can be represented as functions. Dipoles (two charges) are the simplest type of multipole; more complex multipoles include quadrupoles and octupoles.

Nanoframeworks: Three-dimensional nanomaterials that consist of repeating arrays of nanoscale pores separated by walls. Nano-frameworks can be organic, inorganic, or organometallic (metal-organic frameworks, or MOFs).

Nanoparticles: Spherical particles made up of hundreds or thousands of atoms.

Nanoscience: Nanoscience is study of systems with components on the scale of a billionth of a meter, including large molecules (mac-romolecules), assemblies of molecules, molecules confined to within small regions of space, solid materials with small structural features, or combinations of any or all of these systems.

Nanotechnology: The application of nanoscience toward developing new technologies via engineering or manipulation of functional systems at the molecular scale.

Nanowire: A nanostructure that has a diameter on the scale of a nano-meter ($10-9$ m) and an unrestricted length.

Nanolithography: Refers to a wide variety of nanoscale surface manu-facturing techniques for creating patterned surfaces, such as those found in semiconductor circuits.

Nodal planes: Planes where the probability density of a quantum system is zero.

Node: A point where the probability density of a quantum system is zero.

Noncovalent interactions: Interactions that can hold molecules or functional groups within a molecule together that do not involve formation of a covalent bond (sharing of electrons). Examples of noncovalent interactions include electrostatic interactions and dispersion interactions.

Nonlinear optical: A term that indicates the dielectric polarization of the media responds in a nonlinear manner to the electric field of light. Nonlinear optical processes involve multiple photons/electric fields.

Normalized (quantum mechanics): A normalized function is a function multiplied by a scaling constant such that its value is 1. The constant required to normalize a function is known as the **normalization constant**.

Oligonucleotide: A short nucleic acid polymer comprised of fewer than 20 bases.

Onsager model: A model for relating the dipole moment of polar molecules to the dielectric constant of a bulk system that assumes that dipoles only interact with the average field of other dipoles within a system and not with other specific molecules.

Open system: A system that exchanges both energy and matter with its surroundings.

Operator: A mathematical operation that is applied to one or more numbers or functions.

Overlap repulsion: Repulsion between electrons in different atoms or molecules due to both electrostatic and quantum mechanical effects. Overlap repulsion prevents materials from collapsing into single points.

p-Polarized light: Electric field vector polarized along the plane perpendicular to the plane of incidence.

Particle-in-a-box model: A model quantum mechanical system in which electrons are confined in a region of zero potential energy and between barriers of infinite potential energy. Also known as the infinite potential well model. The corresponding model in solid-state physics for bulk materials is called the **free-electron model**.

Path function: A function that depends on intermediate values, such that different paths between the end points of the function will have different values. Mechanical work and reaction rates are examples of path functions.

Path length: The distance in which light moves through a material.

Pauli exclusion principle: States that no two electrons can have the same four quantum numbers.

Phonon: A quantum mechanical description of a special type of vibration, known as the normal modes in classical physics. Phonons are the discrete amounts of energy crystalline structures that these modes of vibration can absorb. In this aspect of quantized energy, the phonon is analogous to the photon.

Phosphorescence: A relaxation process when an excited electron undergoes nonradiative intersystem-crossing to a slightly more stable excited state, called a triplet state, before relaxing and releasing a photon of light.

Photobleaching (FRAP): Photobleaching is the process of destroying a fluorophore with intense light. FRAP, or fluorescence recovery after photobleaching, is the method quantifying lateral diffusion rates by monitoring fluorescence under a microscope and measuring the time it takes for a photobleached segment of a fluorophore film to reorganize. Fluorophores reorganize, intact ones replacing the inactivated ones. The hole in fluorescence created by photobleaching disappears at a rate determined by the lateral diffusion.

Photodynamic therapy: A medical technique that involves killing targeted cells by exciting dyes and/or nanoparticles that are bound to those cells.

Photorelaxation: Process of an excited state chromophore dissipating energy through release of a photon.

Photovoltaic cell: A semiconductor device that captures energy from light to produce an electric current, also known as a solar cell.

Physisorption: Process of exploiting weaker molecular forces to deposit molecules on the surface. Generally weaker and less specific than chemisorption.

Piezoelectric effect: An effect found in certain types of materials in which applying mechanical force to the material (e.g., pushing or pulling) induces an electric field in the material. The **converse**

piezoelectric effect also occurs in piezoelectric materials and involves the material oscillating due to an applied electric field.

Pi bonding (or π-bonding): Bond resulting from the overlap of atomic orbitals that are in contact through two areas of overlap above and below the internuclear axis. For example, in sp^2 hybridized carbon atoms, the unhybridized 2p orbitals on neighboring carbons overlap to form the π-bond.

Pi–pi stacking (or π–π stacking): Packing of large, often aromatic molecules whereby favorable overlap across pi systems of different molecules occurs.

Plasma oscillations: Rapid oscillations of the electron density in a conducting medium (e.g., a metal).

Plasmon: The smallest unit of plasma oscillations; used for quantization.

Poisson–Boltzmann equation: Calculates the actual distribution of the counterions at equilibrium.

Polar molecule: A molecule with a nonzero dipole moment. Water, HCl, and acetonitrile are examples of polar molecules.

Polarizability: Defined as the susceptibility of an atom or molecule's electron cloud to distortion by an external electric field.

Poling: Orientation of polar molecules or polar moieties in a macromolecule/polymer using an externally applied electric field. Can also refer to orientation of magnetic materials using an externally applied magnetic field.

Polyelectrolyte: Polyelectrolytes are polymers, or chains of molecules, which contain free ions that make them electrically conductive. Soluble in water, polyelectrolytes become charged when in solution and are often countered by a salt ion of opposite charge. The amount of charge on a polyelectrolyte determines whether it is classified as strong or weak. Strong polyelectrolytes are fully soluble, whereas weak polyelectrolytes, with fractional charge, are only partially soluble.

Polymer cushioned phospholipid bilayers: Allow bilayers to effectively be decoupled from the surface by minimizing interactions with the underlying substrate.

Potential energy (V): Energy stored within a system due to its position in a force field.

Profilometry: A mechanical method for measuring the thickness of a film by measuring motion of a very sharp stylus across a substrate.

Pseudo-first-order approximation: An approximation that assumes that the concentrations of all but one reacting species in a reaction are nearly constant such that the rate law can be written as a function of only one species.

Pyroelectric effect: An effect found in certain materials where heating the material induces an electric field in the material.

Rate of reaction: The change in the concentration of a reacting species as a function of time during a reaction (e.g., the first derivative of the concentration of a reactant or product with respect to time).

Rate-limiting (determining) step: The slowest elementary step in a reaction, which determines the rate of the entire reaction.

Reaction mechanism: The series of elementary steps that leads from reactants to products for a specific reaction. The sum of the elementary steps yields the stoichiometric equation.

QCM: Quartz crystal microbalance.

Quantization of energy: The quantum mechanical principal that within a confined system, only certain values of energy are allowed for the system.

Quantum confinement: The change in the properties of a system due to the electrons in a system being trapped in a nanoscale region. The dimensionality of quantum confinement is defined by the number of directions in which the movement of electrons is not confined.

Quantum dots: Quantum dots are inorganic semiconductor nanoparticles, typically 2–10 nm in size. The excitons (electron-hole pairs) of quantum dots are confined to three dimensions. Quantum dots typically consist of a core, shell, and final coat, as shown in Figure 5.10. They are characterized by the nature of each of these layers, their size and aspect ratio, their quantum efficiency in optical materials, and their coercivity in magnetic materials. Properties of an ensemble of quantum dots are additionally determined by particle size distribution and differences in morphology within the ensemble.

Quantum mechanical tunneling: A phenomenon in which the wavefunction of a particle extends through a finite energy barrier, such that the particle has a finite probability of being found in regions where it could not be found under classical mechanics. Tunneling allows particles to pass between quantum confined regions.

Quantum mechanics: A system of physical laws developed since the early twentieth century in which observable physical properties of a system such as energy and momentum are limited to discrete values. Quantum mechanics is necessary for describing many nanoscale systems due to their small size limiting the number of available quantum states.

Quantum number: A value that defines which eigenstate a quantum mechanical system is in. Quantum mechanical systems may have one or more quantum mechanical numbers depending on their boundary conditions.

Raman scattering: A scattering process in which the light increases or decreases in energy, often due to a change in vibrational energy of the scattering molecule. Raman scattering is measured using Raman spectroscopy, a vibrational spectroscopy technique.

Rate law: A mathematical relationship between the concentrations of species in a reaction and the rate of the reaction. The constant of proportionality is called the rate constant.

Rayleigh scattering: A process in which photons interact with a molecule or material without being absorbed or transferring any energy and then continue in a random direction. The degree of scattering is related to the wavelength of light being scattered and the electric polarizability of the particles doing the scattering.

Reduced mass: A mathematical method for representing the mass of a multibody system (e.g., two atoms in a diatomic molecule) as a single mass.

Refractive index: The ratio of the speed of light in vacuum versus the speed of light in vacuum, related to the dielectric constant at high frequencies (e.g., visible light). If generalized as a complex number (the **complex refractive index**), the imaginary component describes the extent to which light is absorbed at a given frequency.

Resonant frequency: The frequency at which the local amplitude of oscillation is a maximum.

Reversible reaction: A reaction in which the rate constant for the reverse of a reaction is of similar magnitude to the forward rate constant for a reaction. In contrast, **irreversible** reactions either have products that leave the system or very small reverse rate constants such that the reverse reaction cannot occur at any significant frequency.

s-Polarized light: Electric field vector polarized along the plane of incidence.

Sauerbrey equation: The simplest mathematical relation used for analyzing QCM data, in which the change in mass adsorbed to the surface of a QCM sensor is linearly proportional to the shift in resonance frequency of the quartz crystal in the sensor.

Scanning tunneling microscopy (STM): A sensitive imaging technique that exploits a small tunneling current between a surface and a tip to detect minute changes in surface topology.

Scatchard equation: An equation that relates the equilibrium constant of a reaction involving many small molecules (ligands) binding to a large molecule (macromolecule) with many potential binding sites to the ligand and the concentration of binding sites, under the assumption that all of the binding sites are identical and independent (binding at one site does not affect binding at others, or noncooperative binding).

Schrödinger wave equation: The fundamental equation of quantum mechanics, which represents the state of a system in terms of a wavefunction and a Hamiltonian (energy operator).

Second-harmonic generation: A nonlinear optical effect in which the frequency of light is doubled when passed through a material.

Selection rules: Quantum mechanical rules that describe whether or not an absorption or emission process can occur.

Self-assembled monolayer (SAM): A surface coverage that forms when molecules are spontaneously attracted to a functional surface, often by electrostatic or van der Waals forces. SAMs can also result through chemisorption.

Self-assembly: The process whereby molecules in the bulk phase spontaneously organize either in the bulk phase or on a surface.

Semiconductor: A material with properties between an insulator and a conductor, in which the energy difference between the highest occupied state in the valence band (HOMO) and the lowest unoccupied state in the conduction band (LUMO) is small. Semiconductors can conduct electricity under certain conditions and absorb visible light at some wavelengths.

Sigma bonding (σ bonding): A strong covalent bond characterized by symmetry with respect to rotation about the bond axis.

Small unilamellar vesicles (SUVs): Tiny (50–200 nm), single-walled vesicles of uniform diameter.

Soft sphere model: More realistic way of modeling an atom, this model assumes atoms are compressible to some degree and do not have completely rigid boundaries.

Solvatochromism: Shift in the wavelength of light absorbed or emitted from a molecule or nanomaterial due to the polarity of the surrounding solvent having more or less favorable interactions with the ground and excited states of a system. A solvatochromic shift that reduces the energy of a transition is called a **bathochromic shift**, and a shift that increases the energy of a transition is called a **hypsochromic shift**.

Spectroscopy: The study of the absorption of transmission of light through a medium as a function of either wavelength or frequency.

Spot density: Number of spots that can be placed in a given area of a DNA microarray, which varies greatly among in situ, noncontact, and contact printing methods.

State function: A function that depends only on the initial and final values of the function regardless of path taken. Temperature and Gibbs energy are examples of state functions.

Steady-state approximation: An approximation in multistep reactions where the concentration of an intermediate remains constant due to it being produced and consumed at the same rate.

Steric hindrance: Stress in a molecular structure or assembly that arises when oversized functional groups are forced into too small a space such that exchange repulsion greatly reduces or exceeds attractive interactions. Such stress can result in torsional strain.

Stokes shift: A decrease in energy between an absorbed photon and an emitted photon (or scattered, in the case of Raman spectroscopy) due to a simultaneous transition to a higher vibrational state and the energy from the vibrational transition being subtracted from the photon. Most materials exhibit a Stokes shift when emitting photons.

Sum-frequency generation (SFG): A nonlinear optical effect when two photons of different frequencies combine to produce a photon of the sum frequency when passed through a material.

Superhydrophobic surface: Highly hydrophobic surface with water contact angles exceeding 150°.

Supramolecular chemistry: The study of chemical compounds comprised of a distinct number of cooperating molecules. Supramolecular machines are comprised of a "cooperating set" of molecules and exploit intra- and intermolecular forces to complete a specialized task.

Surface functionalization: Deliberate attachment of specific molecules to a surface in order to allow for the specific binding of a subsequent adsorbate.

Surface tension: The work done in increasing the area of a surface by transporting a molecule from the bulk phase to that surface. A measurement of the cohesion of like molecules at an interface. The tendency of a surface to contract due to energetic factors.

Surfactant: Compounds that lower the surface tension of a liquid and form micelles above the critical micelle concentration (CMC).

Surroundings (thermodynamic): When a system is defined, the surroundings are everything else in the universe that is not part of the system.

System (thermodynamic): The specific region of interest (often the material sample) when analyzing a physical or chemical process.

Thio-: A prefix that indicates an oxygen atom in the common compound is instead sulfur. For example, ether has a general structure ROR′ and a thioether would have a general structure RSR′.

Thiol: Thiols are compounds having the structure RSH (where R ≠ H).

Top-down synthesis: In this approach, structures are synthesized starting from a bulk substrate and chiseled down until only a field of desired assemblies remains.

Total intermolecular potential: A function that represents all of the interactions between a set of molecules, summed over electrostatic, dispersion, and overlap repulsion terms.

Transducer: A device that converts one type of energy to another.

Transition dipole: A short-lived dipole formed by the movement of electron density during an absorption or emission process. Most absorption or emission processes must have a nonzero transition dipole to occur.

Transition states: Transition states are local maxima in the energy of a system during the course of a reaction. While existing for a negligible amount of time, a system must have enough energy to reach a transition state for a reaction to occur.

Transmittance: The ratio of the intensity of a light beam after passing through a sample to the original intensity of the light beam.

Valence band: In band theory, the highest energy-filled band, analogous to the HOMO in a molecular system.

van der Waals interactions: Attractive or repulsive noncovalent interactions between molecules involving charges, permanent or induced dipoles, or higher multipoles.

Variable angle spectroscopic ellipsometry: An advanced ellipsometry technique that uses measurements at many wavelengths and angles to determine the thickness and wavelength-dependent complex refractive index for multilayer films.

Vibrational modes: Oscillations of atoms in a molecule or material in response to being promoted to an excited vibrational state, typically with energies corresponding to infrared light. Vibrational modes have characteristic energies corresponding with different types of bonds (e.g., C-H) stretch, and may range from simple motions of two atoms to complex motions of many atoms in a molecule or material.

Viscoelasticity: The property of materials to both resist flow when a stress is applied and rapidly return to their original state once the stress is removed.

Wavefunction: The wavefunction of a system is a mathematical function that describes the particles in a system (e.g., electrons) as a wave. The square of the wavefunction is the probability density of a particle's location in space and time.

Waveguide: An object composed of multiple layers of materials with different refractive indices that confines and directs waves by means of total reflection of the wave within itself.

Wenzel wetting: A wetting state in which water rests on a surface whose morphology has been altered so that in a given area, water is in contact with more surface than if the surface were completely flat.

Index

Page numbers followed by f and t indicate figures and tables, respectively.